Fluid Flow and Heat Transfer: Science and Engineering

Fluid Flow and Heat Transfer: Science and Engineering

Edited by Emilia Lloyd

Clanrye International
www.clanryeinternational.com

Clanrye International,
750 Third Avenue, 9th Floor,
New York, NY 10017, USA

ISBN: 978-1-64726-678-3

Cataloging-in-Publication Data

Fluid flow and heat transfer : science and engineering / edited by Emilia Lloyd.
p. cm.
Includes bibliographical references and index.
ISBN 978-1-64726-678-3
1. Fluid dynamics. 2. Heat--Transmission. 3. Fluid mechanics. 4. Energy transfer. I. Lloyd, Emilia.
QC151 .F58 2023
532.05--dc23

For information on all Clanrye International publications
visit our website at www.clanryeinternational.com

CLANRYE INTERNATIONAL

Contents

Preface

Fluid flow and heat transfer processes play a significant role in various fields of engineering and science from microscopic to the planetary scale. Heat transfer can be improved by using nanofluids. The problem with fluid flow occurs because of a continuously moving surface in an ambient fluid, which differs from the fluid that flows past a fixed surface. Fluid flow problems in the absence of chemical reactions and electromagnetic forces involve thermal and mechanical variables only. There are three methods or techniques that are utilized to address problems in fluid mechanics and heat transfer, namely, numerical methods, experimental methods, and theoretical methods. This book aims to shed light on the science and engineering of fluid flow and heat transfer. It consists of contributions made by international experts. This book will serve as a reference to a broad spectrum of readers.

All of the data presented henceforth, was collaborated in the wake of recent advancements in the field. The aim of this book is to present the diversified developments from across the globe in a comprehensible manner. The opinions expressed in each chapter belong solely to the contributing authors. Their interpretations of the topics are the integral part of this book, which I have carefully compiled for a better understanding of the readers.

At the end, I would like to thank all those who dedicated their time and efforts for the successful completion of this book. I also wish to convey my gratitude towards my friends and family who supported me at every step.

Editor

1

Numerical Solution of Biomagnetic Power-Law Fluid Flow and Heat Transfer in a Channel

Adrian S. Halifi [1,†,‡], Sharidan Shafie [1] and Norsarahaida S. Amin [1,2,*,‡]

1 Department of Mathematical Sciences, Faculty of Science, Universiti Teknologi Malaysia, Johor Bahru 81310, Malaysia; adriansyah@utm.my (A.S.H.); sharidan@utm.my (S.S.)
2 Department of Mathematics, Faculty of Science and Technology, Universitas Airlangga, East Java 60115, Indonesia
* Correspondence: norsarahaida@utm.my
† Current address: Centre for Degree and Foundation Studies, UTMSPACE, Universiti Teknologi Malaysia, 81310 Johor Bahru, Malaysia.
‡ These authors contributed equally to this work.

Abstract: The effect of non-Newtonian biomagnetic power-law fluid in a channel undergoing external localised magnetic fields is investigated. The governing equations are derived by considering both effects of Ferrohydrodynamics (FHD) and Magnetohydrodynamics (MHD). These governing equations are difficult to solve due to the inclusion of source term from magnetic equation and the nonlinearity of the power-law model. Numerical scheme of Constrained Interpolation Profile (CIP) is developed to solve the governing equations numerically. Extensive results carried out show that this method is efficient on studying the biomagnetic and non-Newtonian power-law flow. New results show that the inclusion of power-law model affects the vortex formation, skin friction and heat transfer parameter significantly. Regardless of the power-law index, the vortex formation length increases when Magnetic number increases. The effect of this vortex however decreases with the inclusion of power-law where in the shear thinning case, the arising vortex is more pronounced than in the shear thickening case. Furthermore, increasing of power-law index from shear thinning to shear thickening, decreases the wall shear stress and heat transfer parameters. However for high Magnetic number, the wall shear stress and heat transfer parameters increase especially near the location of the magnetic source. The results can be used as a guide on assessing the potential effects of radiofrequency fields (RF) from electromagnetic fields (EMF) exposure on blood vessel.

Keywords: biomagnetic; FHD; MHD; power-law; non-Newtonian; CIP

---

1. Introduction

Biological fluid such as blood includes protein, cell membrane and, haemoglobin molecules that contribute to its paramagnetic behaviour, while the content of hydrogen, oxygen, nitrogen and carbon atoms in vessel tissues contribute to its diamagnetic property [1,2]. Biofluid that has magnetic characteristics and able to interact with an applied magnetic field is called biomagnetic [3]. When an external magnetic field is imposed on the biomagnetic fluid, there are two body forces acting on the fluid, magnetisation force and Lorentz force. These two forces correspond to the principle of Ferrohydrodinamics (FHD) and Magnetohydrodynamics (MHD) respectively. The interplay between its hydrodynamics and magnetic properties when external magnetic field gradient is applied has been the topic of great interest [4]. Motivation in these studies are driven by the biomedic application such as in magnetic drug targeting in tumor treatment, hyperthermia, reduction of bleeding during surgeries, and magnetic separation of cells [5–11]. However, at the same time, there are some drawbacks of magnetic fields toward human bodies. Being exposed to electromagnetic field (EMF) have deleterious effects to living organism [12]. In the last decade, over three billions people are exposed to EMF daily from wireless phone, wi-fi, transmitting towers and base stations [13]. Moreover, there are 150 million patients annually used Magnetic Resonance Imaging(MRI) for clinical examinations [14]. Magnetic field intensity from phone, refrigerator, wireless etc are measured up to 1 T (Tesla) [15], while the MRI is at 1.5 T, 3 T, 7 T [16] and about to be boost up to 14 T [17]. The EMF has biological impact on thermal effect occurs via an alteration in temperature from radiofrequency (RF) fields. There is a possibility that every interaction between RF fields and living tissues causes an energy transfer resulting in a rise in temperature [12]. Therefore, more studies need to be done in order to understand the thermal effects associated with RF exposure.

Several existing studies have examined the effect of applying high magnetic field intensity on the cellular components of the biological fluids. Ichioka et al. [18] reported that the static magnetic field of 8 T significantly decreased the skin blood flow and the body temperature in anaesthetised rat. Similar results have been reported by Haik et al. [19] who studied the effect of magnetic field up to 10 T on human blood. They found that the Reynolds number is decreased when the magnetic field is increased. Motta et al. [20] investigated the effect of magnetic field up to 18 T on light absorption of blood and reported that there is an increase on light absorption level due to paramagnetic orientation of heme group. Higashi et al. [21] studied the effect of applying high static magnetic field up to 8 T on the orientation of normal erythrocytes and found that even at 1 T, the magnetic field influenced the erythrocytes to be oriented with their disk plane parallel to the magnetic field direction.

Mathematical model for biomagnetic fluid flow with an external magnetic field has been introduced by Haik et al. [19], where the arising force due to magnetisation depends on the existence of a spatially varying magnetic field and the biofluid is considered to be electrically poor conductors. This mathematical model is known as Biomagnetic Fluid Dynamics (BFD) [22]. Tzirtzilakis and Loukopoulos [2,23,24] extended this model by considering that blood also exhibit strong electrical conductivity which eventually interacts with the magnetic fields when it flows and give rise to Lorentz force. The mathematical model developed for the biofluid flow under the influence of an applied magnetic field is consistent with the principles of FHD and MHD of ferrofluid [25]. In addition, ferrofluid can exhibit non-Newtonian behaviour [26–29]. However, most of the available works reported only on the effect of FHD on Newtonian fluid [2,23,30–43]. Even if the magnetic field effect is included in power-law fluid flow, the works only used magnetic field effect of MHD [44–49]. Research on the magnetic fluid flow undergoing external magnetic field are extensive. For the research specifically on flow in a channel, Loukopoulos and Tzirtzilakis [23] studied the biomagnetic channel flow where the magnetic field is spatially varying linearly with temperature and magnetic intensity and found that a vortex is arising near the lower plate below of which the magnetic source is placed. This vortex increased in the size depending on the magnitude of applied the magnetic field. Goharkhah and Ashjaee [30] reported the works of using alternating nonuniform magnetic field on ferrofluid flow and heat transfer in a channel. The works on uniform localised magnetic field effect in a channel has been studied by Tzirtzilakis and Loukopolous [2] in laminar flow and Tzirtzilakis et al. [31] in turbulent flow. They found that the effect of magnetic field is decreased in the turbulent flow. Siddiqa et al. [32] studied the effect of including thermal radiation on energy equation in the biomagnetic fluid flow in a channel. A great deal of works have been done on the magnetic field undergoing external magnetic field in geometry other than straight channel. Biomagnetic fluid flow in stenosed channel has been studied by Tzirtzilakis [33], Xenos and Tzirtzilakis [34], and Türk et al. [35]. Türk et al. [36] studied the biomagnetic fluid flow in a channel with multiple stenoses. Flow in aneurysm channel has been studied by Tzirtzilakis [37], and recently by Sharifi et al. [38] on drug targeting using magnetic particles. Tzirakis [39] studied biomagnetic fluid flow in circular ducts. For the works on magnetic fluid flow in lid driven cavity, it has been studied by Tzirtzilakis and Xenos [40], Marioni et al. [41], Bozkaya et al. [42] and by Senel and Tezer-Sezgin [43].

However, most of the works mentioned are concerned with biomagnetic of Newtonian fluid. When blood flows through larger arteries, it behaves according to Newtonian model, but in small blood vessels such as narrow arteries, it behaves according to non-Newtonian flow model at low shear rates [50–54]. Unlike Newtonian fluid, non-Newtonian are a broad class of fluid where the relation between the shear rate and shear stress is nonlinear. Hence, no single constitutive equation can be used to single handedly predict all the flow of non-Newtonian models in various condition. There are several models characterizing flow of blood such as Walburn-Schneck, Casson, Carreau-Yasuda, Quemada, and power-law, where each has its own importance in theoretical and experimental of flow phenomena [55–59]. The power-law model is the simplest form and commonly used for modelling shear-thinning or shear-thickening behaviour of non-Newtonian fluids on certain range.

On the study of purely hydrodynamics power-law flow, Bell and Surana [60] investigated the flow of non-Newtonian power-law in lid driven cavity using p-version least squares finite element formulation and provided the benchmark results for shear thinning and shear thickening flow in lid driven cavity. Their results on power-law lid driven cavity were validated by Neofytou [61], who also carried out simulations for Bingham plastics, Casson and Quemada fluid flow. There are many works that have been reported on magnetic field effect on power-law fluid flow. Kefayati [44] studied the effect of MHD magnetic field on power-law in cavity with two-square concentric duct. Kefayati [45] studied the effect of magnetic field on mixed convection on shear thinning fluids in a cavity with sinusoidal boundary conditions. Adıgüzel and Atalık [46] investigated the effect of MHD magnetic field on power-law ferrofluid flow past a circular cylinder. Jahanbakshi [47] reported the MHD magnetic field effects on natural convection using power-law fluid in an L-shaped enclosure. Ikbal et al. [48] and Sankar et al. [49] investigated the effect of MHD magnetic field on power-law in a stenosed artery. All of these works conclude that the vortex formation decreases under the action of MHD magnetic field. A number of investigations reported works on MHD effect on non-Newtonian flow other than power-law model. Misra et al. [62] studied the works of MHD viscoelastic fluid in a channel with stretching walls and reported that increasing the strength of the magnetic field will have the fluid velocity profile decreases but the temperature profile increases. Liu [63] performed an analysis of electrically conducting fluid of second grade in a porous medium over a stretching sheet subject to a transverse magnetic field and found that when the Eckert number is large, the heat flow transferred from the fluid to the wall rather than from the wall to the fluid when Eckert number is small.

The works that combine both biomagnetic and power-law fluid flow has not been investigated. The reason is that the mathematical model of Navier Stokes with FHD and MHD is difficult to solve due to the magnetisation term FHD introduced in the governing equation. This source term can be very large and will result in instability and divergence of the numerical method without the use of a proper technique. Additionally, in contrast to the Newtonian case, the non-Newtonian fluid flow is generally more complicated to be solved due to its complexity contributed from the constitutive equation to the Navier-Stokes equations. This is due to the fact that the effective viscosity diverges to infinite shear rates and zero shear rates for shear thinning and shear thickening fluid. When solving non-Newtonian flow numerically, it may result in significantly time step constraint when explicit schemes are used, and disrupt the solution technique usually used to solve implicitly the viscous terms in the momentum equation [64]. Several methods have been used to solve the problem analogous in this work. Stream function-vorticity finite difference method (FDM) with stretched grid has been used in [33,65]. FDM central difference has also been used in [48]. In [44,45], finite difference lattice Boltzmann method (FDLBM) is used, while in [35,36] finite element method (FEM) has been applied. In [42,43], dual reciprocity boundary element method (DRBEM) has been used. Lattice Boltzmann method (LBM) has been used in [66,67] while control volume finite element method (CVFEM) is applied to solve the flow problem in [68–70]. More detailed review of numerical method on solving FHD and MHD problems are given by Rashidi et al. [71].

An explicit CIP (Constrained Interpolation Profile) with FDM scheme was developed in this work. This scheme is a very stable finite difference technique that can solve generalised hyperbolic equations with third order accuracy in space. Unlike the conventional schemes such as Lax Wendroff and Upwind, this scheme approximated the values of the advection term with its spatial derivative by cubic polynomials interpolation and attains high-accuracy by being less diffusive [72]. The CIP method was introduced by T.Yabe et al. [73,74] and has since been used to solve many various problems [75–77]. This method can be combined with other methods [78] as well since CIP is efficient in solving the hyperbolic equation.

As far as we know, no previous research has investigated the works on biomagnetic and non-Newtonian fluid flow using CIP scheme. Therefore the aim of this work is to study the extent of using non-Newtonian power-law on biomagnetic flow in terms of heat transfer, skin friction and flow characteristics where the governing equations are solved numerically using CIP scheme. For non-Newtonian power-law fluid, the viscosity $\bar{\mu}_a$ is assumed to be a shear dependent fluid given as

$$\bar{\mu}_a = m(I_2)^{(n-1)/2}, \tag{1}$$

where m is the fluid consistency, I_2 is the second invariant of the rate of strain tensor and n is the power-law index. For two-dimensional flow, I_2 is defined as

$$I_2 = 2\left[\left(\frac{\partial \bar{u}}{\partial \bar{x}}\right)^2 + \left(\frac{\partial \bar{v}}{\partial \bar{y}}\right)^2\right] + \left(\frac{\partial \bar{u}}{\partial \bar{y}} + \frac{\partial \bar{v}}{\partial \bar{x}}\right)^2, \tag{2}$$

where $\bar{u}$ is the velocity along $\bar{x}$-coordinate, and $\bar{v}$ is the velocity along $\bar{y}$-coordinate. For values of $n < 1$, the viscosity is shear thinning, while for $n > 1$, the viscosity is shear thickening and $n = 1$ refers to the Newtonian viscosity model where $\bar{\mu}_a = m$. Biological fluids have shear thinning properties as blood has a power-law index at $n = 0.6$ and synovial fluid has a power-law index at $n = 0.2$ [46]. In addition, this study may give valuable information on the thermal and hydrodynamics effects on blood associated with RF exposure.

2. Problem Formulation

The continuity, momentum and energy equations for biomagnetic fluid flow are given as [24,71]

$$\nabla \cdot \vec{v} = 0, \tag{3}$$

$$\rho\left(\frac{\partial \vec{v}}{\partial \bar{t}} + \vec{v} \cdot \nabla \vec{v}\right) = -\nabla \bar{p} + \nabla \cdot \vec{\tau} + (\vec{J} \times \bar{B}) + \bar{\mu}_0(\bar{M}.\nabla)\bar{H}, \tag{4}$$

$$\rho c_p\left(\frac{\partial \bar{T}}{\partial \bar{t}} + \vec{v} \cdot \nabla \bar{T}\right) = \kappa \nabla^2 \bar{T} + \vec{\tau} \cdot \nabla \vec{v} - \bar{\mu}_0 \bar{T}\frac{\partial \bar{M}}{\partial \bar{T}}\vec{v} \cdot \nabla \bar{H} + \frac{\vec{J} \cdot \vec{J}}{\bar{\sigma}}, \tag{5}$$

where $\vec{v}$ is the dimensional velocity vector, t is the dimensional time, ρ is the fluid density, p is the dimensional pressure, $\vec{\tau}$ is the stress tensor, $\vec{J}$ is the current density, $\bar{B}$ is the magnetic field, $\bar{\mu}_0$ is magnetic permeability, $\bar{H}$ is the magnetic field intensity, $\bar{T}$ is the dimensional temperature, c_p is the specific heat at constant pressure, κ is the thermal conductivity, and $\bar{\sigma}$ is the electrical conductivity.

The fluid is considered to be magnetic non-Newtonian and the flow is two-dimensional, laminar, incompressible and under the influence of an applied magnetic field. The body force terms incorporated are from the principles of Magnetohydrodynamics and Ferrohydrodynamics. The flow is considered to be taking place in between two walls. It is assumed that the flow is dominated by the viscous force and the convection is due to temperature gradient, hence the effect of gravity is insignificant. The flow at the entrance is assumed to be fully developed and the walls are kept at a constant temperature $\bar{T}_w$, while the fluid is at temperature $\bar{T}_f$ such that $\bar{T}_w < \bar{T}_f$. The configuration of the flow is shown in Figure 1.

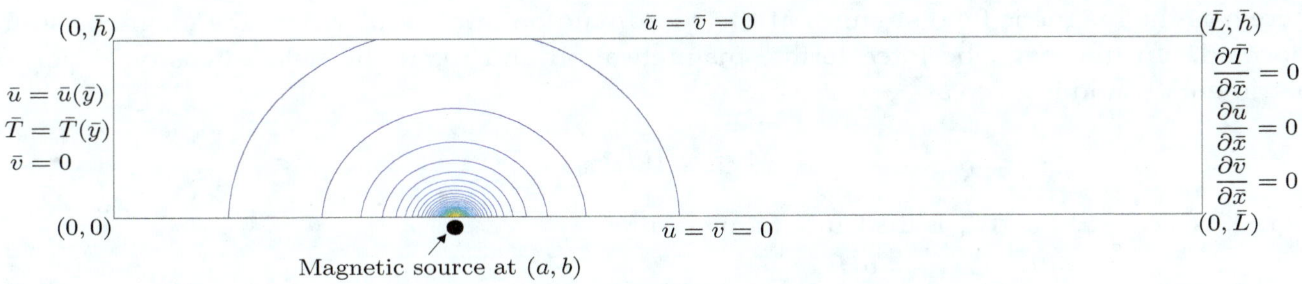

Figure 1. Flow domain and contours of magnetic field strength H.

With these assumptions, the governing equations in Cartesian form are given as

$$\frac{\partial \bar{u}}{\partial \bar{x}} + \frac{\partial \bar{v}}{\partial \bar{y}} = 0, \tag{6}$$

$$\begin{aligned}\bar{\rho}\left(\frac{\partial \bar{u}}{\partial \bar{t}} + \bar{u}\frac{\partial \bar{u}}{\partial \bar{x}} + \bar{v}\frac{\partial \bar{u}}{\partial \bar{y}}\right) &= -\frac{\partial \bar{p}}{\partial \bar{x}} + \bar{\mu}_a\left(\frac{\partial^2 \bar{u}}{\partial \bar{x}^2} + \frac{\partial^2 \bar{u}}{\partial \bar{y}^2}\right) \\ &+ 2\frac{\partial \bar{\mu}_a}{\partial \bar{x}}\frac{\partial \bar{u}}{\partial \bar{x}} + \frac{\partial \bar{\mu}_a}{\partial \bar{y}}\left(\frac{\partial \bar{v}}{\partial \bar{x}} + \frac{\partial \bar{u}}{\partial \bar{y}}\right) \\ &+ \bar{\mu}_0 \bar{M}\frac{\partial \bar{H}}{\partial \bar{x}} - \bar{\sigma}\bar{B}_y^2 \bar{u} + \bar{\sigma}\bar{B}_x \bar{B}_y \bar{v},\end{aligned} \tag{7}$$

$$\begin{aligned}\bar{\rho}\left(\frac{\partial \bar{v}}{\partial \bar{t}} + \bar{u}\frac{\partial \bar{v}}{\partial \bar{x}} + \bar{v}\frac{\partial \bar{v}}{\partial \bar{y}}\right) &= -\frac{\partial \bar{p}}{\partial \bar{y}} + \bar{\mu}_a\left(\frac{\partial^2 \bar{v}}{\partial \bar{x}^2} + \frac{\partial^2 \bar{v}}{\partial \bar{y}^2}\right) \\ &+ 2\frac{\partial \bar{\mu}_a}{\partial \bar{y}}\frac{\partial \bar{v}}{\partial \bar{y}} + \frac{\partial \bar{\mu}_a}{\partial \bar{x}}\left(\frac{\partial \bar{v}}{\partial \bar{x}} + \frac{\partial \bar{u}}{\partial \bar{y}}\right) \\ &+ \bar{\mu}_0 \bar{M}\frac{\partial \bar{H}}{\partial \bar{y}} - \bar{\sigma}\bar{B}_x^2 \bar{v} + \bar{\sigma}\bar{B}_x \bar{B}_y \bar{u},\end{aligned} \tag{8}$$

$$\begin{aligned}\bar{\rho}c_p\left(\frac{\partial \bar{T}}{\partial \bar{t}} + \bar{u}\frac{\partial \bar{T}}{\partial \bar{x}} + \bar{v}\frac{\partial \bar{T}}{\partial \bar{y}}\right) &= \kappa\left(\frac{\partial^2 \bar{T}}{\partial \bar{x}^2} + \frac{\partial^2 \bar{T}}{\partial \bar{y}^2}\right) \\ &+ \bar{\mu}_a\left(2\left[\left(\frac{\partial \bar{u}}{\partial \bar{x}}\right)^2 + \left(\frac{\partial \bar{v}}{\partial \bar{y}}\right)^2\right] + \left[\frac{\partial \bar{u}}{\partial \bar{y}} + \frac{\partial \bar{v}}{\partial \bar{x}}\right]^2\right) \\ &- \bar{\mu}_0 \bar{T}\frac{\partial \bar{M}}{\partial \bar{T}}\left(\bar{u}\frac{\partial \bar{H}}{\partial \bar{x}} + \bar{v}\frac{\partial \bar{H}}{\partial \bar{y}}\right) + \bar{\sigma}(\bar{u}\bar{B}_y - \bar{v}\bar{B}_x)^2.\end{aligned} \tag{9}$$

In the above equations, $\vec{v} = (\bar{u}, \bar{v})$ is the velocity, $\bar{p}$ is the pressure, $\bar{T}$ is the temperature, $\bar{u}(\bar{y})$ and $\bar{T}(\bar{y})$ is the parabolic velocity profile and temperature profile respectively for fully developed flow, $\bar{\sigma}$ is the electrical conductivity, $\bar{\mu}_0$ is the magnetic permeability vacuum, c_p is the specific heat at constant pressure, κ is the thermal conductivity, $\bar{H} = (\bar{H}_x, \bar{H}_y)$ is the magnetic field strength, $\bar{M}$ represents the magnetisation, and $\bar{B}$ is the magnetic induction where $\bar{B} = \bar{\mu}_0 \bar{H} \Rightarrow (\bar{B}_x, \bar{B}_y) = \mu_0(\bar{H}_x, \bar{H}_y)$. The bar symbol on u, v, T, t, x and y denote the dimensional form of these quantities.

The components of magnetic field intensity $\bar{H}_x$ and $\bar{H}_y$ along the $\bar{x}$ and $\bar{y}$ coordinates $\vec{H} = (\bar{H}_x, \bar{H}_y)$ are given by [40]

$$\bar{H}_x(\bar{x}, \bar{y}) = \frac{\gamma}{2\pi}\frac{\bar{x} - a}{\sqrt{(\bar{x} - a)^2 + (\bar{y} - b)^2}}, \qquad \bar{H}_y(\bar{x}, \bar{y}) = -\frac{\gamma}{2\pi}\frac{\bar{y} - b}{\sqrt{(\bar{x} - a)^2 + (\bar{y} - b)^2}}, \tag{10}$$

and the magnitude $\bar{H}$ is given by

$$\bar{H}(\bar{x}, \bar{y}) = \sqrt{\bar{H}_x^2 + \bar{H}_y^2} = \frac{\gamma}{2\pi}\frac{1}{\sqrt{(\bar{x} - a)^2 + (\bar{y} - b)^2}}, \tag{11}$$

where γ is the magnetic field strength at the coordinate point of (a,b) where the magnetic source is located. In the magnetic force terms, magnetisation and magnetic field intensity $\bar{H}$ derived experimentally and is given as

$$\bar{M} = \bar{K}\bar{H}(\bar{T}_c - \bar{T}), \tag{12}$$

where $\bar{K}$ is a constant and $\bar{T}_c$ is the Curie temperature.

The terms $\bar{\mu}_0\bar{M}\frac{\partial \bar{H}}{\partial \bar{x}}$ and $\bar{\mu}_0\bar{M}\frac{\partial \bar{H}}{\partial \bar{y}}$ in Equations (7) and (8) respectively represent the component of magnetic force per unit volume and dependent on the existence of magnetic gradient. The term $\bar{\mu}_0\bar{T}\frac{\partial \bar{M}}{\partial \bar{t}}\left(u\frac{\partial \bar{H}}{\partial \bar{x}} + v\frac{\partial \bar{H}}{\partial \bar{y}}\right)$ in Equation (9) refers to the thermal power unit per unit volume due to the magnetocaloric effect. These three terms arise due to the principles of FHD. The terms $\bar{\sigma}\bar{B}_y^2\bar{u} + \bar{\sigma}\bar{B}_x\bar{B}_y\bar{v}$ and $\bar{\sigma}\bar{B}_x^2\bar{v} + \bar{\sigma}\bar{B}_x\bar{B}_y\bar{u}$ in Equations (7) and (8) respectively represent the Lorentz force per unit volume and included due to the electrical conductivity of the fluid. The term $\bar{\sigma}(\bar{u}\bar{B}_y - \bar{v}\bar{B}_x)^2$ in Equation (9) represents the Joule heating. These three terms arise due to the principles of MHD. Therefore the principle of FHD and MHD are combined in the mathematical model with the application of non-Newtonian fluid.

The non-Newtonian fluid considered here can be represented by the Oswald-de Waele power-law model. The relationship between velocity field and shear stress may be expressed in terms of the effective viscosity as

$$\bar{\mu}_a = m\left(2\left[\left(\frac{\partial \bar{u}}{\partial \bar{x}}\right)^2 + \left(\frac{\partial \bar{v}}{\partial \bar{y}}\right)^2\right] + \left[\frac{\partial \bar{u}}{\partial \bar{y}} + \frac{\partial \bar{v}}{\partial \bar{x}}\right]^2\right)^{\frac{n-1}{2}}, \tag{13}$$

where m is the fluid consistency and n is the power-law index.

The boundary conditions of the problem are

$$\begin{aligned}
&\text{Inflow} && \bar{u}(0,\bar{y}) = \bar{u}(\bar{y}), && \bar{v}(0,\bar{y}) = 0, && \bar{T} = \bar{T}(y),\\
&\text{Outflow} && \frac{\partial \bar{u}(\bar{L},\bar{y})}{\partial \bar{x}} = 0, && \frac{\partial \bar{v}(\bar{L},\bar{y})}{\partial \bar{x}} = 0, && \frac{\partial \bar{T}(\bar{L},\bar{y})}{\partial \bar{x}} = 0,\\
&\text{Lower wall} && \bar{u}(\bar{x},0) = 0, && \bar{v}(\bar{x},0) = 0, && \bar{T}(\bar{x},0) = \bar{T}_w,\\
&\text{Upper wall} && \bar{u}(\bar{x},\bar{h}) = 0, && \bar{v}(\bar{x},\bar{h}) = 0, && \bar{T}(\bar{x},h) = \bar{T}_w.
\end{aligned} \tag{14}$$

3. Numerical Formulation and Solution Procedure

3.1. Non-Dimensionalisation and Transformation to Stream Function-Vorticity

Equations (7)–(9), with the boundary conditions of Equation (14) are transformed into dimensionless form using the following dimensionless variables

$$\begin{aligned}
&x = \frac{\bar{x}}{\bar{L}}, \quad y = \frac{\bar{y}}{\bar{L}}, \quad t = \frac{\bar{t}\bar{u}}{\bar{L}} \quad u = \frac{\bar{u}}{\bar{u}_r}, \quad v = \frac{\bar{v}}{\bar{u}_r},\\
&p = \frac{\bar{p}}{\bar{\rho}\bar{u}_r^2}, \quad H_x = \frac{\bar{H}_x}{\bar{H}_0}, \quad H_y = \frac{\bar{H}_y}{\bar{H}_0}, \quad \mu_a = \frac{\bar{\mu}_a L^{n-1}}{m u_r^{n-1}}, \quad T = \frac{\bar{T} - \bar{T}_w}{\bar{T}_f - \bar{T}_w},
\end{aligned} \tag{15}$$

where $\bar{H}_0 = \bar{H}(a,0)$ and u_r is the maximum velocity at the inflow.

To solve the governing equation, stream function-vorticity approach is used and the velocity $\vec{v} = (u,v)$ is transformed into dimensionless stream function $\psi(x,y)$ and vorticity $w(x,y)$ by the relation

$$w = \frac{\partial v}{\partial x} - \frac{\partial u}{\partial y}, \tag{16}$$

$$u = \frac{\partial \psi}{\partial y}, \quad v = -\frac{\partial \psi}{\partial x}. \tag{17}$$

Using these transformation, the pressure p is eliminated, and Equation (6) is automatically satisfied and hence only the nondimensional variables ψ, w and T need to be solved numerically. The transformation results to the following system of equations

$$\frac{\partial^2 \psi}{\partial x^2} + \frac{\partial^2 \psi}{\partial y^2} = -w, \tag{18}$$

$$\begin{aligned} \frac{\partial w}{\partial t} + u\frac{\partial w}{\partial x} + v\frac{\partial w}{\partial y} &= \frac{\mu_a}{Re}\left(\frac{\partial^2 w}{\partial x^2} + \frac{\partial^2 w}{\partial y^2}\right) + \frac{1}{Re}\left(\frac{\partial^2 \mu_a}{\partial x^2} - \frac{\partial^2 \mu_a}{\partial y^2}\right)\left(\frac{\partial v}{\partial x} + \frac{\partial u}{\partial y}\right) \\ &- \frac{2}{Re}\frac{\partial^2 \mu_a}{\partial x \partial y}\left(\frac{\partial u}{\partial x} - \frac{\partial v}{\partial y}\right) + \frac{2}{Re}\left(\frac{\partial \mu_a}{\partial x}\frac{\partial w}{\partial x} + \frac{\partial \mu_a}{\partial y}\frac{\partial w}{\partial y}\right) \\ &+ Mn_F\left(\frac{\partial H}{\partial x}\frac{\partial T}{\partial y} - \frac{\partial H}{\partial y}\frac{\partial T}{\partial x}\right) \\ &+ N\left(\frac{\partial}{\partial y}(H_y^2 u - H_x H_y v) - \frac{\partial}{\partial x}(H_x^2 v - H_x H_y u)\right), \end{aligned} \tag{19}$$

$$\begin{aligned} \frac{\partial T}{\partial t} + u\frac{\partial T}{\partial x} + v\frac{\partial T}{\partial y} &= \frac{1}{RePr}\left(\frac{\partial^2 T}{\partial x^2} + \frac{\partial^2 T}{\partial y^2}\right) \\ &+ \frac{Ec\mu_a}{Re}\left(2\left[\left(\frac{\partial u}{\partial x}\right)^2 + \left(\frac{\partial v}{\partial y}\right)^2\right] + \left[\frac{\partial u}{\partial y} + \frac{\partial v}{\partial x}\right]^2\right) \\ &+ Mn_F EcH(\epsilon + T)\left(u\frac{\partial H}{\partial x} + v\frac{\partial H}{\partial y}\right) \\ &+ EcN(uH_x - vH_y)^2. \end{aligned} \tag{20}$$

The effective viscosity is nondimensionalised to be

$$\mu_a = \left(2\left[\left(\frac{\partial u}{\partial x}\right)^2 + \left(\frac{\partial v}{\partial y}\right)^2\right] + \left[\frac{\partial u}{\partial y} + \frac{\partial v}{\partial x}\right]^2\right)^{\frac{n-1}{2}}. \tag{21}$$

The magnetic field intensity is also nondimensionalised by using Equation (15) and is given by

$$H(x,y) = \frac{b}{\sqrt{(x-a)^2 + (y-b)^2}}. \tag{22}$$

The boundary configuration for the flow problem is also nondimensionalised using Equation (15) where the length of the horizontal wall is considered to be 10 times the length of the vertical inlet and outlet. The boundary conditions are non-dimensionalised and transformed according to Equations (16) and (17), and given as

$$\begin{aligned} &\text{Inflow} && (x = 0,\, 0 \le y \le 1), && \psi = 2y^2 - \frac{4}{3}y^3, \quad w = 8y - 4, \quad T = 4y(1-y), \\ &\text{Outflow} && (x = 10,\, 0 \le y \le 1) && \frac{\partial \psi}{\partial x} = \frac{\partial w}{\partial x} = \frac{\partial T}{\partial x} = 0, \\ &\text{Lower wall} && (y = 0,\, 0 \le x \le 10) && \psi = \frac{2}{3}, \quad -w = \frac{\partial^2 \psi}{\partial y^2}, \quad T = 0, \\ &\text{Upper wall} && (y = 1,\, 0 \le x \le 10) && \psi = 0, \quad -w = \frac{\partial^2 \psi}{\partial y^2}, \quad T = 0. \end{aligned} \tag{23}$$

3.2. Constrained Interpolated Profile (CIP) Based Finite Difference Method

To solve Equations (18)–(23) numerically, the Constrained Interpolated Profile (CIP) method is implemented. In this method, a cubic polynomial is used to interpolate the spatial profile of the value f and its spatial derivative ∇f. In this case, f is $f = (w, T)$. Once the values of f and ∇f are calculated, the same values for the next time step are simply obtained by shifting the cubic polynomial.

To implement the CIP algorithm, the governing equations of Equations (19) and (20) are split into two categories [79] which are non-advection phase and advection phase.

Advection Phase:

$$\frac{\partial w}{\partial t} + u\frac{\partial w}{\partial x} + v\frac{\partial w}{\partial y} = 0, \tag{24}$$

$$\frac{\partial T}{\partial t} + u\frac{\partial T}{\partial x} + v\frac{\partial T}{\partial y} = 0. \tag{25}$$

Non-advection Phase:

$$\begin{aligned}\frac{\partial w^*}{\partial t} = {} & \frac{\mu_a}{Re}\left(\frac{\partial^2 w}{\partial x^2} + \frac{\partial^2 w}{\partial y^2}\right) + \frac{1}{Re}\left(\frac{\partial^2 \mu_a}{\partial x^2} - \frac{\partial^2 \mu_a}{\partial y^2}\right)\left(\frac{\partial v}{\partial x} + \frac{\partial u}{\partial y}\right) \\ & - \frac{2}{Re}\frac{\partial^2 \mu_a}{\partial x \partial y}\left(\frac{\partial u}{\partial x} - \frac{\partial v}{\partial y}\right) + \frac{2}{Re}\left(\frac{\partial \mu_a}{\partial x}\frac{\partial w}{\partial x} + \frac{\partial \mu_a}{\partial y}\frac{\partial w}{\partial y}\right) \\ & + Mn_F\left(\frac{\partial H}{\partial x}\frac{\partial T}{\partial y} - \frac{\partial H}{\partial y}\frac{\partial T}{\partial x}\right) \\ & + N\left(\frac{\partial}{\partial y}(H_y^2 u - H_x H_y v) - \frac{\partial}{\partial x}(H_x^2 v - H_x H_y u)\right),\end{aligned} \tag{26}$$

$$\begin{aligned}\frac{\partial T^*}{\partial t} = {} & \frac{1}{RePr}\left(\frac{\partial^2 T}{\partial x^2} + \frac{\partial^2 T}{\partial y^2}\right) \\ & + \frac{Ec\mu_a}{Re}\left(2\left[\left(\frac{\partial u}{\partial x}\right)^2 + \left(\frac{\partial v}{\partial y}\right)^2\right] + \left[\frac{\partial u}{\partial y} + \frac{\partial v}{\partial x}\right]^2\right) \\ & + Mn_F EcH(\epsilon + T)\left(u\frac{\partial H}{\partial x} + v\frac{\partial H}{\partial y}\right) \\ & + EcN(uH_x - vH_y)^2.\end{aligned} \tag{27}$$

The computation of advection phase, Equations (24) and (25) is conducted by the CIP method while the non-advection phase, labeled with superscript (*) is solved numerically using finite difference method. In order to adopt this method, the Poisson equation of Equation (18) need to be solved first, then followed by the non-advection phase and finally the advection phase.

Equation (18) is solved by using Successive Overrelaxation (SOR). The iterative procedure is carried out until the residue between two consecutive iteration converges to desired tolerance. The tolerance for SOR is suggested to be smaller than the tolerance for the steady state solution in which $|\psi^{new} - \psi^{old}| < 10^{-7}$, and the relaxation factor used is 1.7.

For the non-advection phase, the explicit scheme is used. The central finite difference discretisation is implemented on first order and second order spatial derivatives terms. The magnetic terms $\frac{\partial H}{\partial x}$, $\frac{\partial H}{\partial y}$, H_x and H_y are computed analytically from Equation (22). It should be noted that the method has also been tested to be solved using Crank-Nicolson scheme and no significant difference is observed. However using the implicit scheme will require solving the two matrices equation in every time step for each Equations (19) and (20), and inadvertently it is time consuming due to exhaustive computer power required.

Once the non-advection phase are computed, the spatial derivative are calculated for both w^* and T^*. Hence by denoting $f = (w^*, T^*)$, the spatial derivative $\frac{\partial f^*}{\partial x} = \partial_x f^*$ and $\frac{\partial f^*}{\partial x} = \partial_y f^*$ can be calculated as

$$\begin{aligned}\frac{\partial_x f^*_{i,j} - \partial_x f^k_{i,j}}{\Delta t} &= \frac{f^*_{i+1,j} - f^*_{i-1,j}}{2\Delta x} - \frac{f^k_{i+1,j} - f^k_{i-1,j}}{2\Delta x} \\ &\quad - \frac{u_{i+1,j} - u_{i-1,j}}{2\Delta x} \cdot \partial_x f^k_{i,j} - \frac{v_{i+1,j} - v_{i-1,j}}{2\Delta x} \cdot \partial_y f^k_{i,j}, \qquad (28)\\ \frac{\partial_y f^*_{i,j} - \partial_y f^k_{i,j}}{\Delta t} &= \frac{f^*_{i,j+1} - f^*_{i,j-1}}{2\Delta y} - \frac{f^k_{i,j+1} - f^k_{i,j-1}}{2\Delta y} \\ &\quad - \frac{u_{i,j+1} - u_{i,j-1}}{2\Delta y} \cdot \partial_x f^k_{i,j} - \frac{v_{i,j+1} - v_{i,j-1}}{2\Delta y} \cdot \partial_y f^k_{i,j}. \qquad (29)\end{aligned}$$

The terms f^k, $\partial_x f^k_{i,j}$, and $\partial_y f^k_{i,j}$ refer to the values of previous time step. The spatial derivatives of u and v in Equations (28) and (29) are approximated by using central order finite difference.

Once $f^* = (w^*, T^*)$ and their respective spatial derivative are obtained, they will be used in CIP algorithm. Hence $f^{k+1} = (w^{k+1}, T^{k+1})$ are advanced to the new time level after the interpolation of CIP via shifting such that $f^{k+1} = f(x_i - u\Delta t, y_j - v\Delta t)$. Similarly in non-advection phase, the spatial gradient for the new time iteration is also required. The CIP scheme to solve Equation (24) and (25) is given as

$$\begin{aligned}f^{k+1}_{i,j} =& C_{30}\Delta x^3 + C_{20}\Delta x^2 + \frac{\partial f^*_{i,j}}{\partial x}\Delta x + f^*_{i,j} + C_{03}\Delta y^3 + C_{02}\Delta y^2 \\ &+ \frac{\partial f^*}{\partial y}\Delta y + C_{21}\Delta x^2\Delta y + C_{12}\Delta x\Delta y^2 + C_{11}\Delta x\Delta y, \qquad (30)\\ \frac{\partial f^{k+1}_{i,j}}{\partial x} &= 3C_{30}\Delta x^2 + 2C_{20}\Delta x + \frac{\partial f^*_{i,j}}{\partial x} + 2C_{21}\Delta x\Delta y + C_{12}\Delta y^2 + C_{11}\Delta y, \qquad (31)\\ \frac{\partial f^{k+1}_{i,j}}{\partial y} &= 3C_{03}\Delta y^2 + 2C_{02}\Delta y + \frac{\partial f^*}{\partial y} + C_{21}\Delta x^2 + 2C_{12}\Delta x\Delta y + C_{11}\Delta x. \qquad (32)\end{aligned}$$

The term C is the coefficient for the interpolation function Equation (30), and for ease of reference, the subscript on C denotes the power of its multiplication with Δx and Δy respectively. Since $\frac{\partial f^*_{i,j}}{\partial x}$ and $\frac{f^*_{i,j}}{\partial y}$ are already known from the non-advection phase at all grid points, only the coefficients C are needed to complete the calculation of Equations (30)–(32), and they can be determined as [79]

$$\begin{aligned}C_{30} &= \left(\Delta x[\partial_x f^k_{i+1,j} + \partial_x f^k_{i,j}] - 2[f^k_{i+1,j} - f^k_{i,j}]\right)/\Delta x^3,\\ C_{20} &= \left(-\Delta x[\partial_x f^k_{i+1,j} + 2\partial_x f^k_{i,j}] + 3[f^k_{i+1,j} - f^k_{i,j}]\right)/\Delta x^2,\\ C_{03} &= \left(\Delta y[\partial_y f^k_{i,j+1} + \partial_y f^k_{i,j}] - 2[f^k_{i,j+1} - f^k_{i,j}]\right)/\Delta y^3,\\ C_{02} &= \left(\Delta y[\partial_y f^k_{i,j+1} + 2\partial_y f^k_{i,j}] + 3[f^k_{i,j+1} - f^k_{i,j}]\right)/\Delta y^2,\\ C_{21} &= \left(\partial_x f^k_{i+1,j+1} - \partial_x f^k_{i+1,j} - \partial_x f^k_{i,j+1} + \partial_x f^k_{i,j}\right)/(2\Delta x^2\Delta y),\\ C_{12} &= \left(\partial_y f^k_{i+1,j+1} - \partial_y f^k_{i+1,j} - \partial_y f^k_{i,j+1} + \partial_y f^k_{i,j}\right)/(2\Delta x\Delta y^2),\\ C_{11} &= \left(f^k_{i+1,j+1} - f^k_{i+1,j} - f^k_{i,j+1} + f^k_{i,j}\right)/(\Delta x\Delta y) - C_{21}\Delta x - C_{12}\Delta\, y.\end{aligned}$$

The above equations are only for $u < 0$ and $v < 0$. For the case of $u \geq 0$, the pointer $i+1$ is to be changed to $i-1$ and Δx is changed to $-\Delta x$. Similarly, when $v \geq 0$, $j-1$ and $-\Delta y$ are used instead of $j+1$ and Δy. The calculated $f^{k+1} = (w^{k+1}, T^{k+1})$ and their respective spatial gradients are then updated to be f^k and will be used for the new time level. The whole procedure is repeated until a steady state flow is achieved in which the tolerance is set to be

$$\max_{i,j} |\phi_{i,j}^{k+1} - \phi_{i,j}^{k}| < 10^{-6}, \tag{33}$$

where ϕ^k is the estimation of $\phi^k = (\psi^k, w^k, T^k)$ at iteration k.

3.3. Boundary Conditions

For the numerical solutions for the system of Equations (18)–(20), boundary conditions for flow variables w, ψ and T are discretised according to Equation (23). The boundary conditions for w, ψ, and T on the entrance and the exit are easily implemented. On the walls, the boundary conditions for ψ and T are also implemented according to Equation (23). However, the boundary condition for w on the walls are not straightforward.

The boundary conditions for the velocities on the lower and upper walls are set as

$$u_{i,J} = 0, \qquad v_{i,J} = 0, \tag{34}$$

where J refers to the index of the lower and upper walls. By using boundary conditions of Equation (34) and central differences of Equation (17), we obtained

$$u_{i,J} = 0 = \left.\frac{\partial \psi}{\partial y}\right|_{(i,J)} = \frac{\psi_{i,J+1} - \psi_{i,J-1}}{2\Delta y} + O\left(\Delta y^2\right) \Rightarrow \psi_{i,J+1} = \psi_{i,J-1} + O\left(\Delta y^3\right), \tag{35}$$

$$v_{i,J} = 0 = -\left.\frac{\partial \psi}{\partial x}\right|_{(i,J)} = \frac{\psi_{i-1,J} - \psi_{i+1,J}}{2\Delta x} + O\left(\Delta x^2\right) \Rightarrow \psi_{i+1,J} = \psi_{i-1,J} + O\left(\Delta x^3\right). \tag{36}$$

Equation (18) is then discsretised using second order central difference, and substituting the relations from Equations (35) and (36), the boundary conditions for the vorticity on the lower and upper walls are then proved to be [33]

$$w_{i,J} = -\frac{2(\psi_{i-1,J} - \psi_{i,J})}{\Delta x^2} + \frac{2(\psi_{i,J} - \psi_{i,J-1})}{\Delta y^2} + O\left(\Delta x^2\right) + O\left(\Delta y^2\right), \tag{37}$$

where fictitious points are to be used accordingly. The discretisation of boundary conditions used for w, ψ, and T during the advection phase and non-advection phase are similar.

The procedure for CIP-Finite Difference method is summarised as follows:

- Set the initial and boundary conditions according to Equation (23).
- Compute Equation (18) by using SOR.
- Calculate the velocities u and v by using Equation (17).
- Using the new u and v, calculate effective viscosity Equation (21).
- Solve numerically the non-advection phase of Equations (26) and (27).
- Calculate the spatial gradient of the non-advection phase from Equations (28) and (29).
- Advance the time iteration by solving numerically the advection phase and their spatial gradient of Equations (24) and (25) by using the CIP algorithm from Equations (30)–(32).

- Repeat the above steps until the steady state is achieved by using the convergence criterion in Equation (33).

3.4. Dimensionless Parameters

To implement the numerical method, it is necessary to assign the values of parameters in the governing equation. Other than the power-law index n, the nondimensional parameters appear in the governing equations are

$$
\begin{aligned}
Re &= \frac{\rho u_r^{2-n} L^n}{m} && \text{(Reynolds number)},\\
Pr &= \frac{c_p m}{\kappa}\left(\frac{u_r}{L}\right)^{n-1} && \text{(Prandtl number)},\\
Mn_F &= \frac{\bar{\mu}_0 \bar{H}_0 \bar{K}(\bar{T}_f - \bar{T}_w)}{\rho u_r} && \text{(Magnetic number arising from FHD)},\\
N &= \frac{\bar{\sigma}\mu_0^2 \bar{H}_0^2 L}{\rho u_r} && \text{(Stuart number)},\\
Ec &= \frac{u_r^2}{c_p(\bar{T}_f - \bar{T}_w)} && \text{(Eckert number)},\\
\epsilon &= \frac{\bar{T}_w}{(\bar{T}_f - \bar{T}_w)} && \text{(Temperature number)}.
\end{aligned}
$$

The special definition of parameters Pr and Re above can be converted to conventional Newtonian fluid by taking the power-law index $n = 1$. The parameter Mn_F is the Magnetic number that arises due to FHD, whereas the parameter N is the Stuart number, defined as the ratio of magnetic to inertia effects.

As in the works of [23,33,65], the magnetic fluid is considered to be blood as blood can be paramagnetic, diamagnetic and non-Newtonian fluid. The parameters are calculated by considering that $\rho = 1050$ kg m^{-3}, $m = \mu_a = 3.2 \times 10^{-3}$ kg m^{-1} s^{-1}, with maximum velocity $u_r = 1.524 \times 10^{-2}$ ms^{-1}. The distance of the plates are $L = 2.0 \times 10^{-2}$ m. Hence, the Reynolds number is at $Re = 100$. The location of the magnetic wire for applying the localised magnetic field is set to be $(a, b) = (2.5, -0.05)$, where the maximum intensity occurred on $H_0 = H(2.5, 0)$. The temperatures of the plate is assumed to be $\bar{T}_w = 37\,°\text{C}$ while the temperature of the fluid is $\bar{T}_f = 41\,°\text{C}$, in which making the temperature number ϵ to be 77.5. The Prandtl number is considered constant at $Pr = 25$ by taking $C_p = 3.9 \times 10^3$ J kg^{-1} and $\kappa = 0.5$ J m^{-1} s^{-1}. From these values, the Eckert number is set to be $Ec = 1.49 \times 10^{-8}$.

The important dimensionless parameter for this work is the Magnetic number Mn_F and it can be written as

$$
\begin{aligned}
Mn_F &= \frac{\bar{\mu}_0 \bar{H}_0 \bar{K}(\bar{T}_f - \bar{T}_w)}{\rho u_r},\\
&= \frac{\bar{B}_0 \bar{M}_0}{\rho u_r^2},\\
&= \frac{\bar{B}_0 \bar{M}_0 L^2 \rho}{\bar{\mu}_a^2 Re^2},
\end{aligned} \tag{38}
$$

where $\bar{B}_0$ and $\bar{M}_0 = 40$ A m^{-1} are the magnetic field induction and the magnetisation at the point $(a, 0)$ respectively. The Stuart number N is calculated subsequently using the parameters assigned as in Mn_F, hence

$$N = \frac{\bar{\sigma}\mu_0^2\bar{H}_0^2 L}{\rho u_r} = \frac{\bar{\mu}_a}{L\rho u_r}\frac{L^2\bar{\sigma}\bar{B}_0^2}{\bar{\mu}_a} = \frac{L^2\bar{\sigma}\bar{B}_0^2}{Re\bar{\mu}_a}. \tag{39}$$

Equations (38) and (39) show the relationship of Mn_F and N with Re, where increasing the magnitude of Magnetic number is the same as lowering the Reynolds number. Hence, the Reynolds number Re is fixed to be $Re = 100$ throughout the numerical experiment, as varying the value of Mn_F and N are sufficient. The parameter for Mn_F and N assigned are dependent on the reference magnetic field induction of $\bar{B}_0$ in Tesla (T) unit. The values of $\bar{B}_0$ are chosen to be 1 T, 4 T and 8 T and the corresponding values for Mn_F and N are given in Table 1.

Table 1. Value of reference magnetic field $\bar{B}_0$ and the corresponding values of Mn_F and N for $Re = 100$.

$\bar{B}_0$ (Tesla)	Mn_F	N
0	0	0
1	164	0.0008
4	656	0.014
8	1316	0.056

This variation of Tesla number was implemented to study the effect of applied magnetic field on power-law flows, in the case of shear thinning ($n < 1$), Newtonian ($n = 1$) and shear thickening ($n > 1$). Another important parameter in this study is the power-law index, that determine whether the fluid is Newtonian or non-Newtonian. The power-law index n is varied from $n = 0.6$ for shear thinning fluid, to $n = 1$ and $n = 1.4$ for shear thickening fluid. This scenario is apt for the magnetic fluid such as blood since its power-law index is varied in the human biological system. This can be also interpreted as using different class of non-Newtonian magnetic fluid for the flow in a channel.

Flow and heat characteristics that are being compared in this study are the local skin friction and the local rate of heat transfer coefficients. These quantities are defined as

$$C_f = \left.\frac{2\psi''(x,y)}{Re}\right|_{y=0,1}, \qquad Nu = \left.\frac{\partial T}{\partial y}\right|_{y=0,1}, \tag{40}$$

where C_f is the local skin friction coefficient and Nu is the Nusselt number that describes the local rate of heat transfer coefficient. These coefficients are calculated both on the lower wall ($y = 0$) and upper wall ($y = 1$) with different Magnetic number Mn_F and power-law index n number.

3.5. Grid Independence and Method Validation

Numerical solution of the system of Equations (18)–(20) has been developed using Matlab®. An extensive grid independence test is performed to check for the grid independent solution. The test is based on the u-velocity profile along the various x location of the channel, where the focus is emphasised near the location on the magnetic source. Five different size of grids configuration have been performed on the case flow problem $n = 1$, $Mn_F = 1312$ and $N = 0.056$. The magnetic source is located at $(a, b) = (2.5, -0.05)$. As shown in Figure 2, when the grid size is refined from 400×40 to 500×100, there is a significant change on velocity profile, particularly near the magnetic source and in the interval of the vortex size. The numerical simulation is sensitive on the vertical grid, while no significant change occurred when refining the horizontal grid. For the temperature profile, the local heat transfer on the lower wall of the channel has been used as a reference to test the grid independence. Only the lower wall is shown since most of the significant changes occurred near the magnetic source. The same grid independence test result is observed as seen in Figure 3, where the grids size have reached grid independence for temperature profile. These grids sizes have also been tested with smaller time step and no significant changes is accounted. It is observed that the grid size configuration of 400 × 80 is sufficient to capture the velocity profile change near the magnetic location.

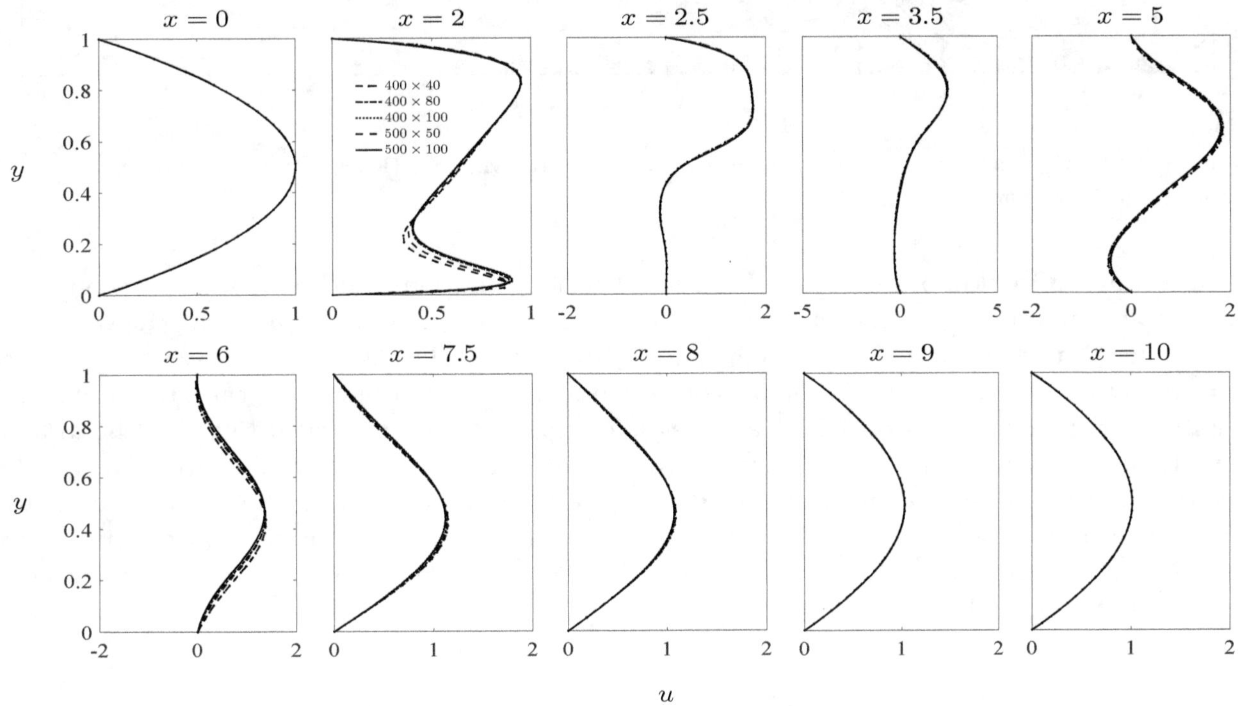

Figure 2. Grid independence test for u-velocity profile at different locations x of the channel.

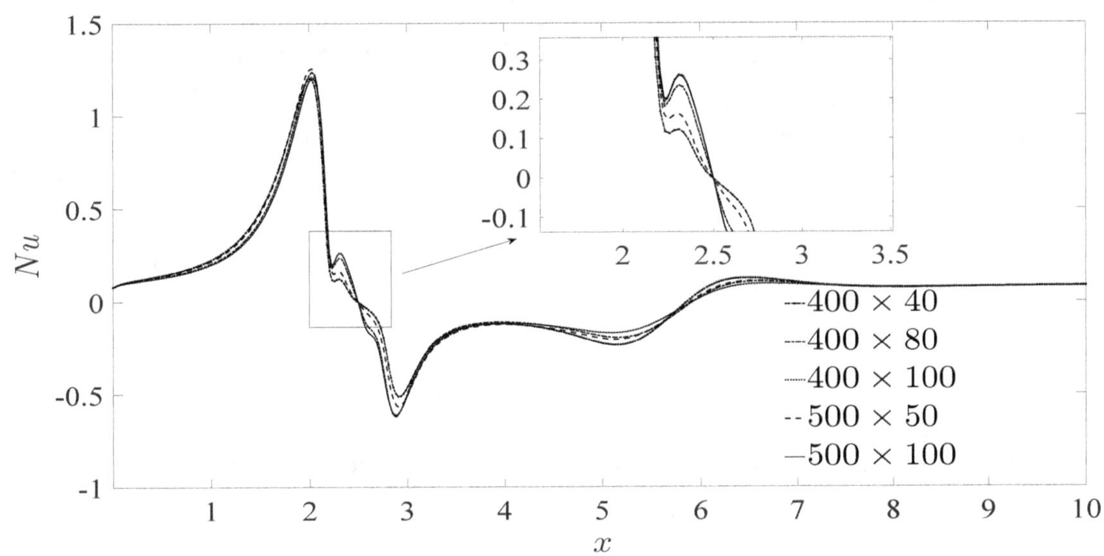

Figure 3. Grid independence test of local heat transfer parameter on the channel's lower wall.

The performance and the validation of present numerical method, are done separately on the benchmark cases for non-Newtonian flow in lid driven cavity, and for the flow in a channel undergoing the localised magnetic field. The results validation regarding the flow in a channel under the action of localised magnetic field is performed as in [33]. From Figure 4, it can be seen that the present numerical method developed is in good agreement with the compared studies.

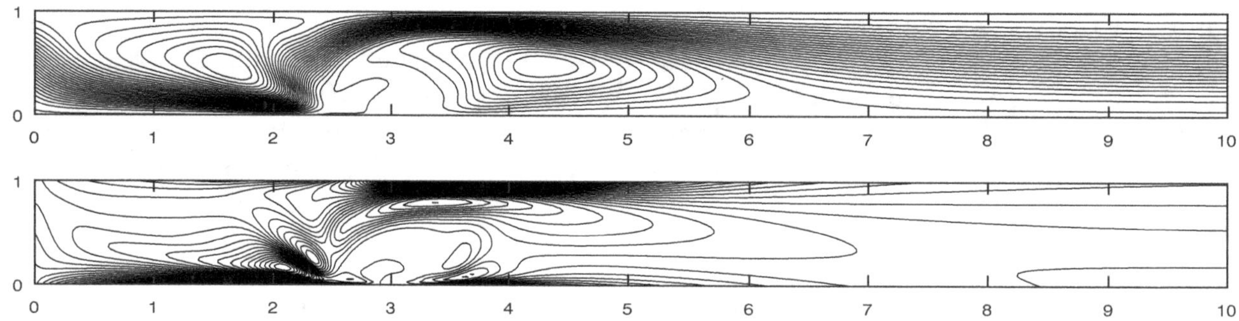

Figure 4. Stream function and vorticity contour for $Re = 10$, $Mn_F = 131{,}250$ obtained by using present method and comparable with [33].

The second validation of the code is based on the non-Newtonian comparison. The computation involving non-Newtonian of power-law model in a lid driven cavity is validated with the works of Bell and Surana [60] and Neofytou [61]. In accordance with these works, simulations have been performed for $Re = 100$ with different values of the power-law index. For Newtonian flow case, the comparison is made with the benchmark works of Ghia et al. [80]. The grid size used is 150×150, with the dimension of the wall is 1×1 where the top wall is moving with velocity $u = 1$. Figure 5a shows the results for u-velocity along the vertical centreline and the Figure 5b v-velocity along the horizontal centreline. The non-Newtonian cases for $n = 0.5$ and $n = 1.5$ show very good agreement when compared with the results by Bell and Surana [60] and Neofytou [61]. For $n = 1$, the result is in very good agreement with the result by Ghia et al. [80].

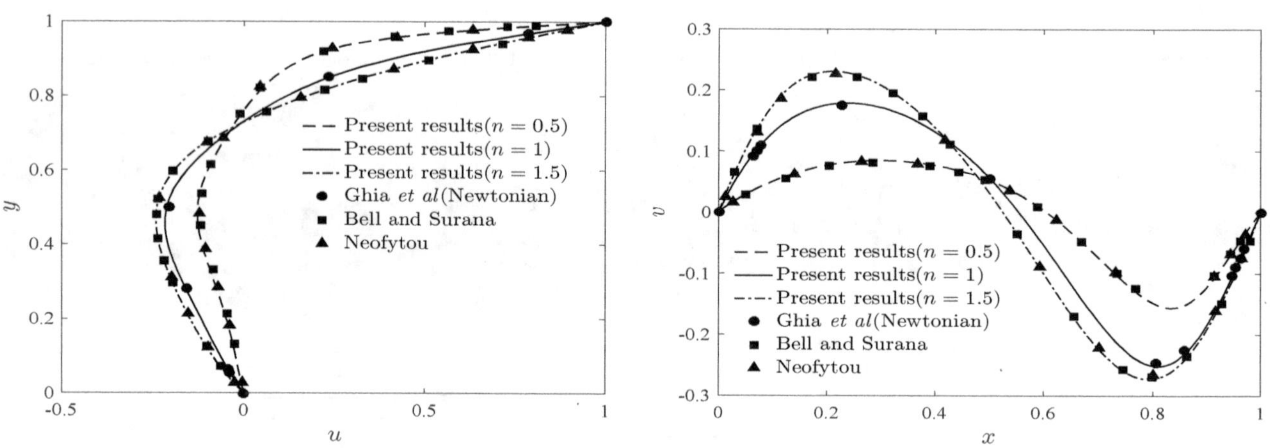

(a) (b)

Figure 5. Comparison between present results and the results by Bell and Surana [60], Neofytou [61] for non-Newtonian power-law at various values of power-law index n, and Newtonian fluid flow by Ghia et al. [80]. (**a**) u-velocity profiles along the vertical centreline of the cavity. (**b**) v-velocity profiles along the horizontal centreline of the cavity for $Re = 100$.

4. Results and Discussion

The time step is chosen to be $\Delta t = 10^{-4}$ for Newtonian and shear thickening cases while for shear thinning cases, the value of $\Delta t = 10^{-5}$ is required for the numerical method to converge. For shear thinning fluid the lowest number of power-law index in this work that converged is $n = 0.6$ while the shear thickening cases were tested until $n = 1.5$, although only the results for $n = 0.6$, $n = 1$ and $n = 1.4$ are shown.

The flow is purely FHD by setting $N = 0$, and will be purely MHD if $Mn_F = 0$. When both Mn_F and N are zero, there is no magnetic source and the flow is purely hydrodynamics hence the flow in the channel has horizontal streamline corresponds to the parabolic profile for u-velocity. When the magnetic source is applied, a vortex arises near the area of the magnetic source. This phenomenon of how the flow is affected by the applied magnetic source with the variation of the power-law index is the main interest of this work. When the flow is tested on pure MHD, the effect of the magnetic field is not significant as the value of Stuart number N that is required to induced the vortex in the area of magnetic source has to be in the order of hundreds and the change is negligible even when N is increased into the order of thousands.

Figure 6 shows the streamlines contours for variations of power-law index $n = 0.6, 1.0$ and $n = 1.4$, with variation of reference magnetic field $\bar{B}_0 = 1$ T, $\bar{B}_0 = 4$ T, and $\bar{B}_0 = 8$ T. The notable effect of applying magnetic field on the flow is the formation of arising vortex near the location of magnetic field. As discussed in previous literature, this vortex extension increases with the increment of magnetic field from 1 T to 8 T. Interesting results are observed when the power-law index is increased from $n = 0.6$ to 1.4, the disturbance in the flow field is also affected. For the case of shear thinning fluid $n = 0.6$ it can be seen that the extension of the vortex is large and extended towards the downstream, compared to the case of when $n = 1.0$ and $n = 1.4$. Additionally when $n = 1.4$, the vortex formation is smaller although its size increases when the Magnetic number is increased. The measured range of sizes of the vortex for each cases summarised in Table 2.

Table 2. Range of vortex for different Magnetic numbers at various power-law index.

n	$\bar{B}_0 = 1$ T	$\bar{B}_0 = 4$ T	$\bar{B}_0 = 8$ T
$n = 0.6$	$2 < x < 5.66$	$2 < x < 7.61$	$2 < x < 8.42$
$n = 1.0$	$2 < x < 4.69$	$2 < x < 5.59$	$2 < x < 5.94$
$n = 1.4$	$2 < x < 4.04$	$2 < x < 4.61$	$2 < x < 4.87$

Figure 7 shows the corresponding temperature contours for variations of power-law with variation of reference magnetic corresponds to the case presented in Figure 6. It is shown that in the area of magnetic source, the maximum temperature is displaced towards the upper plate. The effect is more pronounced when the Magnetic number is increased, regardless the power-law index. The region of the lower vortex is shown to be kept cooler and this region is extended towards the downstream when the Magnetic number is increased for all the cases of different power-law index.

From the previous discussion, it can be seen that the influence of increasing power-law index number n has the opposite effect to the increasing Magnetic numbers on the flow field. Table 3 shows the minimum velocity for different Magnetic numbers at various power-law index. The negative symbol indicates the direction of the velocity and the reverse flow of the vortex region near the magnetic source. From Magnetic number 1 T to 8 T the minimum u-velocity is increased for every n. Overall, the shear thickening flow has the highest velocity followed by Newtonian and shear thinning flow.

Table 3. Minimum u-velocity for different Magnetic numbers at various power-law index.

	$\bar{B}_0 = 1$ T	$\bar{B}_0 = 4$ T	$\bar{B}_0 = 8$ T
$n = 0.6$	$u_{min} = -0.1792$	$u_{min} = -0.2831$	$u_{min} = -0.4106$
$n = 1.0$	$u_{min} = -0.2258$	$u_{min} = -0.3783$	$u_{min} = -0.4848$
$n = 1.4$	$u_{min} = -0.2357$	$u_{min} = -0.4068$	$u_{min} = -0.5093$

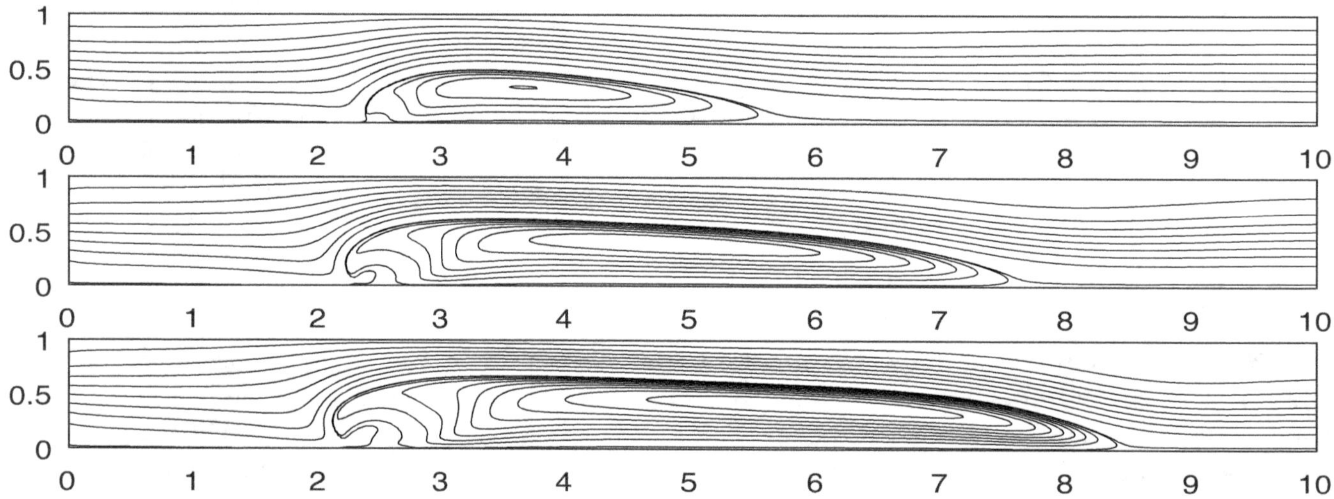

(**a**) Streamline for $\bar{B}_0 = 1$ T (top) $\bar{B}_0 = 4$ T (middle), $\bar{B}_0 = 8$ T (bottom)

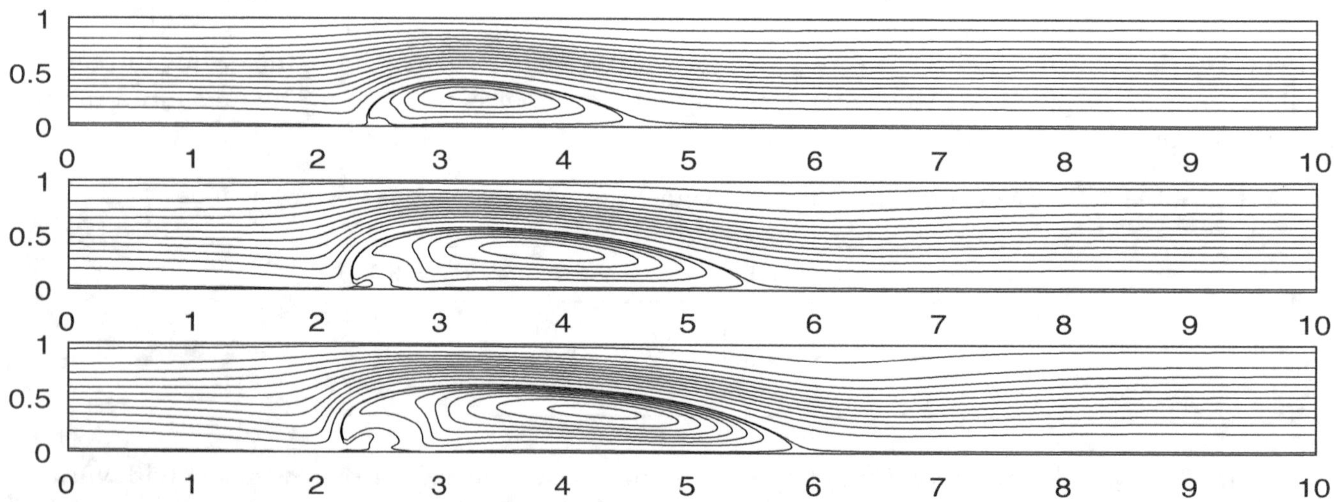

(**b**) Streamline for $\bar{B}_0 = 1$ T (top), $\bar{B}_0 = 4$ T (middle), $\bar{B}_0 = 8$ T (bottom)

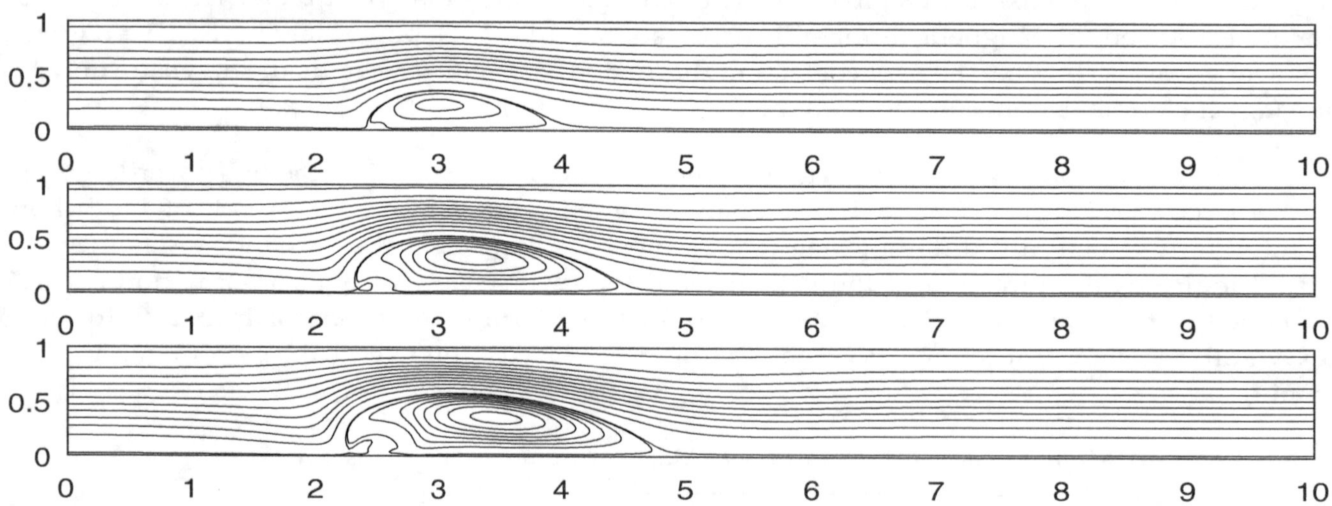

(**c**) Streamline for $B_0 = 1$ T (top), $B_0 = 4$ T (middle), $B_0 = 8$ T (bottom)

Figure 6. Streamline plot for power-law index: (**a**) $n = 0.6$ (shear thinning). (**b**) $n = 1.0$ (Newtonian). (**c**) $n = 1.4$ (shear thickening).

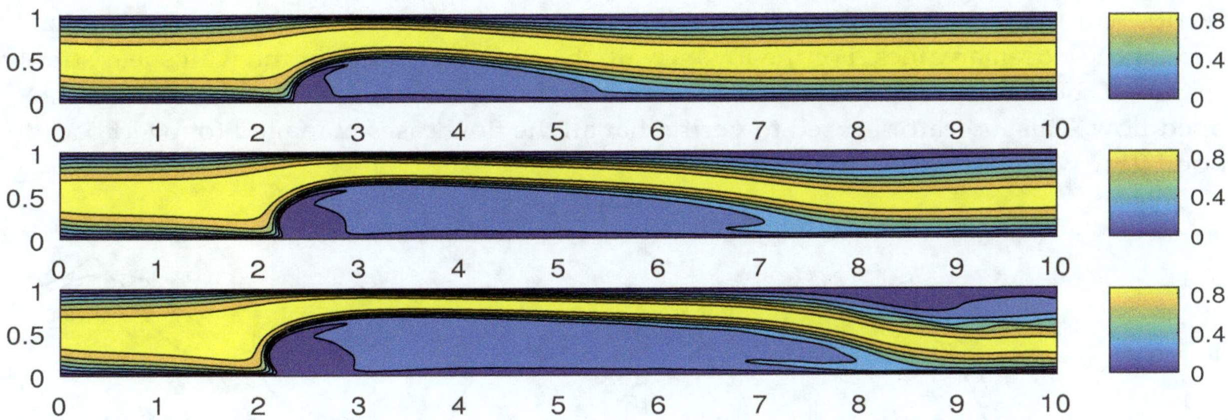

(**a**) Temperature for $\bar{B}_0 = 1$ T (top) $\bar{B}_0 = 4$ T (middle), $\bar{B}_0 = 8$ T (bottom)

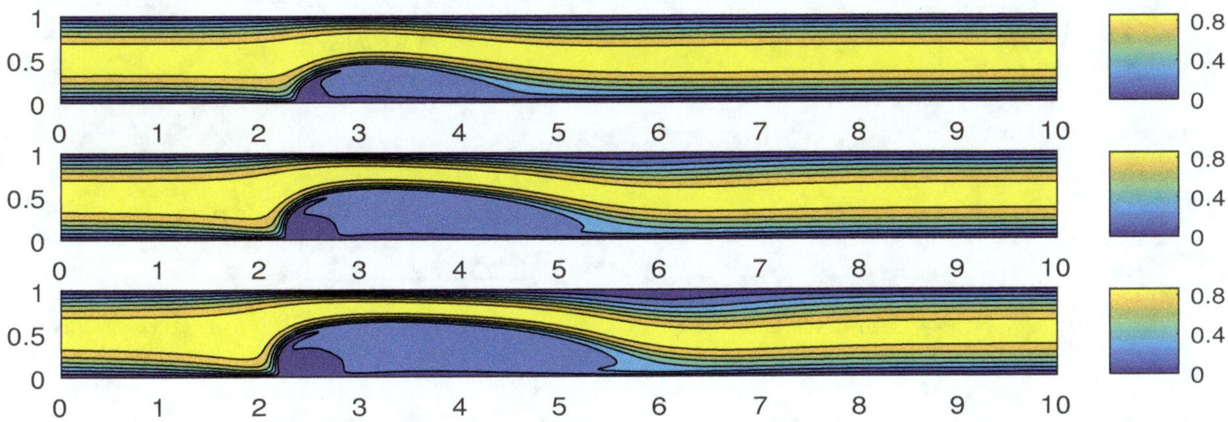

(**b**) Temperature for $\bar{B}_0 = 1$ T (top), $\bar{B}_0 = 4$ T (middle), $\bar{B}_0 = 8$ T (bottom)

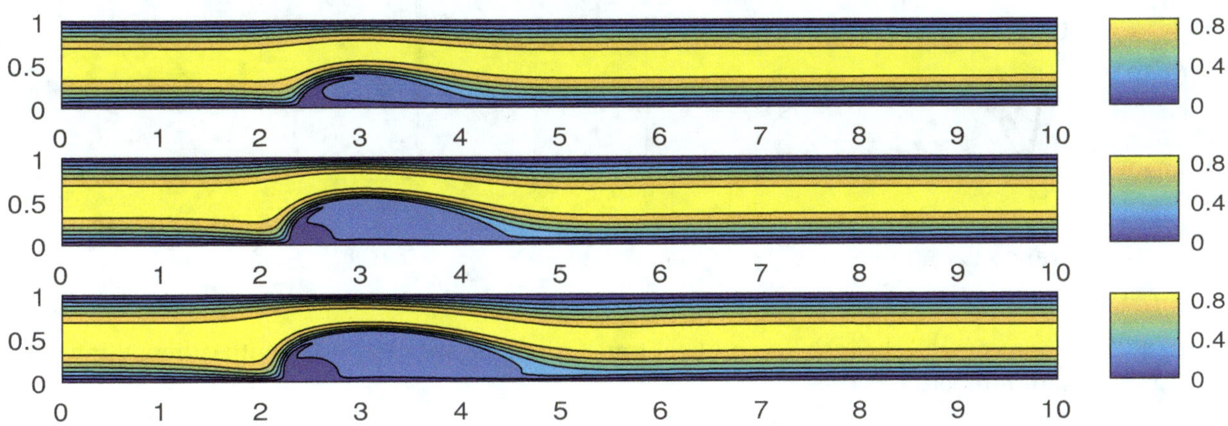

(**c**) Temperature for $\bar{B}_0 = 1$ T (top), $\bar{B}_0 = 4$ T (middle), $\bar{B}_0 = 8$ T (bottom)

Figure 7. Temperature plot for power-law index: (**a**) $n = 0.6$ (shear thinning). (**b**) $n = 1.0$ (Newtonian). (**c**) $n = 1.4$ (shear thickening).

The variation of wall shear stress and heat transfer parameters of various Magnetic numbers for $n = 0.6$, $n = 1.0$ and $n = 1.4$ are shown in Figures 8–10 respectively. The wall shear stress are shown to have more influence on the lower plate, particularly at the location of the magnetic source. The variation of this parameter is qualitatively the same as magnetic field are varied from 1 T to 8 T, where the higher magnetic field results in greater variation of the wall shear stress irrespective the value of n. The variation of wall shear for the upper wall (Figures 8a, 9a and 10a) for all cases changes smoothly from increasing to decreasing near the area of the magnetic source, and its sign does not change. For the upper wall, the lower values of this parameter indicates the higher rates of skin friction. For the case of the lower wall (Figures 8c, 9c and 10c), the rapid changes of this parameter is largest for the case of shear thinning flow, followed by Newtonian and shear thickening flow as shown by the difference of the extrema values. For the lower wall, the negative values of the wall shear stress refers to inversion of the flow, and when wall shear stress reach its original value at $x = 0$, it refers to fully developed flow. This indicator is used to verify that all the flow cases managed to revert back to fully developed flow when it reaches downstream.

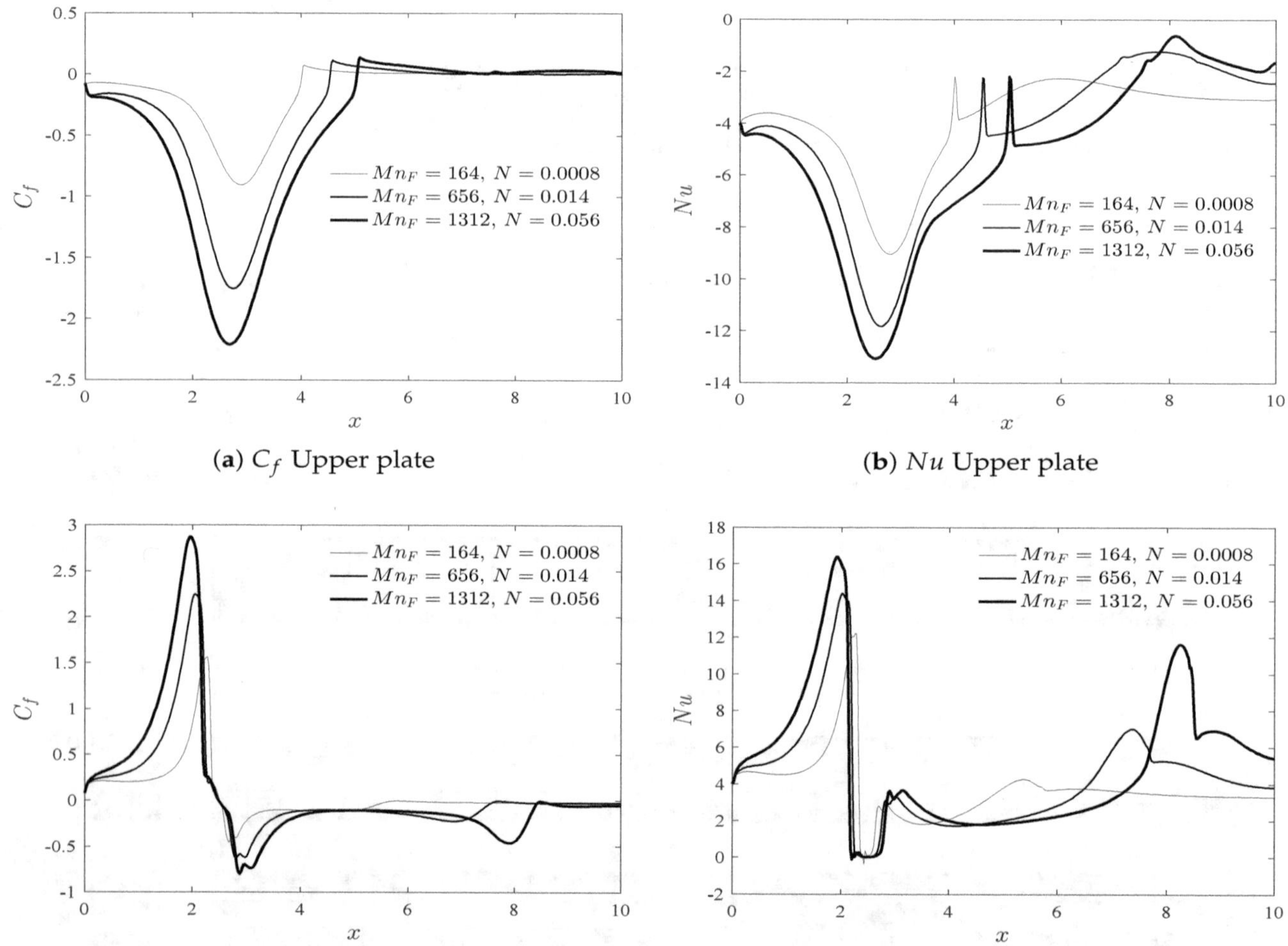

(**a**) C_f Upper plate (**b**) Nu Upper plate

(**c**) C_f Lower plate (**d**) Nu Lower plate

Figure 8. Variations of wall shear stress and heat transfer parameter for $n = 0.6$ with different intensity of magnetic field number.

The heat transfer parameters for the upper wall (Figures 8b, 9b and 10b) and lower wall (Figures 8d, 9d and 10d) similarly shows the same qualitative variation when magnetic fields are varied from 1 T to 8 T, for each case flow. On the lower wall, it is shown that the rate of heat transfer increases when Magnetic number increases, regardless the power-law index. For the case of $n = 0.6$, the overall heat transfer to lower wall is greater than the case of $n = 1.0$ and $n = 1.4$. The heat transfer coefficient is relatively high near the location of magnetic source, and another second region reaching the downstream due to the disturbance of the temperature field is extended far downstream. These second location of high heat transfer coefficient are observed at region $5.5 < x < 8.5$, $4.5 < x < 6$ and $4 < x < 5$ for $n = 0.6$, $n = 1.0$ and $n = 1.4$ respectively. On the upper wall, the lower values of this coefficient refers to higher rates of heat transfer to the wall and it exhibits a similarity with the case of the lower wall.

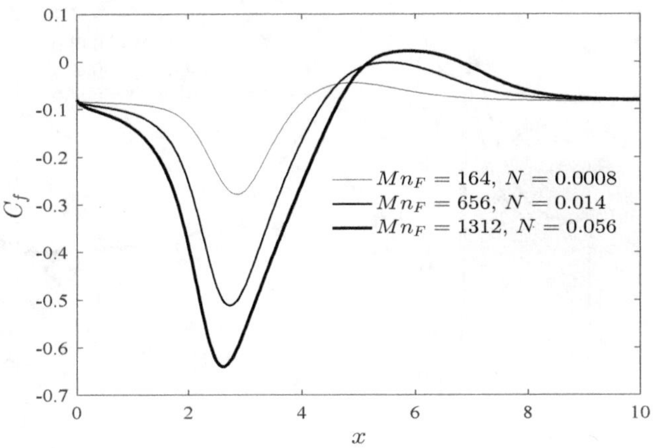

(**a**) C_f Upper plate

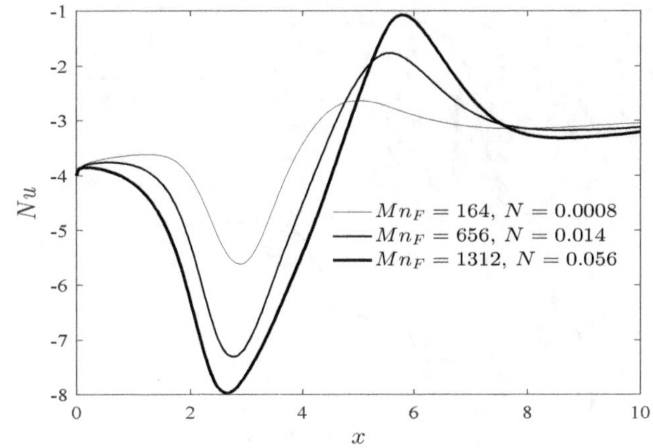

(**b**) Nu Upper plate

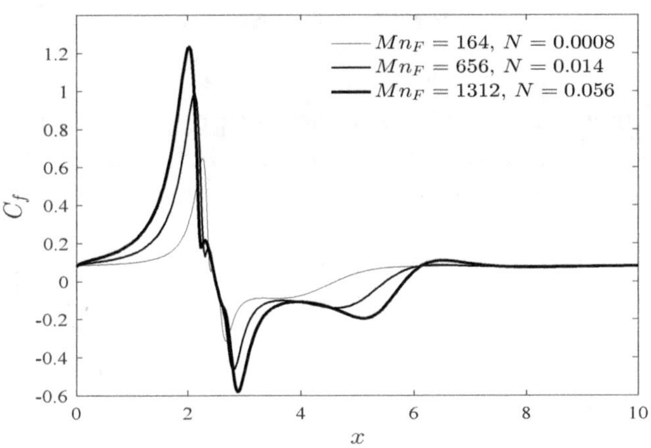

(**c**) C_f Lower plate

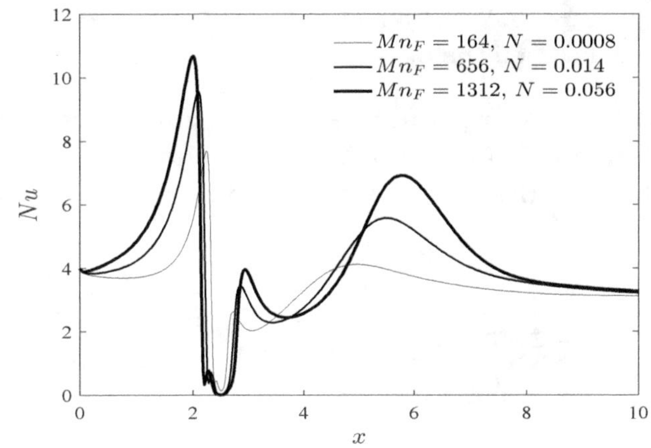

(**d**) Nu Lower plate

Figure 9. Variations of wall shear stress and heat transfer parameter for $n = 1$ with different intensity of magnetic field.

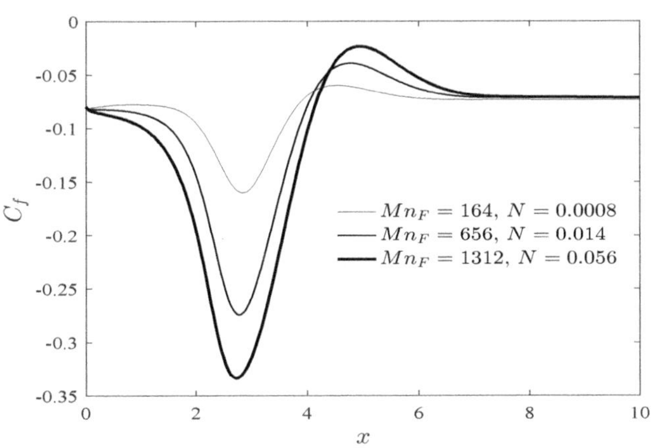

(**a**) C_f Upper plate

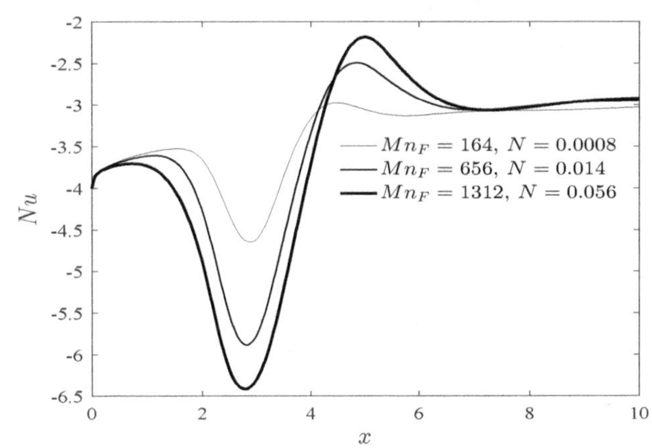

(**b**) Nu Upper plate

Figure 10. *Cont.*

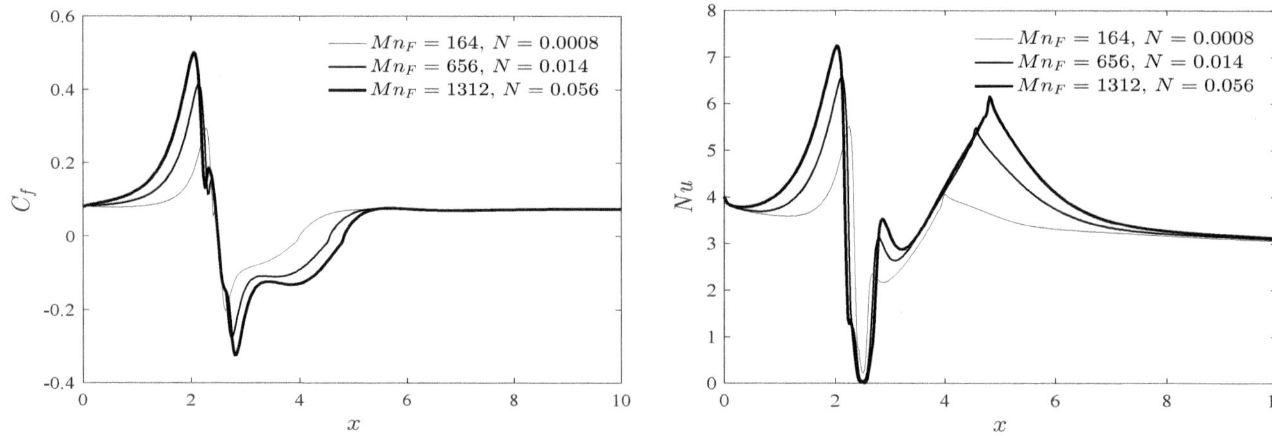

(c) C_f Lower plate (d) Nu Lower plate

Figure 10. Variations of wall shear stress and heat transfer parameter for $n = 1.4$ with different intensity of magnetic field.

Another quantitative comparison can be made from Figure 11, for different power-law index n and magnetic field at 8 T, where it shows that the wall shear stress and heat transfer parameters are highest for the case of shear thinning flow, followed by Newtonian and shear thickening.

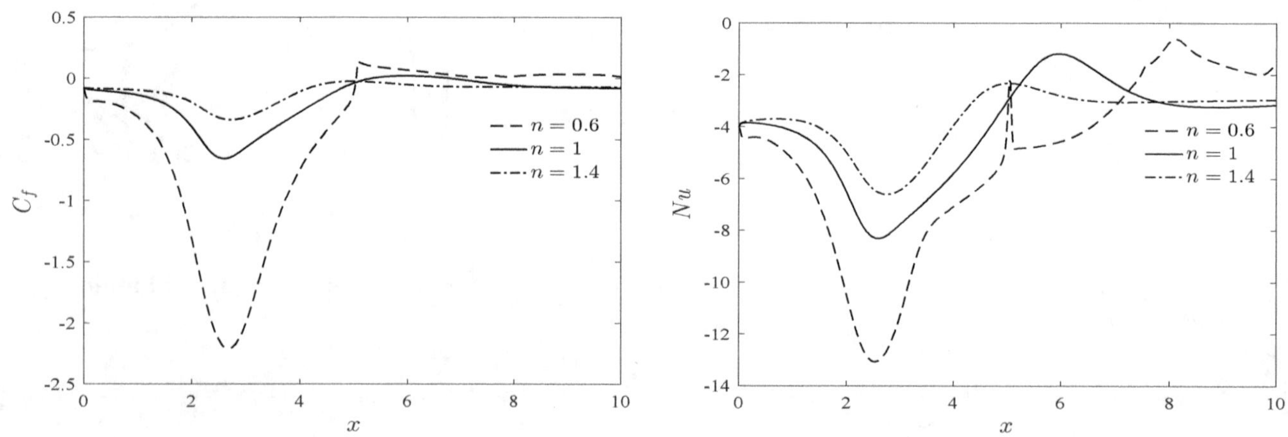

(a) C_f Upper plate (b) Nu Upper plate

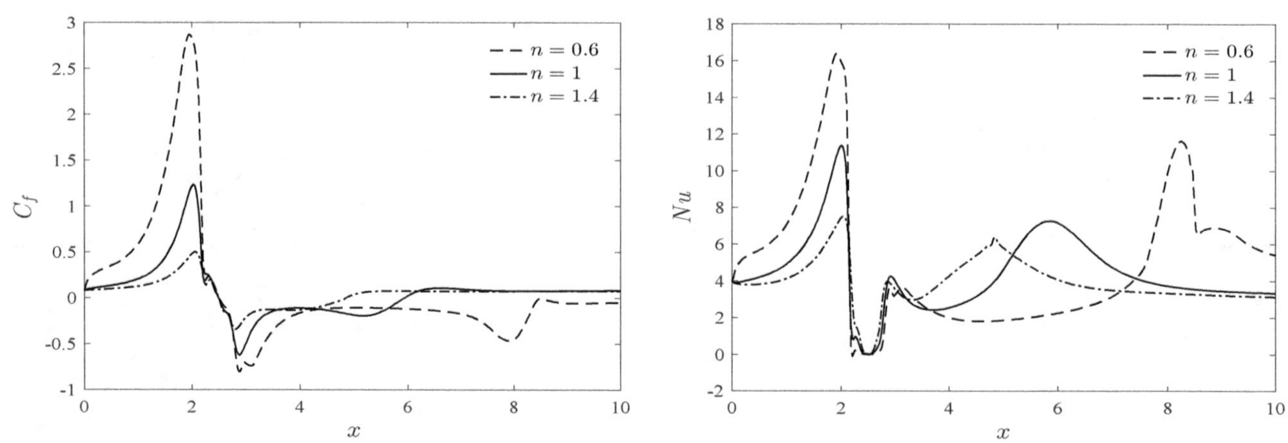

(c) C_f Lower plate (d) Nu Lower plate

Figure 11. Variations of wall shear stress parameter C_f and heat transfer parameter Nu for $Mn_F = 1312$, $N = 0.056$ for different power-law index n.

5. Conclusions

In this paper, the effect of magnetic field on non-Newtonian biomagnetic power-law flow in a channel has been studied numerically by Constrained Interpolated Profile Finite Difference Method (CIP-FDM). This study has been conducted for the parameters $\bar{B}_0 = 1$ T, $\bar{B}_0 = 4$ T and $\bar{B}_0 = 8$ T for power-law index $n = 0.6$ (shear thinning), $n = 1$ (Newtonian) and $n = 1.4$ (shear thickening). The investigation is performed based on the study of the streamline, temperature, vorticity contours, wall shear stress and heat transfer parameters. A good agreement validated with previous numerical investigation demonstrates that the CIP-FDM is an appropriate method for the power-law flow in a channel under an applied magnetic field. The enhancement of Magnetic number increases the length of vortex formation regardless the power-law index. The consideration of power-law index, causes the size of vortex formation to expand in shear thinning flow, and shrink in shear thickening flow. The amplification of Magnetic number increases both the wall shear stress and the heat transfer parameters dominantly near the location of magnetic source. The opposite effect is observed when the power-law is included as the wall shear stress and heat transfer parameters are decreased. The work reported here shows the severe drawbacks associated with RF radiation to blood vessel locally on the displacement of the high temperature and the blockage due to back flow. For future research, it is recommended to perform an analysis on fluid structure interaction by calculating the elastic deformation of the vessel.

Author Contributions: Conceptualization by A.S.H. and N.S.A.; investigation by A.S.H.; methodology by A.S.H.; supervision by N.S.A. and S.S.; validation by A.S.H. and N.S.A.; visualization by A.S.H.; writing—original draft preparation by A.S.H. and N.S.A.; funding acquisition by S.S. All authors have read and agreed to the published version of the manuscript.

Abbreviations

The following abbreviations are used in this manuscript:

FHD	Ferrohydrodynamics
MHD	Magnetohydrodynamics
EMF	electromagnetic field
MRI	Magnetic Resonance Imaging
T	Tesla
RF	radiofrequency field
CIP	Constrained Interpolation Profile
FDM	Finite Difference method
FEM	Finite Element method
FDLBM	Finite Difference Lattice Boltzmann method
DRBEM	Dual Reciprocity Boundary Element method
LBM	Lattice Boltzmann method
CVFEM	Control Volume Finite Element method
SOR	Successive Overrelaxation
CIP-FDM	Constrained Interpolation Profile Finite Difference Method

nomenclatures

$\bar{\mu}_a$	viscosity
m	fluid consistency
I_2	second invariant of the rate of strain tensor
n	Power-law index
$\vec{v}$	velocity vector
ρ	density
p	pressure
$\vec{\tau}$	stress tensor
$\vec{J}$	current density

M	magnetisation
$\bar{B}$	magnetic field
$\bar{\mu}_0$	magnetic permeability
$\bar{H}$	magnetic field intensity
c_p	specific heat
κ	thermal conductivity
$\bar{\sigma}$	electrical conductivity
$\bar{T}_w$	temperature of the wall
$\bar{T}_f$	temperature of the fluid
x, y	Cartesian coordinates
T	temperature
u	x-velocity
v	y-velocity
t	time
(a, b)	coordinate of magnetic source
ψ	stream function
w	vorticity
Re	Reynolds number
Mn_F	Magnetic number
N	Stuart number
Pr	Prandtl number
Ec	Eckert number
ϵ	temperature number
C_f	local skin friction coefficient, wall shear stress
Nu	Nusselt number, heat transfer coefficient

References

1. Jain, V.; Abdulmalik, O.; Propert, K.J.; Wehrli, F.W. Investigating the magnetic susceptibility properties of fresh human blood for noninvasive oxygen saturation quantification. *Magn. Reson. Med.* **2012**, *68*, 863–867. [CrossRef]
2. Tzirtzilakis, E.; Loukopoulos, V. Biofluid flow in a channel under the action of a uniform localized magnetic field. *Comput. Mech.* **2005**, *36*, 360–374. [CrossRef]
3. Haik, Y.; Chen, C.J.; Chatterjee, J. Numerical simulation of biomagnetic fluid in a channel with thrombus. *J. Vis.* **2002**, *5*, 187–195. [CrossRef]
4. Neuringer, J.L.; Rosensweig, R.E. Ferrohydrodynamics. *Phys. Fluids* **1964**, *7*, 1927–1937. [CrossRef]
5. Voltairas, P.; Fotiadis, D.; Michalis, L. Hydrodynamics of magnetic drug targeting. *J. Biomech.* **2002**, *35*, 813–821. [CrossRef]
6. Ruuge, E.; Rusetski, A. Magnetic fluids as drug carriers: Targeted transport of drugs by a magnetic field. *J. Magn. Magn. Mater.* **1993**, *122*, 335–339. [CrossRef]
7. Fiorentini, G.; Szasz, A. Hyperthermia today: Electric energy, a new opportunity in cancer treatment. *J. Cancer Res. Ther.* **2006**, *2*, 41. [CrossRef]
8. Plavins, J.; Lauva, M. Study of colloidal magnetite-binding erythrocytes: Prospects for cell separation. *J. Magn. Magn. Mater.* **1993**, *122*, 349–353. [CrossRef]
9. Raj, K.; Moskowitz, B.; Tsuda, S. New commercial trends of nanostructured ferrofluids. *Indian J. Eng. Mater. Sci.* **2004**, *11*, 241–252.
10. Scherer, C.; Figueiredo Neto, A.M. Ferrofluids: Properties and applications. *Braz. J. Phys.* **2005**, *35*, 718–727. [CrossRef]
11. Odenbach, S. *Colloidal Magnetic Fluids: Basics, Development and Application of Ferrofluids*; Springer: Berlin/Heidelberg, Germany, 2009; Volume 763. [CrossRef]
12. Kıvrak, E.G.; Yurt, K.K.; Kaplan, A.A.; Alkan, I.; Altun, G. Effects of electromagnetic fields exposure on the antioxidant defense system. *J. Microsc. Ultrastruct.* **2017**, *5*, 167–176. [CrossRef]
13. Fragopoulou, A.F.; Koussoulakos, S.L.; Margaritis, L.H. Cranial and postcranial skeletal variations induced in mouse embryos by mobile phone radiation. *Pathophysiology* **2010**, *17*, 169–177. [CrossRef]

14. Van den Brink, J.S. Thermal Effects Associated with RF Exposures in Diagnostic MRI: Overview of Existing and Emerging Concepts of Protection. *Concepts Magn. Reson. Part B* **2019**, *2019*, 9618680. [CrossRef]
15. Zhang, X.; Yarema, K.; Xu, A. *Biological Effects of Static Magnetic Fields*; Springer: Singapore, 2017.
16. Duan, G.; Zhao, X.; Anderson, S.W.; Zhang, X. Boosting magnetic resonance imaging signal-to-noise ratio using magnetic metamaterials. *Commun. Phys.* **2019**, *2*, 1–8. [CrossRef]
17. Nowogrodzki, A. The world's strongest MRI machines are pushing human imaging to new limits. *Nature* **2018**, *563*, 24–27. [CrossRef]
18. Ichioka, S.; Minegishi, M.; Iwasaka, M.; Shibata, M.; Nakatsuka, T.; Harii, K.; Kamiya, A.; Ueno, S. High-intensity static magnetic fields modulate skin microcirculation and temperature in vivo. *Bioelectromagn. J. Bioelectromagn. Soc.* **2000**, *21*, 183–188. [CrossRef]
19. Haik, Y.; Pai, V.; Chen, C. Biomagnetic fluid dynamics. In *Fluid Dyn. Interfaces;* Shyy, W., Narayanan, R., Eds.; Cambridge University Press: Cambridge, UK, 1999; pp. 439–452.
20. Motta, M.; Haik, Y.; Gandhari, A.; Chen, C.J. High magnetic field effects on human deoxygenated hemoglobin light absorption. *Bioelectrochem. Bioenergy* **1998**, *47*, 297–300. [CrossRef]
21. Higashi, T.; Yamagishi, A.; Takeuchi, T.; Kawaguchi, N.; Sagawa, S.; Onishi, S.; Date, M. Orientation of erythrocytes in a strong static magnetic field. *Blood* **1993**, *82*, 1328–1334. [CrossRef]
22. Khashan, S.A.; Haik, Y. Numerical simulation of biomagnetic fluid downstream an eccentric stenotic orifice. *Phys. Fluids* **2006**, *18*, 113601. [CrossRef]
23. Loukopoulos, V.; Tzirtzilakis, E. Biomagnetic channel flow in spatially varying magnetic field. *Int. J. Eng. Sci.* **2004**, *42*, 571–590. [CrossRef]
24. Tzirtzilakis, E. A mathematical model for blood flow in magnetic field. *Phys. Fluids* **2005**, *17*, 077103. [CrossRef]
25. Rosensweig, R.E. *Ferrohydrodynamics*, Reprinted ed.; Dover Publications: Mineola, NY, USA, 2013.
26. Sahoo, B.C.; Vajjha, R.S.; Ganguli, R.; Chukwu, G.A.; Das, D.K. Determination of rheological behavior of aluminum oxide nanofluid and development of new viscosity correlations. *Pet. Sci. Technol.* **2009**, *27*, 1757–1770. [CrossRef]
27. Zhou, S.Q.; Ni, R.; Funfschilling, D. Effects of shear rate and temperature on viscosity of alumina polyalphaolefins nanofluids. *J. Appl. Phys.* **2010**, *107*, 054317. [CrossRef]
28. Odenbach, S.; Thurm, S. Magnetoviscous effects in ferrofluids. In *Ferrofluids*; Springer: Berlin/Heidelberg, Germany, 2002; pp. 185–201._10. [CrossRef]
29. Odenbach, S. *Ferrofluids: Magnetically Controllable Fluids and Their Applications*; Springer: Berlin/Heidelberg, Germany, 2008; Volume 594. [CrossRef]
30. Goharkhah, M.; Ashjaee, M. Effect of an alternating nonuniform magnetic field on ferrofluid flow and heat transfer in a channel. *J. Magn. Magn. Mater.* **2014**, *362*, 80–89. [CrossRef]
31. Tzirtzilakis, E.; Xenos, M.; Loukopoulos, V.; Kafoussias, N. Turbulent biomagnetic fluid flow in a rectangular channel under the action of a localized magnetic field. *Int. J. Eng. Sci.* **2006**, *44*, 1205–1224. [CrossRef]
32. Siddiqa, S.; Naqvi, S.; Begum, N.; Awan, S.; Hossain, M. Thermal radiation therapy of biomagnetic fluid flow in the presence of localized magnetic field. *Int. J. Therm. Sci.* **2018**, *132*, 457–465. [CrossRef]
33. Tzirtzilakis, E. A simple numerical methodology for BFD problems using stream function vorticity formulation. *Commun. Numer. Methods Eng.* **2008**, *24*, 683–700. [CrossRef]
34. Xenos, M.; Tzirtzilakis, E. MHD effects on blood flow in a stenosis. *Adv. Dyn. Syst. Appl.* **2013**, *8*, 427–437.
35. Türk, Ö.; Tezer-Sezgin, M.; Bozkaya, C. Finite element study of biomagnetic fluid flow in a symmetrically stenosed channel. *J. Comput. Appl. Math.* **2014**, *259*, 760–770. [CrossRef]
36. Türk, Ö.; Bozkaya, C.; Tezer-Sezgin, M. A FEM approach to biomagnetic fluid flow in multiple stenosed channels. *Comput. Fluids* **2014**, *97*, 40–51. [CrossRef]
37. Tzirtzilakis, E. Biomagnetic fluid flow in an aneurysm using ferrohydrodynamics principles. *Phys. Fluids* **2015**, *27*, 061902. [CrossRef]
38. Sharifi, A.; Motlagh, S.Y.; Badfar, H. Numerical investigation of magnetic drug targeting using magnetic nanoparticles to the Aneurysmal Vessel. *J. Magn. Magn. Mater.* **2019**, *474*, 236–245. [CrossRef]
39. Tzirakis, K.; Papaharilaou, Y.; Giordano, D.; Ekaterinaris, J. Numerical investigation of biomagnetic fluids in circular ducts. *Int. J. Numer. Methods Biomed. Eng.* **2014**, *30*, 297–317. [CrossRef] [PubMed]
40. Tzirtzilakis, E.; Xenos, M. Biomagnetic fluid flow in a driven cavity. *Meccanica* **2013**, *48*, 187–200. [CrossRef]
41. Marioni, L.; Bay, F.; Hachem, E. Numerical stability analysis and flow simulation of lid-driven cavity subjected to high magnetic field. *Phys. Fluids* **2016** *28*, 057102. [CrossRef]

42. Bozkaya, N.; Tezer-Sezgin, M. The DRBEM solution of incompressible MHD flow equations. *Int. J. Numer. Methods Fluids* **2011**, *67*, 1264–1282. [CrossRef]
43. Senel, P.; Tezer-Sezgin, M. DRBEM solutions of Stokes and Navier–Stokes equations in cavities under point source magnetic field. *Eng. Anal. Bound. Elem.* **2016**, *64*, 158–175. [CrossRef]
44. Kefayati, G.R. Simulation of magnetic field effect on non-Newtonian blood flow between two-square concentric duct annuli using FDLBM. *J. Taiwan Inst. Chem. Eng.* **2014**, *45*, 1184–1196. [CrossRef]
45. Kefayati, G.R. Simulation of heat transfer and entropy generation of MHD natural convection of non-Newtonian nanofluid in an enclosure. *Int. J. Heat Mass Transf.* **2016**, *92*, 1066–1089. [CrossRef]
46. Adıgüzel, Ö.B.; Atalık, K. Magnetic field effects on Newtonian and non-Newtonian ferrofluid flow past a circular cylinder. *Appl. Math. Model.* **2017**, *42*, 161–174. [CrossRef]
47. Jahanbakhshi, A.; Nadooshan, A.A.; Bayareh, M. Magnetic field effects on natural convection flow of a non-Newtonian fluid in an L-shaped enclosure. *J. Therm. Anal. Calorim.* **2018**, *133*, 1407–1416. [CrossRef]
48. Ikbal, M.A.; Chakravarty, S.; Wong, K.K.; Mazumdar, J.; Mandal, P.K. Unsteady response of non-Newtonian blood flow through a stenosed artery in magnetic field. *J. Comput. Appl. Math.* **2009**, *230*, 243–259. [CrossRef]
49. Sankar, D.; Lee, U. FDM analysis for MHD flow of a non-Newtonian fluid for blood flow in stenosed arteries. *J. Mech. Sci. Technol.* **2011**, *25*, 2573. [CrossRef]
50. Haynes, R.H.; Burton, A.C. Role of the non-Newtonian behavior of blood in hemodynamics. *Am. J. Physiol.-Leg. Content* **1959**, *197*, 943–950. [CrossRef] [PubMed]
51. Chien, S.; Usami, S.; Taylor, H.M.; Lundberg, J.L.; Gregersen, M.I. Effects of hematocrit and plasma proteins on human blood rheology at low shear rates. *J. Appl. Physiol.* **1966**, *21*, 81–87. [CrossRef]
52. Shukla, J.; Parihar, R.; Rao, B. Effects of stenosis on non-Newtonian flow of the blood in an artery. *Bull. Math. Biol.* **1980**, *42*, 283–294. [CrossRef]
53. Pedley, T.J. *The Fluid Mechanics of Large Blood Vessels*; Cambridge University Press: Cambridge, UK, 1980, doi:10.1017/CBO9780511896996. [CrossRef]
54. Berger, S.; Jou, L.D. Flows in stenotic vessels. *Annu. Rev. Fluid Mech.* **2000**, *32*, 347–382. [CrossRef]
55. Easthope, P.; Brooks, D. A comparison of rheological constitutive functions for whole human blood. *Biorheology* **1980**, *17*, 235–247.
56. Walburn, F.J.; Schneck, D.J. A constitutive equation for whole human blood. *Biorheology* **1976**, *13*, 201–210. [CrossRef]
57. Cho, Y.I.; Kensey, K.R. Effects of the non-Newtonian viscosity of blood on flows in a diseased arterial vessel. Part 1: Steady flows. *Biorheology* **1991**, *28*, 241–262. [CrossRef]
58. Quemada, D. A non-linear Maxwell model of biofluids: Application to normal blood. *Biorheology* **1993**, *30*, 253–265. [CrossRef] [PubMed]
59. Boyd, J.; Buick, J.M.; Green, S. Analysis of the Casson and Carreau-Yasuda non-Newtonian blood models in steady and oscillatory flows using the lattice-Boltzmann method. *Phys. Fluids* **2007**, *19*, 093103. [CrossRef]
60. Bell, B.C.; Surana, K.S. P-version least squares finite element formulation for two-dimensional, incompressible, non-Newtonian isothermal and non-isothermal fluid flow. *Int. J. Numer. Methods Fluids* **1994**, *18*, 127–162. [CrossRef]
61. Neofytou, P. A 3rd order upwind finite volume method for generalised Newtonian fluid flows. *Adv. Eng. Softw.* **2005**, *36*, 664–680. [CrossRef]
62. Misra, J.; Shit, G.; Rath, H.J. Flow and heat transfer of a MHD viscoelastic fluid in a channel with stretching walls: Some applications to haemodynamics. *Comput. Fluids* **2008**, *37*, 1–11. [CrossRef]
63. Liu, I.C. Flow and heat transfer of an electrically conducting fluid of second grade in a porous medium over a stretching sheet subject to a transverse magnetic field. *Int. J. Non-Linear Mech.* **2005**, *40*, 465–474. [CrossRef]
64. Rosti, M.E.; Picano, F.; Brandt, L. Numerical approaches to complex fluids. In *Flowing Matter*; Springer: Cham, Switzerland, 2019; pp. 1–34._1. [CrossRef]
65. Tzirtzilakis, E. Biomagnetic fluid flow in a channel with stenosis. *Phys. D Nonlinear Phenom.* **2008**, *237*, 66–81. [CrossRef]
66. Sheikholeslami, M.; Bandpy, M.G.; Ellahi, R.; Zeeshan, A. Simulation of MHD CuO–water nanofluid flow and convective heat transfer considering Lorentz forces. *J. Magn. Magn. Mater.* **2014**, *369*, 69–80. [CrossRef]
67. Sheikholeslami, M.; Ellahi, R. Three dimensional mesoscopic simulation of magnetic field effect on natural convection of nanofluid. *Int. J. Heat Mass Transf.* **2015**, *89*, 799–808. [CrossRef]

68. Sheikholeslami, M.; Ganji, D.D. Ferrohydrodynamic and magnetohydrodynamic effects on ferrofluid flow and convective heat transfer. *Energy* **2014**, *75*, 400–410. [CrossRef]
69. Sheikholeslami, M.; Rashidi, M.; Ganji, D. Effect of non-uniform magnetic field on forced convection heat transfer of Fe3O4–water nanofluid. *Comput. Methods Appl. Mech. Eng.* **2015**, *294*, 299–312. [CrossRef]
70. Sheikholeslami, M.; Vajravelu, K.; Rashidi, M.M. Forced convection heat transfer in a semi annulus under the influence of a variable magnetic field. *Int. J. Heat Mass Transf.* **2016**, *92*, 339–348. [CrossRef]
71. Rashidi, S.; Esfahani, J.A.; Maskaniyan, M. Applications of magnetohydrodynamics in biological systems-a review on the numerical studies. *J. Magn. Magn. Mater.* **2017**, *439*, 358–372. [CrossRef]
72. Yabe, T.; Takizawa, K.; Chino, M.; Imai, M.; Chu, C. Challenge of CIP as a universal solver for solid, liquid and gas. *Int. J. Numer. Methods Fluids* **2005**, *47*, 655–676. [CrossRef]
73. Takewaki, H.; Nishiguchi, A.; Yabe, T. Cubic interpolated pseudo-particle method (CIP) for solving hyperbolic-type equations. *J. Comput. Phys.* **1985**, *61*, 261–268. [CrossRef]
74. Yabe, T.; Aoki, T. A universal solver for hyperbolic equations by cubic-polynomial interpolation I. One-dimensional solver. *Comput. Phys. Commun.* **1991**, *66*, 219–232. [CrossRef]
75. Shi, Y.F.; Xu, B.; Guo, Y. Numerical solution of Korteweg-de Vries-Burgers equation by the compact-type CIP method. *Adv. Differ. Equ.* **2015**, *2015*, 353. [CrossRef]
76. Yabe, T.; Wang, P.Y. Unified numerical procedure for compressible and incompressible fluid. *J. Phys. Soc. Jpn.* **1991**, *60*, 2105–2108. [CrossRef]
77. Kishev, Z.R.; Hu, C.; Kashiwagi, M. Numerical simulation of violent sloshing by a CIP-based method. *J. Mar. Sci. Technol.* **2006**, *11*, 111–122. [CrossRef]
78. Mussa, M.; Abdullah, S.; Azwadi, C.N.; Muhamad, N. Simulation of natural convection heat transfer in an enclosure by the lattice-Boltzmann method. *Comput. Fluids* **2011**, *44*, 162–168. [CrossRef]
79. Yabe, T.; Aoki, T. A universal solver for hyperbolic-equation by cubic-polynomial interpolation. II. Two-and three dimensional solvers. *Comput. Phys. Commun.* **1991**, *66*, 233–242. [CrossRef]
80. Ghia, U.; Ghia, K.N.; Shin, C. High-Re solutions for incompressible flow using the Navier-Stokes equations and a multigrid method. *J. Comput. Phys.* **1982**, *48*, 387–411. [CrossRef]

2

Fluid Mixing Nonequilibrium Processes in Industrial Piping Flows

Mikhail Sukharev

Department of Applied Mathematics and Computer Modeling, National University of Oil and Gas «Gubkin University», 119991 Moscow, Russia; mgsukharev@mail.ru

Abstract: The flow of a multicomponent fluid through a pipeline system of arbitrary configuration is considered. The problem consists in determining the component composition of the fluid for each pipeline of the system based on the values of the concentration of the components throughout the entire set of measuring points, provided that there are no phase transitions. To solve the problem, mathematical models have been developed that, in principle, are suitable for pipeline systems of various functional purposes, the presentation is concretized and carried out in relation to gas transmission systems. The models are stochastic in nature due to measurement errors, which are considered random variables. The solution of the problem is reduced to the optimization of a quadratic function with constraints in the form of equalities and inequalities. The considered mixing processes do not depend on the regime parameters of the fluid flow. The processes are irreversible and non-equilibrium. A criterion is introduced that characterizes the degree of closeness of a multicomponent mixture to an equilibrium state. The criterion is analogous to entropy in thermodynamic processes. A numerical example of calculating the distribution of a three-component mixture is given. The example illustrates the feasibility of the proposed computational procedures and gives an idea of the distribution of the component composition and the change in «entropy» along the directions of pumping of the gas supply system.

Keywords: multi-component flows; gas transmission systems; non-equilibrium processes; mathematical models; maximum likelihood method; calorific value; entropy

---

1. Introduction

Gas enters the Unified Gas Supply System (UGSS) of the Russian Federation from different sources and differs in its composition, although methane remains the main component. The share from Russian fields ranges from 90% to 98%. By way of example, let us consider composition analysis of gas transported via a gas line in the North of Siberia (gas composition in molar fractions, %: methane 93.2; ethane 4.1; propane 1.30; i-butane 0.19; n-butane 0.18; neo-pentane 0.0013; i-pentane 0.030; n-pentane 0.021; hexane 0.0084; carbon dioxide 0.32; nitrogen 0.59; oxygen—less than 0.005; helium—0.0096; hydrogen 0.0018; water vapour—0).

In addition to gas and gas condensate fields, UGSS supply sources include oil fields (associated gas), gas processing plants, underground storage facilities, and import supplies. From plants, oil fields, underground storages in depleted oil fields there arrives rich gas with a greater share of heavy hydrocarbons and, therefore, higher calorific value. In the process of transportation, the concentrations change with the mixing of fluids at the junction points of the pipelines. Composition and calorific value of gas transported via the UGSS are measured by samples taken at gas metering stations (GMS). These indicators can be measured periodically using gas cylinders or continuously if the necessary equipment is available at the measuring installation [1–3]. Along the length of each pipeline, the concentrations/calorific values do not change, so it is natural to associate them with a given pipeline regardless of where the measuring point is. The problem is to estimate the concentration values for each pipeline based on the set of measurements. Any measurements are fraught with random errors and, therefore, the measured concentrations must be random values, so the concentration estimation models must be stochastic.

In the process of operational control, such parameters of the mode as pressure, flow rate, temperature are measured, and their changes and distribution in the directions of pumping are monitored. However, the calorific value turns out to be a not less important parameter, and tracking its distribution throughout the system should also be included in the number of dispatching tasks. Furthermore, in pipeline systems, emergency situations occur, accompanied by the injection of substances that reduce the quality of the transported product. Tracking the spread of the "pollutant" is also one of the problems of monitoring the component composition.

The paper considers a large-scale gas transmission system (GTS) and sets the task of finding for it the most reasonable ways to estimate the component composition (CC) for each direction of pumping. The component composition uniquely determines the calorific value of the gas. Caloric content can be calculated either knowing the CC, or setting this problem as the main one and solving it using the same methods as the CC calculation.

Here is an incomplete list of problems for the solution of which it is necessary to know the distribution of the CC of the fluid through the directions of pumping.

(a) Settlements between suppliers and consumers are based on energy measurements and with regard to supplied gas calorific value. Differences in consumer requirements to gas calorific value make suppliers interested in calorific value management, which can be achieved by optimizing gas system flows.

(b) Necessity of the UGSS metrological support enhancement. With the gradual depletion of reserves in the Nadym-Purtaz region, deeper deposits will be developed, which will increase methane homologues quantity in the total amount of hydrocarbons produced. The problem of rational distribution of rich gas will become more important. It will be necessary to increase the accuracy of determining CC/caloric value of gas supplied to consumers. The economic significance of the models considered in the paper will grow. Such models will be required to compare UGSS metrological equipment development and modernization options.

(c) The problem of placement of gas chemical complexes. Hydrocarbons of the methane homologous series are a more valuable raw material for gas chemistry than methane. It is economically feasible to separate them from natural gas for use in gas chemistry. The choice of gas chemical facilities location presents a serious problem and amounts to two options: whether to place the complexes in proximity to gas fields or to industrial centers. Each of the options has its pros and cons. An analysis of the gas component distribution within the UGSS is indispensable for making the right decision. To make the right decision, it is necessary to analyze the component composition of gas through the UGSS pipelines.

(d) Calculation of dew point temperature for water and hydrocarbons. Failure to comply with regulatory requirements for their parameters is fraught with an increase in the risk of hydrate and condensate formation, and for export deliveries it entails penalties. In our opinion, the ideas of evaluating the component composition will also be suitable for solving these problems, but under complicating circumstances: the need to consider the phase space with the inclusion of pressure and temperature.

(e) Provide some modification to account for technological specifics, and herein the proposed methods are also applicable to other pipeline systems: oil trunk transportation (compounding oils from different fields, monitoring the content of hydrogen sulfide), water supply (compliance with water quality standards), etc.

Thus, studying the CC distributions of natural gases is very important in practice. The significance of the problem was also noted by Tevyashev et al. [4]. A formalized model and methods for its solution for two-component mixtures were first proposed by Sukharev, Kislenko et al. [2,3]. As shown above, the tasks of determining the CC arise both in operational control and in planning the development of the UGSS. Papers [5,6] should be considered the first steps in the study of the problems. In this article, the following steps have been taken in the same direction. First, not two-component, but three-component mixtures are considered. This led not only to an increase in the number of sought variables. The non-equilibrium and irreversibility of mixing processes lead to the introduction into the model of a large number of constraints in the form of inequalities. Some of them turn out to be "needless", that is, they are automatically executed even in solutions when they are not entered into the model. In problems for two-component mixtures, the significance of non-equilibrium constraints was not manifested. In the flows of three-component mixtures (a) the number of unknowns increases (compared to two-component ones), (b) one has to deal with situations where unknowns differ by orders of magnitude, (c) among the non-equilibrium constraints, significant constraints appear as a rule.

To overcome the obstacles that arise, we propose various methods. They are tested in computational experiments. In particular: (a) the nested solutions method—a subject-oriented iterative procedure that allows one to gradually introduce significant non-equilibrium constraints into the model, thereby reducing the dimension of mathematical programming problems at the stages of an iterative procedure; (b) a new approach to solving problems by non-standard introduction of unknowns (differing by orders of magnitude), eliminating (or at least smoothing out) the problem of the appearance of unknowns differing by orders; (c) instead of concentrations, which often turn out to be small quantities of different orders, the unknowns are the jet intensities—quantities that usually differ from each other by no more than an order of magnitude; (d) a criterion for the information content of metrological support, which makes it possible to judge the possibilities of assessing the CC and/or the caloric content of the fluid, depending on the number of measurement points and their location in the pipeline system; (e) quantitative characteristic of the degree of non-equilibrium of the mixing process (this is an analogue of entropy in classical thermodynamics, however, unlike the latter, it is not defined on a continuum, but on a discrete set–cuts of the pipeline system graph).

When determining the component composition/calorific value, it is assumed that the total flow rate of the fluid for each direction of pumping is known, that is, the flow distribution of the fluid—a mixture of gases–is calculated in advance and serves as initial information for determining the flow rates of each component. The inclusion of the flow distribution of the fluid in the initial information essentially means the decomposition of the problem. At the highest level of the hierarchy is the calculation of fluid flow rates by pumping directions. This calculation is carried out by standard methods with a detailed account of technology (structure, operating parameters: pressure and temperature, technical condition of equipment, etc.). An exact knowledge of the distribution of the component composition is not required in this case. The study of the component composition is the next level of the hierarchy, the subject of this work. Decomposition is also possible because the processes associated with phase transitions are not considered. Phase transitions during the operation of gas supply systems are manifested in the form of condensate–water or hydrocarbons–and hydrate formation. In our opinion, the construction of models for the distribution of CC/calorific value taking into account phase transitions is a matter for the near future. However, decomposition as implemented below will not be valid. When assessing the distribution of the CC, it is necessary to take into account the parameters of the gas flow, at least the temperature, as well as the models of condensate precipitation [7–9]. It is very likely that the results obtained in the field of thermodynamics of mixtures [10] can be useful here. Models of distribution of CC/calorific value, along with the developed apparatus of multimodal techniques [11,12], should, in our opinion, be used in the development of computer technologies for monitoring and control of large pipeline systems at the highest level of the hierarchy.

The rest of the manuscript is organized as follows. Section 2 describes the research methods used in the work. In particular, the specific features of irreversible and non-equilibrium processes of gas components mixing during their transportation are discussed (Section 2.1). The mathematical formulization of the problem is presented in two versions: in the first, the unknown quantities are the flow rates of the components (Section 2.2); in the second, the intensity of the "jets" (Section 2.5). The conditions for the non-equilibrium of the mixing processes are derived (Section 2.3). The computational aspects of the procedures for estimating the CC of the fluid are discussed (Sections 2.4 and 2.5). Section 3 (results) provides an example of calculating a system whose graph contains 11 nodes and 13 arcs (transporting directions). An indicator of non-equilibrium of mixing processes is introduced and illustrated by an example. Section 4 contains a discussion of the results.

2. Methods

2.1. Non-Equilibrium Mixing Processes (Technological Aspect)

Successful solution of this task depends largely on how well the computational model is constructed. When calculating large gas supply systems, they often resort to network reduction and replace several parallel gas pipelines with a single arc of the calculation scheme. For this reason, when describing the model, the term transportation direction is more justified, than a gas pipeline, since an element of the calculation scheme can correspond to a pipeline system, and not to a single pipeline. When solving the task of gas mixture flow distribution through the system, the network reduction procedure should be carried out with caution. Questions arise when fluids with different concentrations are mixed at a junction point of the calculation scheme. Can we assume that the concentrations of components in all pipelines diverging from this junction point are equal? Concentration equality would take place if mixing processes were equilibrium ones. In fact, generally speaking, they do not possess the quality of equilibrium. The gas flow rates through pipelines of GTS are quite large and, as observations on large-scale systems show, complete mixing is observed only with a small load in the pipelines. The concentration values along the output lines depend on the local configuration of pipelines at the junctions. It is not possible to take into account such technological details in the aggregated scheme, since the geometric dimensions of the pipeline junction are incomparably smaller than the lengths of the pipelines. (Note that for this reason, attempts to build models of the CC distribution, taking into account the regime parameters of the flow, primarily pressure, are doomed to failure. The only acceptable way is the way we have chosen to take into account all simultaneous measurements of CC). So, to calculate the concentrations, the design scheme should be disaggregated so that the model makes it possible to reflect the difference in concentrations in the arcs emerging from the connection node. Suppose, for example, that two pipelines or two systems of parallel pipelines (pipeline corridors) converge at the junction, and the output is a multi-pipe corridor. Is it possible to represent this corridor as one arc in the aggregated scheme? If gases of different composition enter through the input lines, and it is known from observations of real regimes that the concentrations on the output lines differ, then this is impossible.

The mixing of components occurs for two reasons: due to diffusion and flow turbulence. The latter reason is decisive at the flow rates characteristic of GTS operation. Observations of the GTS regimes show that the processes of mixing natural gas components are isothermal. These processes are irreversible and non-equilibrium. Complete mixing is usually not achieved, which should be considered an experimentally established fact.

Irreversible non-equilibrium processes are studied by thermodynamics, physical chemistry, kinetic theory of gases. Irreversible non-equilibrium processes often take place during the production, transportation and processing of natural gas [13,14]. The range of these processes is expanding due to the emergence of new technologies. In this regard, we indicate several recent publications in the field of pipeline gas transportation. In [15] the problems of transporting a mixture of natural gas with hydrogen are considered, and in [16], those of transporting natural gas mixed with nitrogen.

Gas flow mixing schemes are traditionally considered in theoretical and technical thermodynamics [17–19]. However, usually the main attention is paid to the study of pressure and temperature of mixing flows. Evaluation of working capacity additional loss caused by irreversible heat transfer between the mixing gases or by failure to use difference in flow pressures turns out to be especially important.

The kinetic theory of gases and, in many respects, physical chemistry work not with continual models, but with discrete models operating with the movement of molecules and the atomic structure of matter. They use the apparatus of quantum chemistry, statistical mechanics, and analytical dynamics [20].

In thermodynamics and in scientific disciplines that have arisen on its basis, a powerful apparatus has been developed for describing irreversible and non-equilibrium processes. The physical processes of fluid mixing in industrial pipeline systems considered in this paper are also irreversible and non-equilibrium, but their study does not require the use of this apparatus. First of all, this because the mixing processes are not accompanied by a change in the parameters of the regime: temperature and pressure—and phase transitions. But, nevertheless, to characterize the mixing processes we have proposed an indicator—an analogue of thermodynamic entropy (see Section 4).

2.2. Task Formalization

We will represent the structure of the pipeline system in the form of directed graph $G = (V, E)$, where V is a set of nodes, and E is a set of arcs. The arc in the aggregated scheme corresponds to transportation direction. By m we denote the number of nodes, and by n the number of arcs. (We use the terminology of the classical monographs of Christofides and Berge on graph theory [21,22]). The nodes are divided into 3 groups: inflows V_{in}, outflows V_{out}, junctions V_{joint} (in Figure 1 $V_{in} = \{1, 2, 3, 4\}$, $V_{out} = \{9, 11\}$, $V_{joint} = \{5, 6, 7, 8, 10\}$). We assume that the inflows and outflows are connected to the graph by a single arc, outgoing for inflows (in Figure 1 $1 \equiv (1,2)$, $2 \equiv (2,6)$, $3 \equiv (3,5)$, $4 \equiv (4,5)$) and incoming for outflows (in Figure 1 $11 \equiv (8,9)$, $13 \equiv (10,11)$). A vector $\boldsymbol{\xi}$ is set on the graph G. It represents mass flow rates of fluid consisting of several miscible components (in Figure 1 $\xi = \|170, 90, \ldots, 65, 135\|^{\mathbf{T}}$. This means that the flow distribution is calculated, as is customary in the practice of GTS operation, and the results of the calculation serve as the initial data for determining the CC for each direction of pumping. Naturally, the calculation is carried out taking into account the technical state of the system and the relationship between the regime parameters (pressure and temperature) of the gas flow. Since flow $\boldsymbol{\xi}$ is initially known, all arcs $(i, j) \in E$ can be orientated along the flow, which means $\xi_{ij} \geq 0$. To simplify the notation, we assume that the number of components is 3; all formulas can obviously be rewritten for the case of an arbitrary number of components. $M = I, II, III$ is component number, ξ_{ij} is flow rate of fluid, and η_{ij}^M is component M flow rate along the arc (i, j).

By definition, $\xi_{ij} = \sum\limits_{M=I}^{III} \eta_{ij}$. Values ξ_{ij} satisfy material balance equations in all joint nodes x_k

$$\sum\nolimits_{x_i \in \Gamma^{-1}(x_k)} \xi_{ik} - \sum\nolimits_{x_j \in \Gamma(x_k)} \xi_{kj} = 0, \ x_k \in V_{joint}. \tag{1}$$

Hereinafter, $\Gamma(x_k)$ is a set of nodes into which arcs come from x_k, $\Gamma^{-1}(x_k)$ is a set of nodes from which arcs come into x_k (Figure 2). Flow rates of each component of the mixture is also a vector $\boldsymbol{\eta}^M = \|\eta_{ij}^M\|$, $(i, j) \in E$, $M = I, II, III$, components of that vector must satisfy of material balance conditions analogous to Equation (1)

$$\sum_{x_j \in \Gamma(x_k)} \eta_{kj}^M - \sum_{x_i \in \Gamma^{-1}(x_k)} \eta_{ik}^M = 0, \ x_k \in V_{joint}, \ M = I, II, III. \tag{2}$$

The challenge is to find vectors $\boldsymbol{\eta}^M$, $M = I, II, III$. Instead of values η_{ij}^M, mass concentrations r_{ij}^M can be considered using the relation (in practice, mass, molar, and volumetric concentrations are used, without difficulty they can be mutually converted. In this paper, mass concentrations are used, since for them material balance equations are most naturally written).

$$\eta_{ij}^{M} = r_{ij}^{M}\xi_{ij},\ (i,j) \in E,\ M = I, II, III. \tag{3}$$

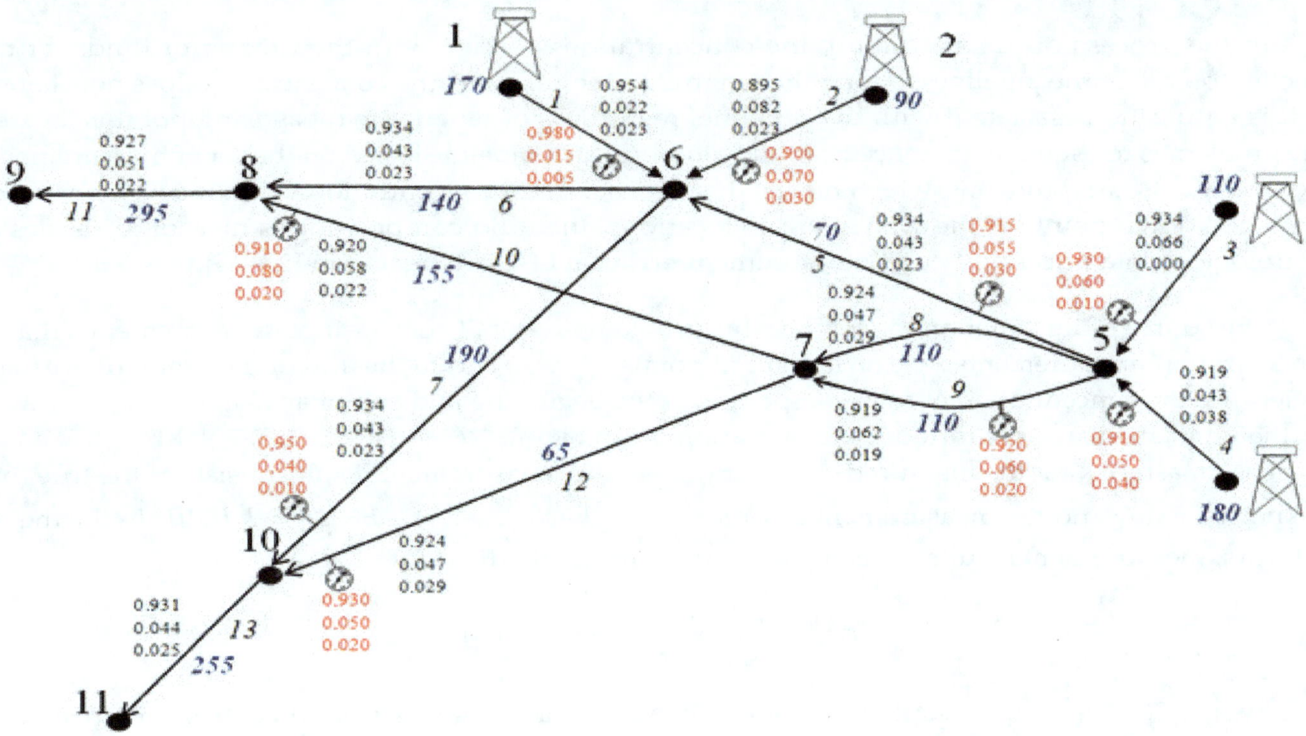

Figure 1. Gas transmission system (GTS) structure, concentration measurements at gas metering stations, calculation results. Designations: **1**—node numbers, *9*—arc numbers, 0.950—component concentration (measurement r_{ij}^{*M}), 0.950—component concentration (calculation $\hat{r}_{ij}^{M}$), *110*—flow rate, MSCMD (million standard cubic meters per day).

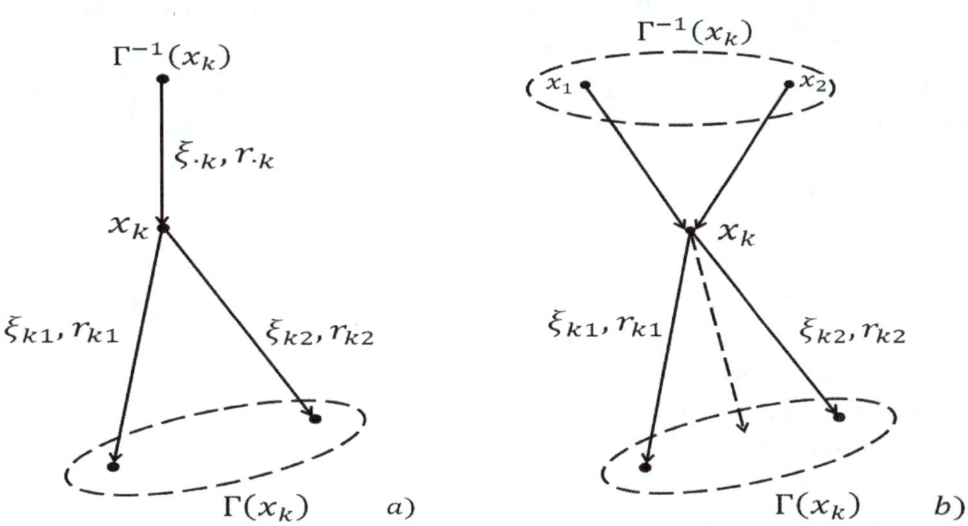

Figure 2. Schemes of separation (**a**) and merging (**b**) in the junction node x_k.

The solution to the problem considered in the paper, like any other problem of operational control, should be based on operational information: current measurements of the parameters of the technological process. The actual location of the GMS at the UGSS has been determined by historically significant factors, but not by the requirements for a reasonable assessment of the distribution of gas CC along transportation directions. In order to increase the degree of reliability of information on distribution of calorific value and/or makeup of gas in the system, the total of measurements at

UGSS facilities should be accounted for. This approach makes allowance for interdependence of the measured operating parameters and requires special methods, adequate mathematical and computer models to be implemented.

In the process of transportation, the concentrations change with the mixing of fluids, at the junction points of the pipelines. Along the length of each pipeline, the concentration does not change, so it is natural to associate it with the pipeline, regardless of where the measuring point is located. The problem is to estimate the concentration values for each pipeline based on the set of measurements. Measurements are erroneous, which means that the measured concentration value is a random value. At some known points of the system (GMS locations) the fluid composition is measured. Issues of metrological data uncertainty and measurement errors at GTS are treated in [23–25].

The source of information for solving this problem is a set of composition measurements, that is, concentration measurements r^*_{ij}. The measured points are assigned to the arcs of the graph (this can be done since the concentration of each component at the beginning and end of any pipeline is the same); for the set of such arcs, we introduce the notation E^* (in Figure 1 $E^* = \{1,2,3,4,5,7,9,10,12\}$). We will also use an asterisk for the measured concentrations. The measurement result consists of the true (but unknown) value and the measurement error $\left(r^M_{ij}\right)^* = r^M_{ij} + \delta r^M_{ij}$, $(i,j) \in E^*$, $M = I, II, III$. In the theory of errors, measurement results are considered to be normally distributed quantities:

$$\delta r^M_{ij} \in N\left(0, \left(\sigma^2_{ij}\right)^M\right),\ M = I, II, III. \tag{4}$$

Symbol $X \in N\left(a, \sigma^2\right)$ means that random variable X has a normal distribution with the mathematical expectation a and the dispersion σ^2. Dispersions σ^2_{ij} characterize the device (or measurement method) error on the arc $(i,j) \in E$. Error $\delta\eta^M_{ij}$ in determining the flow rate of component M also has a normal distribution $\delta\eta^M_{ij} \in N\left(0, \xi^2_{ij}\left(\sigma^2_{ij}\right)^M\right)$, $M = I, II, III$. According to the problem statement, it is required to find such component flow distribution that is most consistent with the entire set of concentration measurements.

The problem under consideration in a certain sense resembles the well-known problem of "unaccounted gas" [26–29]. This problem requires, on the basis of operating parameter measurements, to determine gas balance more precisely (injections in the system, deliveries to consumers, own use), find imbalance reasons, obtain estimates of the flow rates in the directions of pumping, which are most consistent with the current measurements and the dynamics of their change over the past period.

In the stochastic formulation of the problem, it is customary to speak not of determining unknown quantities r_{ij}, but of their estimation. Mathematical statistics for point estimation of an unknown parameter recommends use of the maximum likelihood method (MLM). The estimation of maximum likelihood is the value of the argument at which the likelihood function takes its maximum. The likelihood function in our case is the probability of the totality of all measured values. MLM brings us to the problem of quadratic function conditional minimization [6]:

$$\sum_{M=I}^{III} \sum\nolimits_{(i,j)\in E^*} \left(\left(\eta^{*M}_{ij} - \eta^M_{ij}\right) / \xi_{ij}\sigma^M_{ij}\right)^2 \to min. \tag{5}$$

In fact, not flow rates are measured, but component concentrations. However, using Equation (3), "measurements" η^{*M}_{ij} can be defined through concentration measurements r^{*M}_{ij}, since values ξ_{ij} are given. The minimum of criterion (5) should be sought provided abidance by constraints in the form of equalities and inequalities. Of greatest interest are the restrictions in the form of inequalities arising from mixing process non-equilibrium conditions. These issues are discussed in the next section.

2.3. *Conditions for Non-Equilibrium of Mixing Process*

The processes of mixing natural gas components are non-equilibrium (see Section 2.1). The laws of conservation of mass impose restrictions on the concentrations of the components on the output lines of the joint nodes. The obvious limitations are inequalities:

$$\underline{r}_{in} \leq \underline{r}_{out} \leq \overline{r}_{out} \leq \overline{r}_{in}. \tag{6}$$

The subscript *in* refers to the input lines, the subscript *out* index refers to the output lines, the bar at the top indicates the maximum concentration, the bar at the bottom means the minimum one. Using the notation in Figure 2b, the variables included in inequalities (6) for each component are written in the form $\underline{r}_{in}^{M}(x_k) = \min\limits_{x_i \in \Gamma^{-1}(x_k)} \left(r_{ik}^{M}\right)$, $\overline{r}_{in}^{M} = \max\limits_{x_i \in \Gamma^{-1}(x_k)} \left(r_{ik}^{M}\right)$, $\underline{r}_{out}^{M}(x_k) = \min\limits_{x_i \in \Gamma(x_k)} \left(r_{ki}^{M}\right)$, $\overline{r}_{in}^{M} = \max\limits_{x_i \in \Gamma(x_k)} \left(r_{ki}^{M}\right)$, $M = I, II, III$. Inequalities (6) are directly proved using algebraic calculations. Let us consider the joint node of pipelines with 2 input and 2 output lines (Figure 2b). Fluid flow rates at the input lines are $\xi_{1,in}$, $\xi_{2,in}$. Suppose that the flow through the 1st line is divided into 2 parts: $\xi_{1,in} = \alpha_1 \xi_{1,in} + (1 - \alpha_1)\xi_{1,in}$. The first of them went through the 1st output line, the second through the 2nd output line. In the same way the flow is divided in the second output line. Thus, the fluid flow rates at the output lines equal $\xi_{1,out} = \alpha_1 \xi_{1,in} + \alpha_2 \xi_{2,in}$ $\xi_{2,out} = (1 - \alpha_1)\xi_{1,in} + (1 - \alpha_2)\xi_{2,in}$ accordingly. Consider any component of the mixture. Its concentration in the 1st input line is denoted r_1, and that in the 2nd input line r_2. In our reasoning, one component stands out; all other mixture components play the role of the second component with respective concentrations $1 - r_1$, $1 - r_2$ in the incoming lines, so $\xi_{1,in} = r_1 \xi_{1,in} + (1 - r_1)\xi_{1,in}$, $\xi_{2,in} = r_2 \xi_{2,in} + (1 - r_2)\xi_{2,in}$. Fluid flow rate on the 1st outlet line will consist of 2 summands, each one corresponding to one of the two components $\xi_{1,out} = [\alpha_1 r_1 \xi_{1,in} + \alpha_2 r_2 \xi_{2,in}] + [\alpha_1 (1 - r_1)\xi_{1,in} + \alpha_2 (1 - r_2)\xi_{2,in}]$. The selected component concentration in the 1st output line equals:

$$r_{1,out} = (\alpha_1 r_1 \xi_{1,in} + \alpha_2 r_2 \xi_{2,in}) / (\alpha_1 \xi_{1,in} + \alpha_2 \xi_{2,in}) \tag{7}$$

If, for example, $r_1 < r_2$, that is $\underline{r}_{in} = r_1$, then from relation (7) we directly obtain $\underline{r}_{in} = r_1 < r_{1,out}$. In the same way, inequality $r_{1,out} \leq \overline{r}_{in}$ and similar inequalities for $r_{2,out}$ are proved. Relations (6) are proved for the arbitrarily chosen mixture component; therefore, its record can be generalized as:

$$\underline{r}_{in}^{M} \leq \underline{r}_{out}^{M} \leq \overline{r}_{out}^{M} \leq \overline{r}_{in}^{M}, \; M = I, II, III. \tag{8}$$

This reasoning carried out for the special case of two incoming and two outgoing lines can be applied to a joint node with an arbitrary number of incoming and outgoing lines. The general outline of the proof is preserved. Let us mentally divide the incoming flow along each line into jets, and we obtain an analogue of relation (7) for each outgoing line. Hence the required result follows. Relations (8) are *necessary*, but *not sufficient*. They establish inequalities for the extreme (maximum and minimum) output concentrations. But they do not account for relations of the mixing component quantities at the input and output of the node. Indeed, consider again a certain component and imagine that its maximum concentration on the output lines is equal to the maximum concentration on the input lines $\overline{r}_{out} = \overline{r}_{in}$. For these lines, according to the physics of the process, the flow rate of the fluid in the outlet line cannot exceed the flow rate of the fluid in the inlet line $\overline{\xi}_{out} \leq \overline{\xi}_{in}$. Similar ratios should hold for two, three, etc. output lines.

Let us write down these relations for the general case. The main result is presented by relations (13). The general outline of the reasoning is the same as in the particular case (Figure 2b), but the details of the proof are omitted due to the cumbersomeness. We take a node, denote the number of input lines by N_{in}, and the number of output lines by N_{out}. For analysis, it is not enough to consider only lines with maximum and minimum concentrations. Arranging in decreasing order concentrations in the input lines and in the output lines, we obtain two non- increasing sequences:

$$r_{in}^{(1)} \geq r_{in}^{(2)} \geq \ldots \geq r_{in}^{(N_{in})}. \tag{9}$$

$$r_{out}^{(1)} \geq r_{out}^{(2)} \geq \ldots \geq r_{out}^{(N_{out})}. \tag{10}$$

The superscript in parentheses indicates the rank of the corresponding number in the sequence. It is obvious that $r_{in}^{(1)} = \bar{r}_{in}$, $r_{in}^{(N_{in})} = \underline{r}_{in}$ and $r_{out}^{(1)} = \bar{r}_{out}$, $r_{out}^{(N_{out})} = \underline{r}_{out}$. Using the elements of sequence (9), we construct a line of "limit concentrations". This will be a piecewise smooth curve $r(x) = r_{in}(x)$:

$$r(x) = \begin{cases} r_{in}^{(1)} \text{ provided } 0 \leq x \leq \xi_{in}^{(1)} \\ \frac{r_{in}^{(1)}\xi_{in}^{(1)} + r_{in}^{(2)}\left(x - \xi_{in}^{(1)}\right)}{x} \text{ provided } \xi_{in}^{(1)} < x \leq \xi_{in}^{(1)} + \xi_{in}^{(2)} \\ \ldots \\ \frac{r_{in}^{(N_{in}-1)}\sum_{j=1}^{N_{in}-1}\xi_{in}^{(j)} + r_{in}^{(N_{in})}\left(x - \sum_{j=1}^{N_{in}-1}\xi_{in}^{(j)}\right)}{x} \text{ provided } \sum_{j=1}^{N_{in}-1}\xi_{in}^{(j)} < x \leq \sum_{j=1}^{N_{in}}\xi_{in}^{(j)} \end{cases} \tag{11}$$

Let us proceed to give an interpretation of the point with coordinates $[x;\, r(x)]$ and take by way of example a point lying in the 3rd half-segment of its definition $\xi_{in}^{(1)} + \xi_{in}^{(2)} < x \leq \xi_{in}^{(1)} + \xi_{in}^{(2)} + \xi_{in}^{(3)}$. The inlet flow rate maximum concentration equal to x is obtained if we sum the total gas supply through the inlet lines with concentrations $r_{in}^{(1)}$ and $r_{in}^{(2)}$ and with partial supply through the line with concentration $r_{in}^{(3)}$, provided flow rate $x - \left(\xi_{in}^{(1)} + \xi_{in}^{(2)}\right)$. The maximum concentration for flow rate x is equal to $(1/x)\left[r_{in}^{(1)}\xi_{in}^{(1)} + r_{in}^{(2)}\xi_{in}^{(2)} + r_{in}^{(3)}\left(x - \xi_{in}^{(1)} - \xi_{in}^{(2)}\right)\right]$.

The distribution of flow rates and concentrations in the output lines will be acceptable if all points with coordinates:

$$\left[\xi_{out}^{(1)}; r_{out}^{(1)}\right], \left[\xi_{out}^{(1)} + \xi_{out}^{(2)}; \frac{r_{out}^{(1)}\xi_{out}^{(1)} + r_{out}^{(2)}\xi_{out}^{(2)}}{\xi_{out}^{(1)} + \xi_{out}^{(2)}}\right], \ldots, \left[\sum_{j=1}^{N_{out}}\xi_{out}^{(j)}; \frac{\sum_{j=1}^{N_{out}} r_{out}^{(j)}\xi_{out}^{(j)}}{\sum_{j=1}^{N_{out}}\xi_{out}^{(j)}}\right] \tag{12}$$

lie no higher than curve (11). The first of these points corresponds to the output line with maximum concentration $r_{out}^{(1)}$, the 2nd one to two output lines with concentrations $r_{out}^{(1)}$ and $r_{out}^{(2)}$, etc. The last point with abscissa $\xi_{\Sigma} = \sum_{j=1}^{N_{out}}\xi_{out}^{(j)}$ and ordinate ρ always lies on the curve $r(x)$. Here ρ is the system average value of the considered component. If the remaining points are no higher than $r(x)$, then the concentration distribution (10) is acceptable. If for ordinates of points in the aggregate (9) we introduce the notation $y^{(1)} = r_{out}^{(1)}$, $y^{(2)} = \frac{r_{out}^{(1)}\xi_{out}^{(1)} + r_{out}^{(2)}\xi_{out}^{(2)}}{\xi_{out}^{(1)} + \xi_{out}^{(2)}}, \ldots, y^{(N_{out})} = \frac{\sum_{j=1}^{N_{out}} r_{out}^{(j)}\xi_{out}^{(j)}}{\sum_{j=1}^{N_{out}}\xi_{out}^{(j)}}$, then the admissibility condition is written in the form:

$$y^{(1)} \leq r\left(\xi_{out}^{(1)}\right),\ y^{(2)} \leq r\left(\xi_{out}^{(1)} + \xi_{out}^{(2)}\right), \ldots, y^{(N_{out})} \leq r\left(\sum_{j=1}^{N_{out}}\xi_{out}^{(j)}\right). \tag{13}$$

This condition (13) is necessary and sufficient for non-equilibrium process feasibility. It should be noted that the verification procedure (9)–(13) must be carried out for each fluid component.

Now we can fully formulate the problem of estimation of CC gas mixture distribution in the directions of aggregated scheme transportation, listing all the restrictions that should be accounted for in the optimization problem with criterion function (5).

Constraints in the form of equalities include balance conditions in the joint nodes for each component:

$$\sum_{x_j \in \Gamma(x_k)} \eta_{kj}^{M} - \sum_{x_i \in \Gamma^{-1}(x_k)} \eta_{ik}^{M} = 0,\ x_k \in V_{joint},\ M = I, II, III. \tag{14}$$

Conditions of equality of the fluid flow rate on each arc to the sum of all components flow rates on the arc:

$$\sum_{M=I}^{III} \eta_{ij}^{M} = \xi_{ij},\ (i,j) \in E. \tag{15}$$

Moving to concentrations, we write down this equation in the form:

$$\sum_{M=I}^{III} r_{ij}^{M} = 1,\ (i,j) \in E. \tag{16}$$

Limitations in the form of inequalities include: condition (13) of a non-equilibrium process feasibility, which must be satisfied for each fluid component $M = I, II, III$; non-negativity of all concentrations of the problem:

$$r_{ik}^{M} \geq 0,\ (i,k) \in E,\ M = I, II, III. \tag{17}$$

Thus, determination of fluid composition per transportation direction is reduced to optimization problem (5) under constraints (14), (15) in the form of equalities and constraints (13), (17) in the form of inequalities.

2.4. *Assessment of the Information Content of the Gas Transmission System (GTS) Metrological Equipment*

Before proceeding with solution of the task of estimating concentrations r_{ik}^{M}, it is necessary to find out what can be obtained from the available totality of measurements r_{ik}^{*M}, $(i,k) \in E^{*}, M = I, II, III$. Estimates obtained by solving the problem of conditional minimization will be marked with the caret, for example, concentration estimates we shall notate as $\hat{r}_{ik}^{M}$. The estimation quality depends on the quantity and location of the instruments, instruments accuracy class and the frequency of instruments polling sessions. With an insufficient number of measurements it is impossible to obtain all estimates of all concentrations $\hat{r}_{ik}^{M}$, $(i,k) \in E$. To demonstrate the possibilities of metrological support of the system, let us consider the flow of a two-component mixture on the simplest system - a tee with one input line and two output lines (Figure 2a). By n^{*}, m_{in}, m_{out}, m_{joint} we denote the number of measuring points, sources, outflow and joint nodes, respectively.

In the beginning we analyze the simplified problem (5), (14). It is a quadratic program with n unknowns and m_{joint} linear constraints. Different cases may arise depending on values n^{*}, m_{in}, m_{out}, m_{joint}. For a three-way piece (Figure 2a) $n = 3$, $m_{joint} = 1$ there can be 3 options of measured parameters, respectively $n^{*} = 3, 2, 1$. In option 1 (concentrations are measured in each pipeline) to estimate unknowns r_{ij}, we resort to the problem of minimizing quadratic function of 3 variables with one linear constraint. Using the constraint, we express one of the unknown quantities r_{ij} in terms of the other two. The resulting problem of unconditional extremum of 2 variable quadratic function can be reduced to a system of 2 linear equations with the same number of unknowns. Solving this problem, we obtain estimates $\hat{r}_{ij}$ of unknowns r_{ij} which use the results of all 3 measurements. The estimates $\hat{r}_{ij}$ are better substantiated than each value direct measurements r_{ij}^{*}. In option 2, the concentration is measured only in 2 lines (for example, r_{k1}^{*}, r_{k2}^{*} in Figure 2a). Now $n = 3$, $n^{*} = 2$, $m_{joint} = 1$, we have the problem of minimizing quadratic functions of 2 variables r_{k1}, r_{k2}, provided one linear constraint. The maximum of the likelihood function is found directly, and the estimates are equal to measurements $\hat{r}_{k1} = r_{k1}^{*}$, $\hat{r}_{k2} = r_{k2}^{*}$. The constraint is used to find an estimate of the missing unknown concentration along the arc entering the node x_k. In option 3, only one value is measured, for instance r_{k1}^{*} (Figure 2a). The maximum likelihood method allows one to obtain only a trivial result—an estimate of the concentration along that arc $\hat{r}_{k1} = r_{k1}^{*}$.

The same kind of reasoning is carried out in the general case for any ratio of quantities n, n^{*}, m_{joint}. It helps to reveal what results can in principle be obtained with the existing system of measuring gas composition. Concentration estimates cannot always be obtained and not for all graph edges. It all depends on the number and location of measuring points. For the graph of arbitrary configuration, of greatest importance is the value $d = m_{joint} - (n - n^*)$. When $d > 0$, sought estimates can be refined by accounting for mutual influence of all the measurements. If $d = 0$, then the measurement results directly are concentration estimates. If $d < 0$, then the constraints in the form of equalities are insufficient to estimate all unmeasured concentrations r_{ij}. In the case $d \leq 0$ graph G can contain subgraphs that meet the condition $d > 0$. (Subgraph $G_l(X_l, E_l)$ of the graph $G(X, E)$ is a graph for which $X_l \subset X$ and for each graph node $x_k \in X_l$, $\Gamma_l(x_k) = \Gamma(x_k) \cap X_l$, [21,22]). Consequently, sufficient measurements are made on the arcs of these subgraphs to refine concentration estimates. An algorithmic procedure for identifying such subgraphs has been developed. Estimates are obtained as linear functions of measurements $\hat{r}_{ij} = \sum_{(k,l) \in E^*} \left[a_0^{ij} + a_{kl}^{ij} r_{kl}^*\right]$, $(i, j) \in E$.

The quality of the estimates is characterized by their dispersion. Under assumption (4), dispersion of the estimates resulting from the maximum likelihood method is calculated as $\mathbf{D}\hat{r}_{ij} = \sum_{(k,l) \in E^*} \left[\left(a_{kl}^{ij}\right)^2 \mathbf{D} r_{kl}^*\right]$, $(i, j) \in E$.

2.5. Computational Aspects

The considered problem of the distribution of the CC of multicomponent fluids is thus reduced to a mathematical programming problem with constraints in the form of equalities and inequalities. The number of unknowns and constraints in a task can be large. When solving it numerically, pitfalls can be encountered, the manifestation of which will be the flatness or, conversely, the ravine of the target function. The idea of the presence or absence of these phenomena, negative from a computational point of view, can be obtained by conducting a computational experiment. We conducted experiments, and found non-standard computational techniques and algorithms, which we consider to be proof of the efficiency of the developed technique.

A. Two Approaches to Formalizing the Problem

The first approach (unknown variables–flow rates or concentration of components, see Sections 2.2–2.5): the sought variables are concentrations r_{ij}^M. If in relations (5), (14) and 15), instead of component flow rates η_{ij}^M, using relations (3), we substitute their expressions in terms of concentrations r_{ij}^M, then we will obtain a formalization of the problem of determining gas composition in variables r_{ij}^M (component concentrations on the arcs of calculated graph) or the problem of optimizing quadratic function with constraints in the form of equalities and inequalities. Since the fluid flow rates ξ_{ij} are considered known, we can act differently, taking values $\eta_{ij}^{*M} = \xi_{ij} r_{ij}^{*M}$ as initial information for calculation.

The second approach: the sought variables are coefficients α_{ik}^j. We introduce new unknowns α_{ik}^j, presenting the fluid flow rates in the form

$$\xi_{kj} = \sum_{i \in \Gamma^{-1}(x_k)} \alpha_{ik}^j \xi_{ik}, \; x_k \in V_{joint}. \tag{18}$$

Here, we used the notation in Figure 2b. The physical meaning of relation (18) is as follows. We consider joint node $x_k \in V_{joint}$. Each flow through the incoming line (i, k) (with flow rate ξ_{ik}) is divided into "jets" $\alpha_{ik}^1 \xi_{ik}, \alpha_{ik}^2 \xi_{ik}, \ldots, \alpha_{ik}^{N(k)} \xi_{ik}$; $N(k) = mes\,\Gamma(x_k)$, where α_{ik}^j is the fraction of the flow rate ξ_{ik} entering the output line (k, j), $mes\,\Gamma(x_k)$ is the number of arcs outgoing from node x_k. Obviously, for α_{ik}^j the following relations are true:

$$\sum_{j=1}^{N(k)} \alpha_{ik}^j = 1, \; x_i \in \Gamma^{-1}(x_k); \; \alpha_{ik}^j \geq 0. \tag{19}$$

In accordance with relation (18), we obtain the distribution of the components in the outgoing lines:

$$\eta_{kj}^{M} = \sum_{i \in \Gamma^{-1}(x_k)} \alpha_{ik}^{j} \eta_{ik}^{j,M}, M = I, II, III, \ x_k \in V_{joint}. \tag{20}$$

An obvious fact is used in relations (19). Let us consider the fluid flow along any incoming line. In the joint node, this flow is divided into parts ("jets") along the outgoing lines ($N(k)$ is the number of the jets). Component composition of each of these parts will be the same, the fact on which formula (20) is based. Thus, if values $\eta_{ik}^{M}, M = I, II, III$, are taken as unknowns, then the mathematical model does not require inclusion of inequality type relations (13), following from the conditions of non-equilibrium of the process. These relations will be performed automatically.

The technique of mentally dividing each stream entering the node into "jets" was used above (see Section 2.3, Formula (7)). If values α_{ik}^{j} are determined, then the known concentrations are used to calculate the concentrations on all network arcs. This can be done beginning with concentrations in the sources r_{ij}^{*M}, $x_i \in V_{in}$, $x_j \in \Gamma(x_i)$, $M = I, II, III$. The concentrations are calculated sequentially in the order determined by the following node numbering algorithm.

Algorithm. We number the sources (in random order) with the numbers $1, 2, \ldots, N_{in}$. By numbering a node, we color it and all arcs outgoing from it. We assign the next number to the node, all incoming arcs in which are colored. We continue the procedure until all nodes are numbered.

In the same way, all flow rates η_{ij}^{M}, $(i, j) \in E$, $M = I, II, III$ can be expressed through a set of quantities α_{ik}^{j}, which we designate as vector $\boldsymbol{\alpha} = \|\alpha_{ik}^{j}\|$. In particular, we obtain measured values $\eta_{ij}^{*M} = \eta_{ij}^{*M}(\boldsymbol{\alpha})$, $(i, j) \in E^{*}$, $M = I, II, III$. As a result, the problem is reduced to finding the minimum of criterion function (5) subject to fulfillment of constraints (15) and (19). Note that the calculation procedure automatically satisfies conditions $\eta_{ik}^{M} \geq 0$, $(i, k) \in E$, $M = I, II, III$, and, most importantly, conditions (13) of the mixing process non-equilibrium. The last statement is explained by the fact that when deriving non-equilibrium conditions (13), the method of splitting the flow along each incoming line into "jets" was used (see Section 2.3, derivation of formulae (7), (8)).

In the first and second approaches, either component flow rates η_{ij}^{M} or their concentrations r_{ij}^{M} are used. By virtue of relations (3), these variables are easily expressed through each other. However, the choice of variables can influence the effectiveness of the calculation methods.

B. The First Approach. Calculation Procedure Specificity

Nested solutions method. In its complete form, the problem of mathematical programming (5) with conditions (13)–(16) because of a large number of restrictions and significant difference in the sought variable values can be computationally difficult. In this case, one can resort to the method of finding solution to the problem disregarding some constraints. The obtained quasi-solution should be further checked for compliance with previously ignored constraints. If it does comply, then it is the desired result. If it fails to comply, then those of the ignored constraints that are not satisfied in the quasi-solution are added to the conditions. For example, constraints (13) are rather cumbersome. Instead of them, restrictions (8) can be introduced, that is, only part of restrictions (13). A quasi-solution with constraints (8) can satisfy all constraints (13), then a solution to the problem is reached.

The computational experiment performed proved the method to be very effective, although, of course, it is impossible to guarantee its success in any situation. The possibility of rejecting condition (17) was also tested. In the experiments, after the quasi-solution was obtained, condition $r_{ik}^{M} \geq 0$ was introduced only for those unknown concentrations that turned out to be negative in the quasi-solution. In the solution obtained in this case, all constraints were satisfied.

The larger the dimension of the problem, the more efficient the application of the nested solution method. The application of the method in computer systems for operational control is especially justified. For operational control, it is characteristic that with the next arrival of information about measurements from the GMS, a good initial approximation is usually known to solve the problem of determining the component composition. Therefore, first of all, it is necessary to take into account the constraints in the form of inequalities, which are currently fulfilled as equality.

3. Results

Section 3 uses a double designation of arcs, for example, *1* and (1, 6) (see Figure 1), the 2nd way is more convenient in those cases when it is necessary to emphasize the direction of the arc.

3.1. Example

To illustrate the technique developed, we will give an example of calculating CC distribution in a three-component mixture flow in the graph shown in Figure 1. Structural graph *G* contains 13 arcs (transportation directions), 11 nodes (of which 4 sources, 2 outflows, 5 joints). Measurement points are located on 9 arcs. The criterion for the information content is $d = m_{joint} - (n - n^*) = 5 - (13 - 9) = 1$. Therefore, in this case the technique provides an opportunity to take into account mutual influence of the measurements. The initial measurement data and optimization problem solution results are shown in Table 1 and duplicated in Figure 1. The values of the estimates $\hat{r}_{ij}^M$ of the component concentrations shown in Figure 1, as well as in Table 1 are obtained by computer calculations and rounded to 3 decimal places. Naturally, the estimates $\hat{r}_{ij}^M$ were obtained for all arcs, including those without measuring devices (arcs 6, 8, 11, 13, Figure 1). Due to rounding errors, relation (16) for estimates $\hat{r}_{ij}^M$ may be violated in the last digit. So, for the arc (1, 6) we have $\sum_{M=I}^{III} \hat{r}_{16}^M = 0.924 + 0.022 + 0.023 = 0.999$.

Table 1. Composition measurements r_{ij}^{*M} and estimates $\hat{r}_{ij}^M$ (numerator—measurement, denominator—calculation).

Arc №/(*i*, *j*)	Flow Rate, MSCMD	Concentration of Components		
		M = *I*	*M* = *II*	*M* = *III*
1/(1, 6)	170	0.980/0.954	0.015/0.022	0.005/0.023
2/(2, 6)	90	0.900/0.895	0.070/0.082	0.030/0.023
3/(3, 5)	110	0.930/0.934	0.060/0.066	0.010/0.000
4/(4, 5)	180	0.910/0.919	0.050/0.043	0.040/0.038
5/(5, 6)	70	0.915/0.934	0.055/0.043	0.030/0.023
6/(6, 8)	140	—/0.934	—/0.043	—/0.023
7/(6, 10)	190	0.950/0.934	0.040/0.043	0.010/0.023
8/(5, 7)	110	—/0.924	—/0.047	—/0.029
9/(5, 7)	110	0.920/0.919	0.060/0.062	0.020/0.019
10/(7, 8)	155	0.910/0.920	0.080/0.058	0.010/0.022
11/(8, 9)	295	—/0.927	—/0.051	—/0.022
12/(7, 10)	65	0.930/0.924	0,050/0,047	0.020/0.029
13/(10, 11)	255	—/0.931	—/0.044	—/0.025

The difference between direct measurements r_{ij}^{*M} and estimates $\hat{r}_{ij}^M$ for the entire set of measurements, especially for components with low concentrations, in some cases is very large. For example, $r_{(7,8)}^{*II} = 0.080$, $\hat{r}_{(7,8)}^{II} = 0.058$, $r_{1,6}^{*III} = 0.005$, $\hat{r}_{1,6}^{III} = 0.023$, $r_{(3,5)}^{*III} = 0.010$, $\hat{r}_{(3,5)}^{III} = 0.000$. Significant discrepancies in the values r_{ij}^{*M} and $\hat{r}_{ij}^M$ may indicate non-fulfillment of relation (4), in the part that refers to the absence of systematic measurement errors. The presence of significant discrepancies should be an incentive for the standardization of measuring instruments.

The distribution of flows ξ is included in the initial data. The flow rates of the fluid in the directions of pumping $\xi_{(1,6)} = 170, \xi_{(2,6)} = 90, \ldots$ are balanced, and they are shown in Figure 1 and duplicated in Table 1.

Despite the small size of the graph, it allows you to illustrate the characteristic features of the problem of assessing the CC of the GTS pipelines. The sample data have been selected to illustrate the methods.

The concentration measurement data is deliberately selected in such a way as to simulate possible errors. For example, the measurements of the concentration of component $M = II$ on arcs emanating from the sources are 0.015, 0.070, 0.060, 0.050, while the concentration $M = II$ measurement on the outgoing arc (7, 8) is 0.080. Naturally, such a situation is impossible for true concentration values due to violation of the non-equilibrium conditions. Nevertheless, the solution of the problem has been obtained, and the calculation results satisfy the conditions of non-equilibrium in both soft (8) and harder (13) forms. This means that the algorithm made it possible to smooth out the measurement errors without going beyond the limits, including (16) and (17).

Attention should be paid to the difference in the component composition of the fluid along parallel lines 8 and 9 (arcs between nodes 5 and 7). Thus, $\hat{r}_8^{II} = 0.047$, $\hat{r}_9^{II} = 0.062$, $\hat{r}_8^{III} = 0.029$, $\hat{r}_9^{III} = 0.019$. This once again indicates that the distribution of the component composition of the fluid is influenced by a set of interrelationships of operating parameters throughout the GTS. Rough-and-ready, local methods for solving the problem can lead to significant errors.

One of the purposes of this example is to test nested decision methods (see Section 1, Section 2.5B). First, the optimization problem was solved without taking into account inequality constraints. In this setting, the problem is reduced to a quadratic program with linear equality constraints. The resulting solution contained 2 negative components and one violated inequality (8). Then, both inequalities from (17) and the violated inequality from (8), which were not fulfilled at the first stage, were introduced into the model. The solution to this corrected problem has satisfied all the constraints.

It is interesting to note that none of the constraints (13), which are not included in the number of inequality constraints (8), were violated either at the 1st or at the 2nd stage of the solution. Apparently, restrictions from (8) are infrequently violated, which indicates the usefulness of the proposed nested solution method.

In this example, both approaches to the formalization of the problem A and B (Section 2.5) are tested. They have been proven capable of working, both lead to virtually identical numerical results.

Comparison of approaches A and B can be, in our opinion, carried out only on the basis of a computational experiment. However, it is clearly premature to draw conclusions based on the study of only one example. Our assumption about the preferability of method B remains a hypothesis that must be verified by a representative computational experiment.

3.2. Mixing Processes Non-Equilibrium Indicator

Initially, the study of irreversible and non-equilibrium processes was the subject of thermodynamics [17–19]. Then, other branches of science that lie at the intersection of physics and chemistry arose to study processes of this kind (see Section 2.1). In connection with the need to determine the directionality of the physical (heat transfer) process the concept of entropy was introduced, which underlies one of the cardinal laws of physical laws—the second law of thermodynamics. Entropy is a physical quantity deduced from mathematical models of energy transfer processes. Entropy is a monotonic function defined in such a way that in irreversible heat transfer processes it inevitably increases. Currently, there are numerous options for extending the concept of entropy to processes not associated with heat transfer. K. Shannon [30] introduced the concept of information entropy as a measure of information, as a characteristic of the amount of information contained in a message. We can say that entropy is a general name for the quantitative characteristics of a wide variety of processes, primarily processes that are irreversible and characterised by non-equilibrium. For the processes of gas mixing in industrial pipelines considered here, the quantitative indicator of non-equilibrium can also be introduced.

To define it, we introduce the concept of an extended graph G'. G' is obtained from the original graph G by adding to it a fictitious source s' and a fictitious outflow t', as well as fictitious arcs connecting s' to the sources of graph G and outflows of the graph G to t'. Fictitious arcs are shown in Figure 3 with a dotted line.

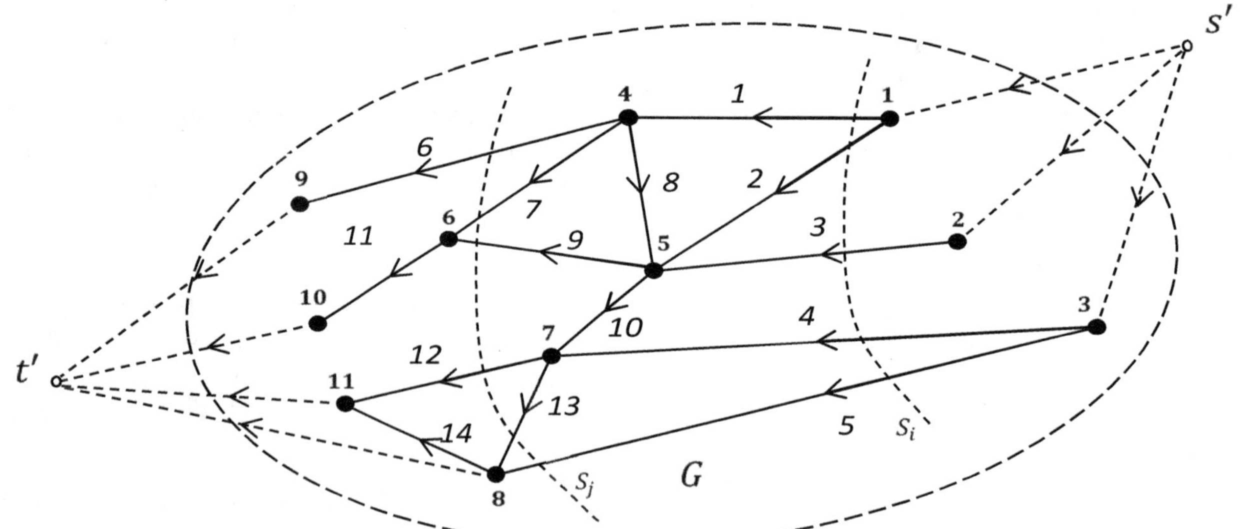

Figure 3. Cuts of the extended graph G'.

In graph G' cuts separating s' from t' are considered. A minimal set of arcs, the removal of which splits the graph into 2 components of connectivity, of which one contains s' and the other t' is called a cut (more precisely, a minimal cut) separating the source s' from the outflow t' [21,22]. The term minimal cut means that when any arc is removed from the set, it ceases to be a cut.

Figure 3 shows 2 cuts S_i, S_j. The fluid flow rate distribution through the cut is determined by population $\alpha_{kj} = \xi_{kj}/\xi_\Sigma, (k,j) \in S_i$. Here $\xi_\Sigma = \sum\limits_{(k,j)\in S_i} \xi_{kj}$ is the total flow rate through the cut. All flow rates ξ_Σ are obviously the same in any cut. Since $\sum\limits_{(k,j)\in S_i} \alpha_{kj} = 1$ it can be argued that the set α_{kj}, r^M_{kj}, $(k,j) \in S_i$ for each component M determines a discrete random variable specified in the cut S_i. The dispersion of this variable:

$$H^M(S_i) = \sum_{(k,j)\in S_i} \left(r^M_{kj} - \overline{r}_M\right)^2 \alpha_{kj}. \tag{21}$$

characterizes the degree of component M scattering around the mean value $\overline{r}^M$, that is, degree of the component flow proximity to the state of complete mixing. The function of the cut:

$$H(S_i) = \sum_{M=I}^{III} H^M(S_i) \tag{22}$$

characterizes the mixture proximity to the equilibrium state, the state of complete mixing. Here, as before, it is assumed that the number of mixture components is III. According to established tradition, entropy as a characteristic of irreversible processes can change in only one direction: it can increase. Let us introduce the concept of "entropy" for the process of non-equilibrium mixing of multicomponent gas mixtures moving through industrial pipeline systems, trying not to break this tradition. Let us define the "entropy" of the cut S_i using the formula:

$$\text{entr}(S_i) = -H(S_i) = -\sum_{M=I}^{III} H^M(S_i). \tag{23}$$

According to definition (23), the gas mixture "entropy" is the sum of the gas mixture component "entropies". In contrast to the overwhelming majority of other applications function $\mathrm{entr}(S_i)$ is defined not for a continuum set, but for a discrete set: the set of cuts of the extended graph G' of pipeline system.

Let us introduce a new concept. Each cut S_i divides extended graph G' into 2 parts, into 2 connected components, subgraphs $G'_{s'}(S_i), G'_{t'}(S_i)$ containing a source and an outflow, respectively, so that $G' = G'_{s'}(S_i) \cup G'_{t'}(S_i) \cup S_i$. We can say that cut S_j includes S_i, and denote it as $S_j \succ S_i$ if $G'_{s'}(S_i) \subset G'_{t'}(S_j)$ (see Figure 3). The necessary condition for the non-equilibrium of component M flow distribution in graph G' is written in the form $H^M(S_i) > H^M(S_j)$.

Let us call the maximum number of arcs in the chain from s' to i as the distance from s' to node i and introduce a sequence of cuts S_k, $k \geq 1$. S_k is a cut that divides extended graph G' into 2 subgraphs, one of which $G'_{s'}(S_k)$ contains all the nodes removed from s' at a distance not exceeding k, and the other subgraph $G'_{t'}(S_k)$ contains the remaining nodes of the graph. By construction, $S_j \succ S_k$, if $j > k$. We denote the distance from s' to t' through K and consider a sequence of cuts $S_1 \prec S_2 \prec \ldots \prec S_K$. The function (22) $H(z)$ is defined for the set of cuts $S_1, S_2, \ldots, S_K$, that is, argument z can be considered a real variable taking values $1, 2, \ldots, K$, or a real n-dimensional vector $\mathbf{z} \in R^n$ which values are determined by the cuts $S_1, S_2, \ldots, S_K$. We can consider, for example, $z_{jk} = \begin{cases} 1, \text{ if } (j,k) \in S_i \\ 0, \text{ if } (j,k) \notin S_i \end{cases}, (j,k) \in E$. The above operation $S_j \prec S_i$ establishes a partial ordering relation in the set of cuts $S_1, S_2, \ldots, S_K$. Cuts S_i, S_j in Figure 3 are related via ratio $S_j \prec S_i$. The "entropy" in these cuts satisfies the inequality $\mathrm{entr}(S_i) \leq \mathrm{entr}(S_j)$.

Table 2 shows values of indicator $\mathrm{entr}(S_i)$ and its additive components $H^M(S_i)$ for 3 cuts of graph G' (Figure 4).

Table 2. Example. Indicator $\mathrm{entr}(S_i)$ values for 3 cuts (Figure 4).

Cut S_i	Arcs of the Cut	$\mathrm{entr}(S_i)$	$H^I(S_i)$	$H^{II}(S_i)$	$H^{III}(S_i)$
S_1	1, 2, 3, 4	-1.0800×10^{-3}	0.431×10^{-3}	0.468×10^{-3}	0.180×10^{-3}
S_2	6, 7, 10, 12	-0.0823×10^{-3}	0.370×10^{-4}	0.415×10^{-4}	0.380×10^{-5}
S_3	11, 13	-0.0170×10^{-3}	0.500×10^{-5}	0.110×10^{-4}	0.110×10^{-5}

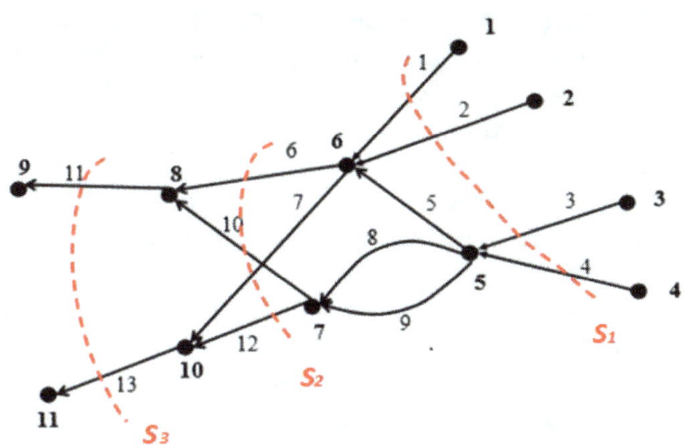

Figure 4. Example. The cuts S_1, S_2, S_3 (see the Table 2).

The indicator of non-equilibrium allows us to gain a general picture of the distribution of CC/caloric content in the GTS. This will help in solving the issues of adjusting the aggregated scheme in the direction of refinement or dividing the graph into subgraphs $G = G_u \cup G_v$, for one of which G_u the non-equilibrium of the mixing process is essential, and for the other G_v it is not essential. In this case, it may be acceptable the switch from the study of CC distribution on G to study on G_u. "Entropy" rapidly decreases with distance from sources and approaching sinks. So $\mathrm{entr}(S_1) : \mathrm{entr}(S_2) = 13.1$, $\mathrm{entr}(S_2) : \mathrm{entr}(S_3) = 4.84$. This indicates a rapid approach of the process to the equilibrium state.

In general, the example testifies to the efficiency of the methods, the efficiency of computational procedures and the acceptability of the apparatus for practical calculations.

4. Conclusions

A technique is proposed that allows calculating the distribution of the component composition of natural gas flows through gas transportation systems of arbitrary configuration. A mathematical model has been developed that takes into account the irreversibility and non-equilibrium of mixing processes, as well as the random nature (instruments errors) of the component concentration measurement. Non-equilibrium conditions are derived for a mixture with an arbitrary number of components. The model takes into account the entire set of measurements and their mode-technological relationships. The possibility of two approaches to the numerical solution of the problem is established. The aim of the research is achieved by solving a mathematical programming problem with constraints in the form of equalities and inequalities. A subject-oriented computational procedure has been developed, which allows one to obtain a result using standard software packages. The efficiency of the technique is demonstrated by a numerical example of calculating the flow distribution of a three-component mixture. A non-equilibrium indicator («entropy» of the mixing process) is introduced, which is defined for a discrete set: the set of cuts of the computational graph.

References

1. *GOST R 57851.1-17. Gas-Condensate Mixture. Part 1. Gas Separation. Determination of the Composition by Gas Chromatography*; Standartinform: Moscow, Russia, 2017; p. 53. (In Russian)
2. ISO. *ISO 39254.2016. Petroleum Products—Determination of Boiling Range Distribution—Gas Chromatography Method*; ISO: Geneva, Switzerland, 2016; p. 25.
3. Kramer, R.; Mickan, B. Gas Measurement and Gas Meters Testing in Practice. *PTB-Mitteilungen* **2009**, *119*, 16–22.
4. Tevyashev, A.D.; Ievleva, C.H.; Pozhidaev, M.V. Mathematical models and methods for reconstructing actual values of the component composition and physicochemical properties of natural gas flows in the GTS nodes. *RI* **2011**, *3*, 81–87. (In Russian)
5. Sukharev, M.G.; Kazak, A.S.; Fomina, E.V. Distribution of gas makeup in gas transmission system pipelines. *Territoriya Neftegaz* **2019**, *9*, 70–79. (In Russian)
6. Kislenko, N.A.; Sukharev, M.G.; Kazak, A.S.; Fomina, E.V. Gas transmission systems multi-component flows studies (flow and measurement processing models). *Energy Syst. Res.* **2019**, *2*, 29–39.
7. Euro-Asian Council for Standardization. *GOST R 53763-2009. Combustible Natural Gases*; Determination of the Dew Point Temperature of Water; Euro-Asian Council for Standardization, Metrology and Certification: Minsk, Belarus, 2009. (In Russian)
8. Euro-Asian Council for Standardization. *GOST 20061-84 Natural Combustible Gases*; Method for Determining the Dew Point Temperature of Hydrocarbons; Euro-Asian Council for Standardization, Metrology and Certification: Minsk, Belarus, 1984. (In Russian)
9. ISO. *ISO 18453 Natural Gas—Correlation between Water Content and Water Dew Point*, 1st ed.; International Standard; ISO: Geneva, Switzerland, 2004.
10. Versaci, M.; Palumbo, A. Magnetorheological Fluids: Qualitative comparison between a mixture model in the Extended Irreversible Thermodynamics framework and an Herschel–Bulkley experimental elastoviscoplastic model. *Int. J. Non-Linear Mech.* **2020**, *118*, 103288. [CrossRef]
11. Yan, D.; Zhang, S.; Jung, U. A variable-selection control chart via penalized likelihood and Gaussian mixture model for multimodal and high-dimensional processes. *Control Eng. Pract.* **2004**, *12*, 745–755. [CrossRef]
12. Tewari, A.; Giering, M.J.; Raghunathan, A. Parametric Characterization of Multimodal Distributions with Non-Gaussian Modes. In Proceedings of the 2011 IEEE 11th International Conference on Data Mining Workshops, Vancouver, BC, Canada, 11 December 2011. [CrossRef]
13. Mokhatab, S.; Poe, W.A.; Mak, J.Y. *Handbook of Natural Gas Transmission and Processing Principles and Practices*; Gulf Professional Publishing: Houston, TX, USA, 2019.
14. Kidnay, A.J.; Parrish, W.R. *Fundamentals of Natural Gas Processing*; Taylor & Fransis Group: Milton Park, UK, 2006; p. 433.

15. Kuczy'nski, S.; Łaciak, M.; Olijnyk, A.; Szurlej, A.; Włodek, T. Thermodynamic and Technical Issues of Hydrogen and Methane-Hydrogen Mixtures Pipeline Transmission. *Energies* **2019**, *12*, 569. [CrossRef]
16. Liu, J.; Dang, M.; Lu, R.; Guan, X.; Yang, J. Research on mixing law of natural gas pipeline replaced by nitrogen. *Adv. Mech. Eng.* **2017**, *9*, 168781401771370. [CrossRef]
17. Gibbs, J.W. *Elementary Principles in Statistical Mechanics, Developed with Especial Reference to the Rational Foundation of Thermodynamics*; Dover Publications: New York, NY, USA, 1960; p. 207.
18. De Groot, S.R.; Mazur, P. *Non-Equilibrium Thermodynamics*; Elsevier: Amsterdam, The Netherlands, 1962.
19. Porshakov, B.P.; Romanov, B.A. *Fundamentals of Thermodynamics and Heat Engineering*; Nedra: Moscow, Russian, 1988. (In Russian)
20. Baykov, V.I.; Pavlyukevich, N.V.; Fedotov, A.K.; Shnip, A.I. *Thermophysics. Nonequilibrium Heat and Mass Transfer Processes*; Higher School: Minsk, Belarus, 2018; p. 476. (In Russian)
21. Christofides, N. *Graph. Theory: An. Algorithmic Approach*; Academic Press: Cambridge, MA, USA, 1975; p. 400.
22. Berge, C. *The Theory of Graphs and Its Applications*; John Wiley & Sons: Hoboken, NJ, USA, 1962; p. 247.
23. ISO. *ISO/IEC Guide 98-3: 2008 (e) Uncertainty of Measurement-Part 3: Guide to the Expression of Uncertainty in Measurement (GUM: 1995)*; ISO: Geneva, Switzerland, 2008.
24. Ficco, G.; Dell'Isola, M.; Vigo, P.; Celenza, L. Uncertainty analysis of energy measurements in natural gas transmission networks. *Flow Meas. Instrum.* **2015**, *42*, 58–68. [CrossRef]
25. Cimerman, F.; Jarm, M.; Širok, B.; Blagojevič, B. Taking in Account Measuring Errors of Volume Conversion Devices in Measuring of the Volume of Natural Gas. *J. Mech. Eng.* **2016**, *62*, 95–104. [CrossRef]
26. Bagajewicz, M.J.; Cabrera, E. Data reconciliation in gas pipeline systems. *Ind. Eng. Chem. Res.* **2003**, *42*, 5596–5606. [CrossRef]
27. Arpino, F.; Dell'Isola, M.; Ficco, G.; Vigo, P. Unaccounted for gas in natural gas transmission networks: Prediction model and analysis of the solutions. *J. Nat. Gas Sci. Eng.* **2014**, *17*, 58–70. [CrossRef]
28. De Oliveira, E.C.; Frota, M.N.; de Oliveira Barreto, G. Use of data reconciliation: A strategy for improving the accuracy in gas flow measurements. *J. Nat. Gas Sci. Eng.* **2015**, *22*, 313–320. [CrossRef]
29. Botev, L.; Johnson, P. Applications of statistical process control in the management of unaccounted for gas. *J. Nat. Gas Sci. Eng.* **2020**, *7*, 103194. [CrossRef]
30. Shannon, C.E. A Mathematical Theory of Communication. *Bell Syst. Tech. J.* **1948**, *27*, 379–423. [CrossRef]

3

Visualization Study on Thermo-Hydrodynamic Behaviors of a Flat Two-Phase Thermosyphon

Chao Wang [1], Feng Yao [1,2,*], Juan Shi [3], Liangyu Wu [1,*] and Mengchen Zhang [3]

1 School of Hydraulic, Energy and Power Engineering, Yangzhou University, Yangzhou 225127, China; cwang@microflows.net

2 Jiangsu Key Laboratory of Micro and Nano Heat Fluid Flow Technology and Energy Application, School of Environmental Science and Engineering, Suzhou University of Science and Technology, Suzhou 215009, China

3 Key Laboratory of Energy Thermal Conversion and Control of Ministry of Education, School of Energy and Environment, Southeast University, Nanjing, Jiangsu 210096, China; 158065072@seu.edu.cn (J.S.); 220130421@seu.edu.cn (M.Z.)

* Correspondence: yaofeng@mail.usts.edu.cn (F.Y.); lywu@yzu.edu.cn (L.W.)

Abstract: The coupled effect of boiling and condensation inside a flat two-phase thermosyphon has a non-negligible influence on the two-phase fluid flow behavior and heat transfer process. Therefore, a flat two-phase thermosyphon with transparent wall was manufactured. Based on this device, a visualization experiment system was developed to study the vapor–liquid two-phase behaviors and thermal performance of the flat two-phase thermosyphon. A cross-shaped wick using copper mesh was embedded into the cavity of two-phase thermosyphon to improve the heat transfer performance. The effects of heat flux density, working medium, and wick structure on the thermal performance are examined and analyzed. The results indicated that a strong liquid disturbance is caused by the bubble motions, leading to the enhancement of both convective boiling and condensation heat transfer. More bubbles are generated as the heat flux increases; therefore, the disturbance of bubble motion on liquid pool and condensation film becomes stronger, resulting in better thermal performance of the flat two-phase thermosyphon. The addition of the wick inside the cavity effectively reduces the temperature oscillation of the evaporator wall. In addition, the wick structure provides backflow paths for the condensate owing to the effect of capillary force and enhances the vapor–liquid phase change heat transfer, resulting in the improvement of thermal performance for the flat two-phase thermosyphon.

Keywords: thermosyphon; phase change; two-phase flow; visualization

1. Introduction

Highly efficient cooling technologies have always been the subject of both scientific and engineering investigation in high-powered electronics circuits. Aiming to achieve highly efficient heat dissipation of high-heat-flux electronics, several advanced cooling technologies, including boiling cooling [1–3], liquid cooling [4,5], functional surface [6,7], microchannels heat sink [8–10], heat pipes [11–13], microfluidic engineering [14,15], metal foam [16], etc., have been introduced and applied in every field. Among these advanced cooling technologies, the heat pipes (such as grooved heat pipes [17], pulsation heat pipes [18], thermosyphons, etc.) are most widely used for the heat dissipation of microelectronic devices under high heat flux density due to high heat transfer capacity, good temperature uniformity, and no power consumption [19]. Among those heat pipes, the flat two-phase thermosyphon possesses higher heat transfer limit and superior temperature uniformity over other heat pipes, due to the operation principle of gravity-driven vapor–liquid flow and boiling phase change heat transfer [20,21]. The flat two-phase thermosyphon is considered a highly efficient heat spreader and shows good application prospects in solving uniform heat dissipation under high-heat-flux electronics [22–26]. Therefore, study on the flat two-phase thermosyphon's thermal properties is of significance and has become a topic of growing technological interest during the past decades.

It should be noted that the flat two-phase thermosyphon is a confined structure, where the vapor–liquid two-phase flow self-circulates between the evaporator and condensation surface. The confinement effect in turn affects the vapor–liquid flow dynamics of the flat heat pipe. The previous investigation indicated that the bubbles generated by the evaporation surface survive on the liquid surface for a while and continue to grow, merge, and burst when the liquid level inside the cavity is low [27]. In addition, the rapid rise of bubbles can promote the liquid convection heat transfer on the evaporator surface [28,29]. Liu et al [30] add a hollow foam metal block into the cavity of a flat two-phase thermosyphon; they observed that the dense bubbles generated in the foam metal fly to the condenser surface and break, and the liquid is caused to splash to the condenser surface by the intense movement of the bubbles under the optimal filling rate. As a result, the convective boiling and film condensation heat transfer are enhanced. In summary, the coupled effect of boiling and condensation inside the flat two-phase thermosyphon has a non-negligible influence on the two-phase fluid flow behavior and heat transfer process.

Although several experimental attempts have been devoted to investigating the thermal performance of flat two-phase thermosyphons, special attention is paid to thermal performance indices, equivalent thermal conductivity, and the maximum heat transfer capacity [31,32]. For example, Lips et al. [33] found that the thermal resistance of the flat two-phase thermosyphon reduces due to the liquid retention induced by the small cavity height. Ju et al. [34] developed a flat two-phase thermosyphon incorporating hybrid wicks, which is composed of a spreading layer with low thermal resistance and a supply structure for the working medium; then, a low thermal resistance and high heat flux limit over large heating areas can be achieved. In addition, the vapor–liquid two-phase flow behaviors are complex due to the interaction between the boiling and condensation. The thermo-hydrodynamic characteristics of the two-phase thermosyphon are still waiting to be explored by visualization experiment. Still, there are few studies have been conducted to visually study the internal vapor–liquid two-phase flow behaviors and thermal performance.

Apart from experimental investigation, limited theoretical attempts on vapor–liquid two-phase flow and phase change heat transfer in the two-phase thermosyphon have been conducted. Wu et al. investigated the co-existing boiling and condensation heat transfer in a confined space using lattice Boltzmann simulation [35]. It is indicated that the interaction of condensation and boiling in a two-phase thermosyphon is embodied in the condensate droplet contact the growing bubble that contributes to bubble detachment or motion.

The heat is mainly transferred by the process of boiling and condensation inside the flat two-phase thermosiphon. However, these two processes coexist in the confined cavity and interact with each other, resulting complex vapor–liquid two-phase flow behaviors. Compared with the study of boiling or condensation in a confined space separately, the research of the coupled effect of boiling and condensation is more helpful for understanding of the thermo-hydrodynamic behaviors inside a confined space. Therefore, an experimental system is developed to visually study the vapor–liquid two-phase flow and phase change heat transfer of the thermosyphon, especially to elucidate the coupled boiling and condensation heat transfer. Through the experiment, the effects of heat flux density, working medium, and wick structure on the vapor–liquid two-phase flow behavior and thermal performance are examined and investigated.

2. Description of Experiment

2.1. Experimental Apparatus

The heat transfer performance test system is shown in Figure 1. As shown, two electric heating rods (6 mm in diameter and 50 mm in length) are embedded inside the bottom side of the copper column (20 mm in diameter and 105 mm in length). The top side of the copper column is closely contacted with the evaporator surface of the flat two-phase thermosyphon. The heating power of the electric heating rods is provided by the DC power supply. The working condition adjustment is realized by adjusting the voltage regulator and the power meter. The electric heating power ranges from 20 W to 90 W in the experiment. The constant-temperature water bath provides cooling water with a constant temperature (±0.1 °C of temperature fluctuation) for the condenser section of the flat two-phase thermosyphon. The flow meter is used to control the flow rate of the circulating water. In the experiment, the temperature of the cooling water was set to room temperature of 25 °C, and the circulating water flow rate was set to 80 mL/min.

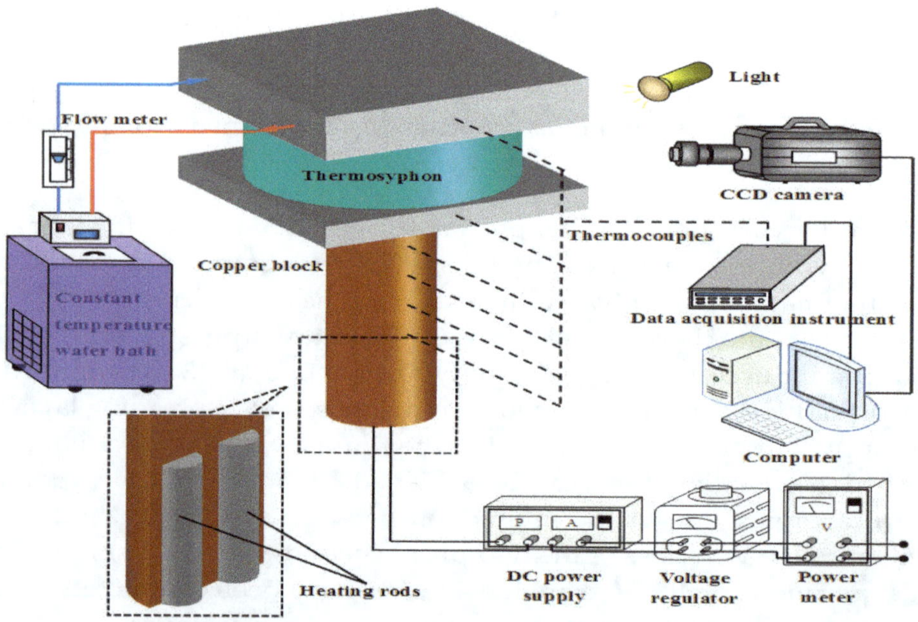

(**a**)

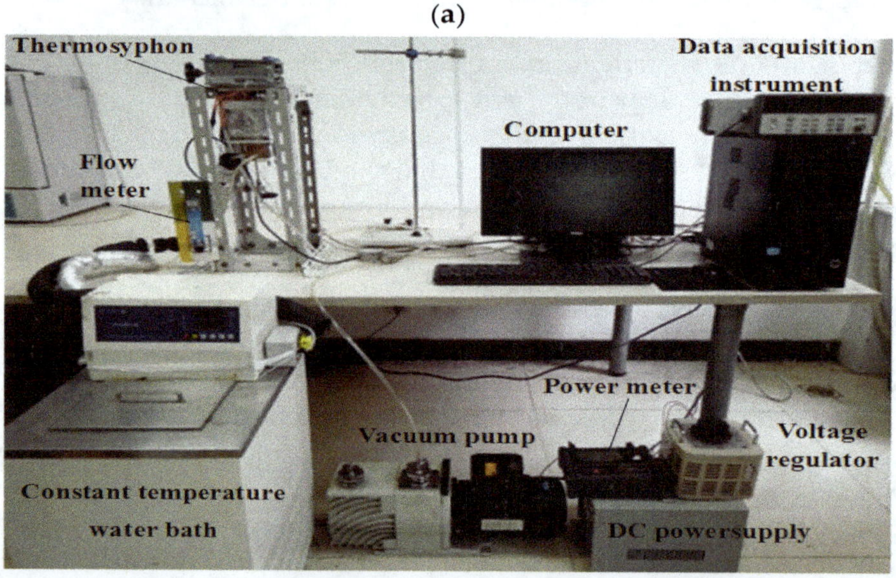

(**b**)

Figure 1. Schematic of experimental setup: (**a**) system diagram; (**b**) experimental rig.

As shown in Figure 2, the flat two-phase thermosyphon is assembled using an evaporator plate, a transparent quartz glass tube, a condenser plate, and a cooling water tank sealing plate. The glass tube is of 50 mm in outer diameter, 44 mm in inner diameter, and 3 mm in thickness. Both the evaporator and condenser plate are made of square brass plates with 54 mm in side length. The thicknesses of the evaporator and condenser plate are 5 mm and 10 mm, respectively. The evaporator and condenser surfaces are a smooth plane with a milled annular groove. Both ends of the glass tube are tightly embedded in the groove on the plate, so a closed cavity is formed. The cooling water tank is set at the back of the condenser plate and sealed by a sealing plate. The working medium is charged into the cavity through a charging hole drilled in the condenser plate. In the experiment, unless specified, deionized water was selected as the working liquid.

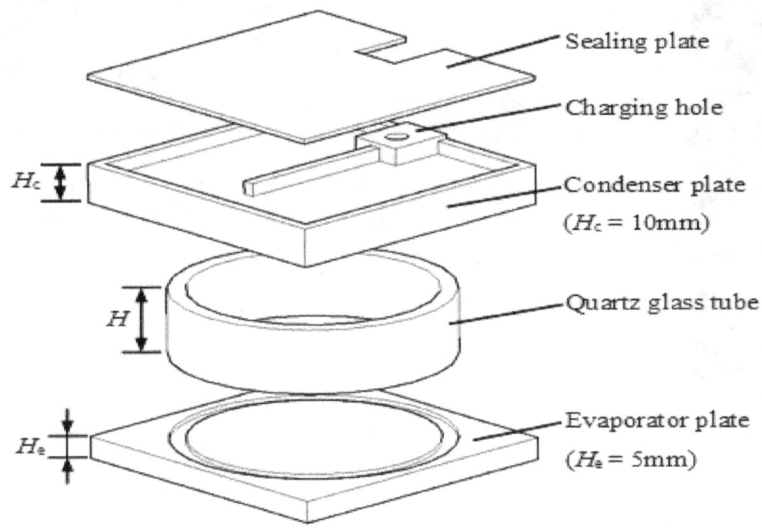

Figure 2. Schematic of the flat two-phase thermosyphon.

The CCD camera and light are used to monitor the vapor–liquid two-phase flow behavior in the cavity, and the thermocouples and data acquisition instrument are used to test the temperature changes during the experiment. Both the images and the tested temperature are recorded on the computer. In order to monitor the working conditions of the testing system, thermocouples are arranged in all parts of the system. Four thermocouple holes with a diameter of 0.5 mm and a depth of 10 mm are arranged in the upper side of the copper column, with respective distances of 5 mm, 20 mm, 35 mm, and 50 mm from the upper surface of the copper column. During operation, the heat load of the copper column imposed on the heat pipe is obtained according to the axial temperature distribution measured by the thermocouples in the copper column. The thermocouples are also arranged at the inlet and outlet of the cooling water tank for measuring the temperature change of the circulating water before and after the condenser section. Through the sensible heat gain of circulation fluid, the heat load input into the heat pipe is also obtained. The experiment test indicated that the difference of the heat load obtained by above two approaches is within 3%. Furthermore, six thermocouple holes have been arranged in the evaporator and condenser plate with the locations shown in Figure 3, so that the temperature changes of the evaporator and condenser surfaces of the flat two-phase thermosyphon can be monitored.

Through the performance test on the above experimental setup, it was indicated that there is an obvious temperature oscillation during the operation of the flat two-phase thermosyphon owing to the intermittent boiling. The presence of temperature oscillation is unfavorable for the application of flat two-phase thermosyphons. In order to eliminate the temperature oscillation, a cross-shaped wick formed by folding 200-mesh copper screen was installed into the cavity of the flat two-phase thermosyphon, as shown in Figure 4. The wick height is 15 mm, which is the same as the cavity height.

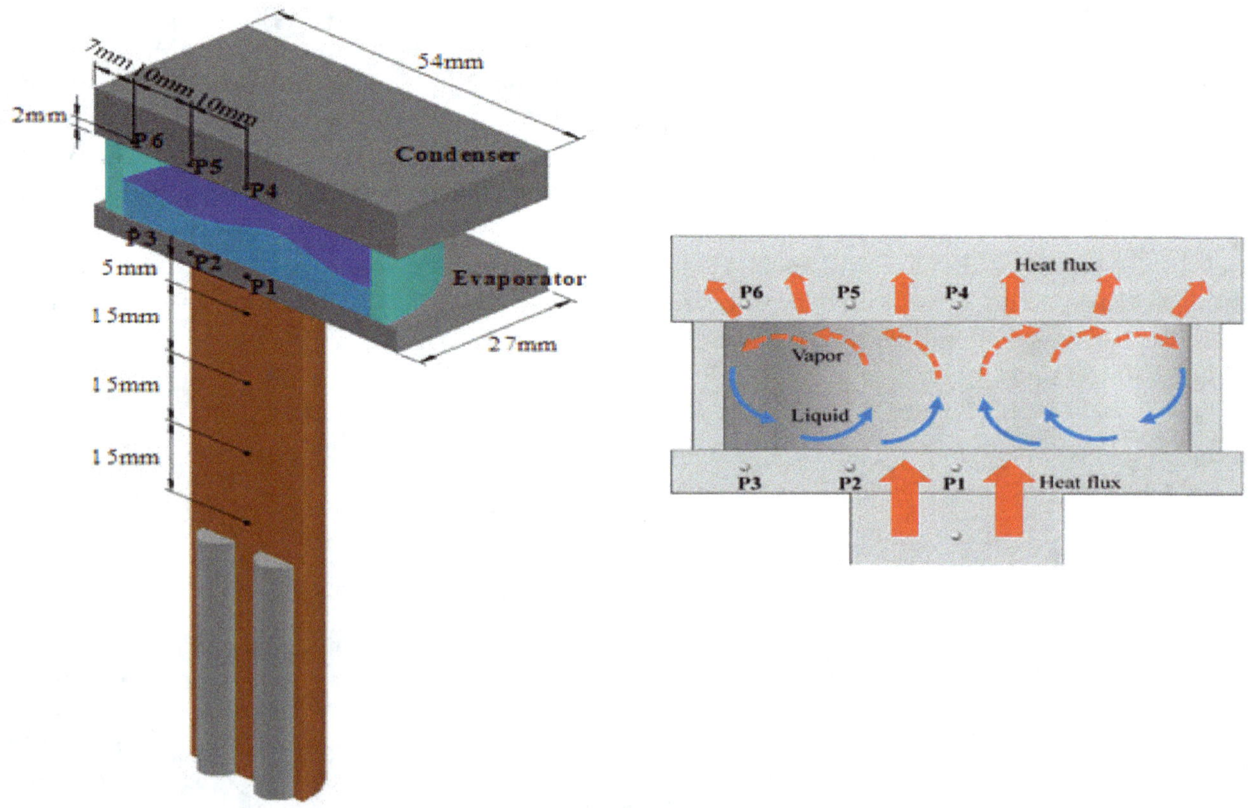

(a) (b)

Figure 3. Schematic of temperature measuring points: (**a**) thermocouple positions; (**b**) diagram of temperature measuring in the two-phase thermosyphon.

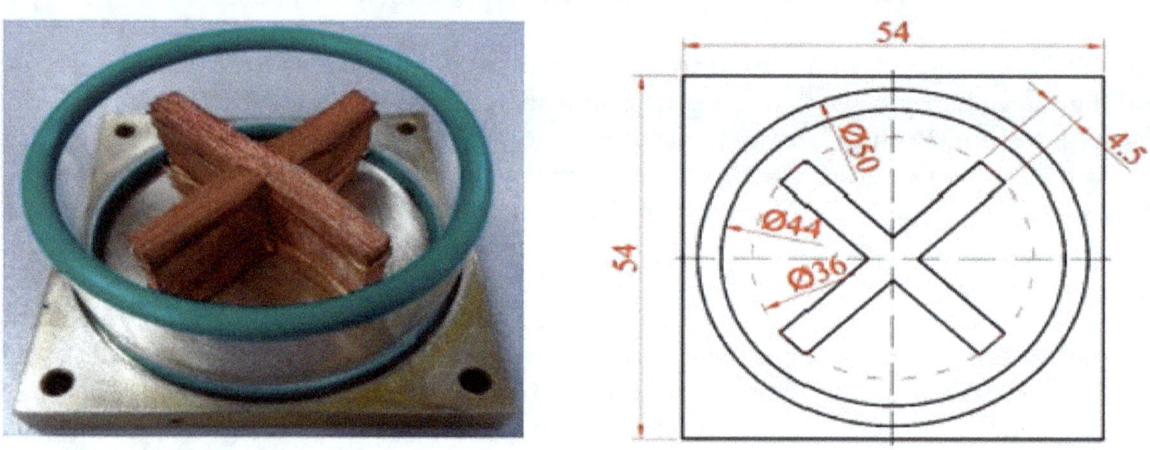

(a) (b)

Figure 4. Structure of the cross-shaped wick: (**a**) picture of cross-shaped wick; (**b**) dimensions of cross-shaped wick.

2.2. Data Reduction

The total thermal resistance of the flat two-phase thermosyphon characterizes the temperature difference required for transporting unit heat power, which is an important parameter for measuring the heat transfer performance of the flat two-phase thermosyphon. In this paper, the thermal resistance is adopted to quantitatively evaluate the heat transfer performance of the flat two-phase thermosyphon, which is defined as

$$R = \Delta\overline{T}/Q \tag{1}$$

where $\Delta\overline{T}$ is the average temperature difference between the evaporator section and the condenser section, as follows:

$$\Delta\overline{T} = \overline{T}_e - \overline{T}_c$$

$$\overline{T}_e = \frac{T_1 + T_2 + T_3}{3},\ \overline{T}_c = \frac{T_4 + T_5 + T_6}{3}$$

where R is the total thermal resistance, and $\overline{T}_e$ and $\overline{T}_c$ represent the average temperatures of the evaporator section and condenser section, respectively. The subscripts 1–6 represent the temperature measurement points of the evaporator and condenser sections (P1–P6, see Figure 3). Q is the heat load. This definition characterizes the resistance of heat transfer from the evaporator surface to the condenser surface, and smaller thermal resistance means better heat transfer performance of the flat two-phase thermosyphon. The heat load can be obtained as $Q = \rho G C_p(T_{\text{out}} - T_{\text{in}})$, where ρ is density, G is the volume flow rate, C_p is specific heat, and T_{out} and T_{in} are the outlet temperature and inlet temperature, respectively, of the cooling water. The uncertainties of $\Delta\overline{T}$, Q, and R are 0.8%, 2.62%, and 2.71% according to the data error analysis.

2.3. Experiment Procedures

Before each experiment, the evaporation surface and the condensation surface were polished with metallographic abrasive paper. Then, in order to remove copper rust and dust, the surfaces were cleaned using deionized water, absolute ethanol, acetone, and deionized water in sequence. After ensuring that the evaporation plate, the condensation plate, and the quartz glass tube were clean, the two-phase thermosyphon was assembled. Then, the two-phase thermosyphon was subjected to vacuuming, liquid filling, and leak detection.

In order to ensure good repeatability of the experimental results, the experiment steps are as follows:

(1) Connect each unit of the experiment system, and then perform circuit and connection checks on every unit to ensure that the experimental system is correct.
(2) Open constant-temperature water bath, and adjust the temperature and flow rate to the preset values. Turn on the high-speed CCD camera to debug the shooting effect.
(3) Turn on the data acquisition instrument, and observe the temperature change until the temperature at each measuring point reaches a steady state.
(4) Maintain the operation of every other piece of equipment, reset the data acquisition instrument, and record the temperature data of the temperature measurement point.
(5) Observe and record the gas–liquid two-phase behavior in the confined space during the start-up and quasi-steady operation in real time. Collect the temperature data. When the change of temperature is less than 0.5 °C over 20 min, the heat balance can be considered to be reached, and the experiment for the current working condition can be finished.
(6) After each working condition is completed, adjust the input voltage to zero, and keep the constant-temperature water bath cooling water circulating until the temperature falls back to the initial temperature, then perform the experiment on the next working condition.

3. Results and Discussion

When there is a heat load imposed on the flat two-phase thermosyphon, the gravity-driven vapor–liquid two-phase self-circulation flow is generated in the cavity, accompanied by sensible heat transport and latent heat transport (evaporation/boiling and condensation). During this vapor–liquid phase change process, the evaporation, boiling, and condensation phase change, vapor–liquid two-phase flow, and phase interface fluctuations are very complex and coupled with each other in the flat two-phase thermosyphon. This paper began an observation of the coupled boiling and condensation phase change inside the confined cavity, and now the effects of heat flux, working medium, and wick structure on two-phase flow behaviors and thermal performance of the flat two-phase thermosyphon are examined and analyzed.

3.1. Coupled Boiling and Condensation Phase Change

The nucleate boiling occurs on the evaporator section as soon as a certain degree of superheat is achieved when the liquid continues to be heated in a two-phase thermosyphon. Figure 5 shows the typical process of nucleate boiling, namely, the bubbles generate on the evaporator surface, grow, and then detach from the surface, accompanying intense gas–liquid two-phase flow and phase change heat transfer in the cavity. As shown in the figure, bubble nucleation is first generated on the heated surface when the energy accumulates to a certain degree. Then, the bubble starts to grow according to the evaporation of the liquid microlayer at the bottom of the bubble and the heat transferring from the superheated liquid around the bubble. When the bubble grows to a certain size on the evaporator surface, it will depart from the surface by the combined effect of surface tension, buoyancy, and inertial force. After that, the bubble continues to grow to a diameter equal to the liquid level. When the bubble rises, the liquid film will be lifted directly against the condenser surface, coursing radial motion of the liquid on the condenser surface by the scour of the gas flow. The rising movement of bubbles causes strong disturbance of the liquid pool, which is beneficial for enhancing convective boiling phase change heat transfer. In addition, the condensation heat transfer can also be increased due to the scour phenomenon caused by the bubble motion on the condenser surface.

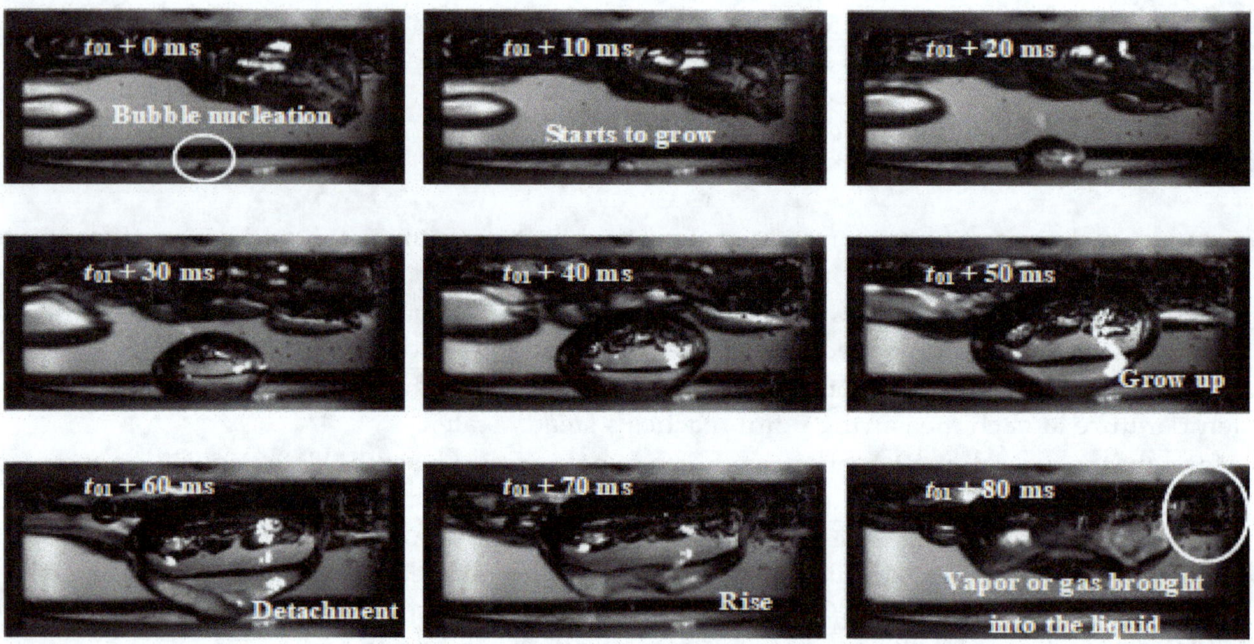

Figure 5. The evolution of gas–liquid two-phase flow pattern in cavity (φ = 80%, q = 17.7 W/cm^2, H = 20 mm).

It is conceivable that during the operation of the flat two-phase thermosyphon, the boiling and condensation phase change heat transfer processes coexist and occur simultaneously inside the cavity, accompanied by complex gas–liquid two-phase flow behavior, such as bubble generation, growth, movement, coalescence, rupture, condensation, and fluctuation of the gas–liquid interface, etc. Therefore, these complex thermo-hydrodynamic behaviors play an important role in the operation and heat transfer of the flat two-phase thermosiphon.

3.2. Effect of Heat Flux

The heat flux directly determines the intensity of vapor–liquid phase change inside a confined cavity. Figure 6 shows the variation of temperature on the center point of the evaporator surface over time under different heat inputs. Figure 7 plots the total thermal resistance as a function of heat flux, and the corresponding vapor–liquid two-phase flow behaviors affected by heat flux are illustrated in Figure 7. It can be seen from the figures that, with the increase of heat flux, the thermal resistance of the flat two-phase thermosyphon shows a decreasing trend (see Figure 7), and the boiling heat transfer intensifies as indicated by increasing bubble number (see Figure 8).

When the heat flux increases, the temperature on the center point of the evaporator surface becomes higher when it reaches steady state, as shown in Figure 6. The superheat degree of the evaporator surface temperature increases, so more nucleation sites are activated, leading to generation of more bubbles. In this case, the liquid disturbance is intensified and, hence, the convective heat transfer between the liquid and solid surface is improved. As a result, the boiling heat transfer is enhanced. In addition, the condensate backflow into the liquid pool is promoted by the bubble rupture as well as the liquid flushing on the condenser surface. Induced by the stronger disturbance of the bubble by a larger heat load, the thickness of the condensate film is thinned, so the condensation heat transfer is enhanced. Therefore, as the heat load increases, the thermal resistance of the flat two-phase thermosyphon gradually becomes smaller.

It also can be seen from Figure 7 that the thermal resistance of the two-phase thermosyphon charged with water is smaller than that charged with ethanol as the working medium. This can be explained by the fact that the viscous resistance of ethanol flowing through the cavity is greater because of its higher viscosity, which causes weaker liquid disturbance and condensate backflow. Furthermore, the heat transport capacity of the ethanol is smaller than that of water due to its lower latent heat of vaporization.

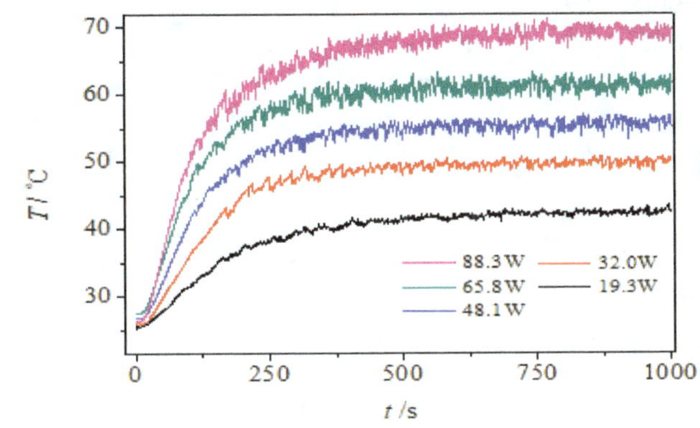

Figure 6. Effect of heat input on the temperature of the center point on evaporator surface.

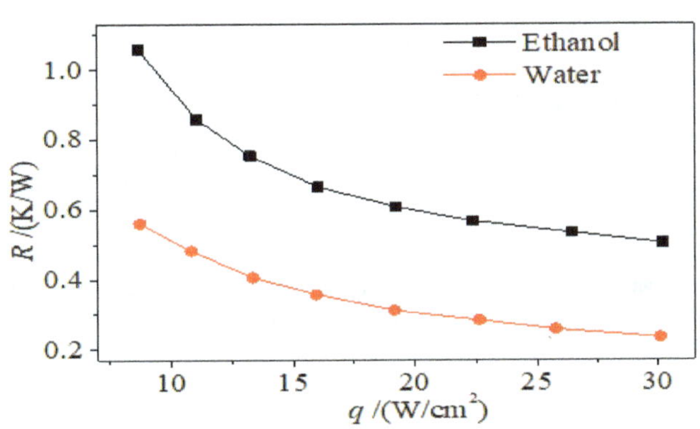

Figure 7. Effect of heat input on thermal resistance (φ = 65%).

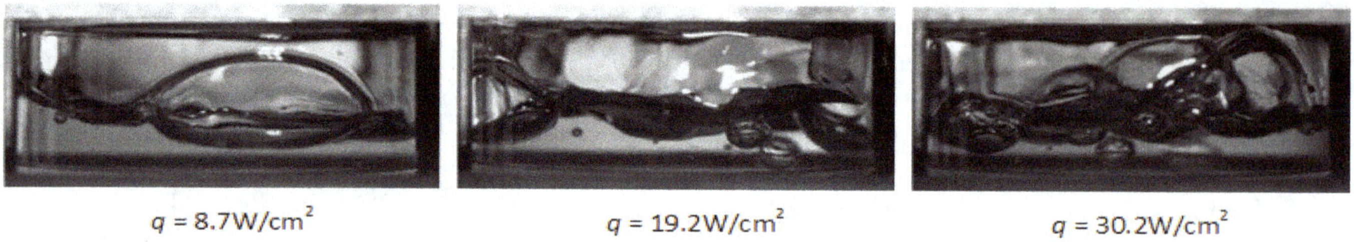

Figure 8. Effect of heat flux on two-phase flow behavior (working medium: water).

3.3. Effect of Wick Presence

The experimental results indicated that the temperature of the evaporator wall oscillates with time when the flat two-phase thermosyphon operates. As shown in Figure 9a, there is large temperature oscillation during the quasi-steady state process, and the temperature oscillation magnitude is about 5 °C when the heat flux is 10.9 W/cm^2. Note that the temperature mainly arises from the intermittent boiling when the heat flux imposed on the heat pipe is not sufficiently large. Considering the unfavorable temperature oscillation, the cross-shaped wick is introduced in this paper to eliminate the temperature oscillation. As expected, the temperature oscillation is obviously reduced once the cross-shaped wick using copper mesh is embedded into the cavity of the two-phase thermosyphon, as shown in Figure 9b. Under the case with the wick, the temperature oscillation magnitude is only about 0.15 °C, indicating the feasibility of the wick structure to enhance the operation stability of the flat two-phase thermosyphon.

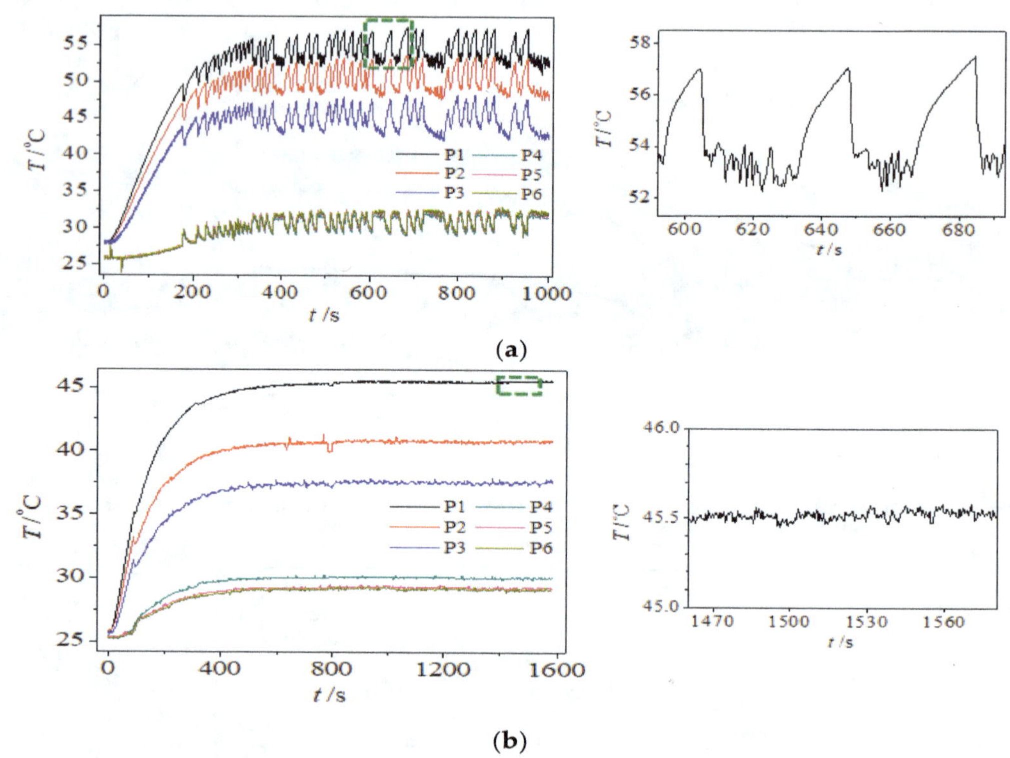

Figure 9. Effect of wick presence on evaporator wall temperature evolution: (**a**) no wick (φ = 50%, q = 10.9 W/cm^2, H = 15 mm); (**b**) cross-shaped wick (φ = 50%, q = 9.9 W/cm^2, H = 15 mm).

Apart from the reduction of temperature oscillation, the presence of the wick structure can also improve the thermal performance of the flat two-phase thermosyphon. Figure 10 describes the thermal resistance of the flat two-phase thermosyphon with a cross-shaped wick as a function of heat flux.

In order to highlight the influence of the wick structure, the thermal resistance of the flat two-phase thermosyphon without a wick is also included in the figure. It can be seen from the figure that the thermal performance of the flat two-phase thermosyphon is better than that with no wick structure. This is due to the reason that the heat is more easily transferred from the evaporator heat spot to the upper and side liquid owing to the high thermal conduction of the metal mesh wick. As shown in Figure 11, the liquid disturbance is increased as the detachment frequency and number of generated bubbles on the evaporator surface are significantly increased, so the convective boiling heat transfer of the liquid pool is improved. Besides, since the wick top contacts with the condenser surface, a shortcut of condensate backflow is provided by the wick, i.e., the condensate backflow can be promoted by the capillary force provided by the wick.

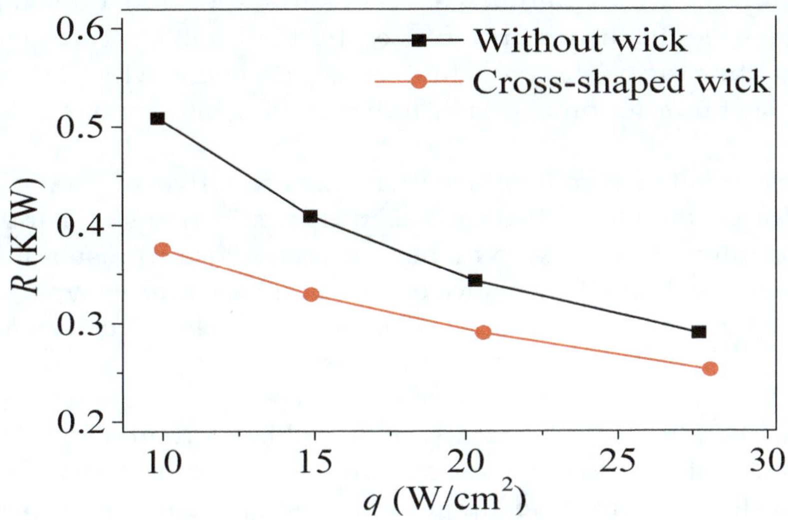

Figure 10. Effect of wick presence on thermal resistance (H = 15 mm, φ = 50%).

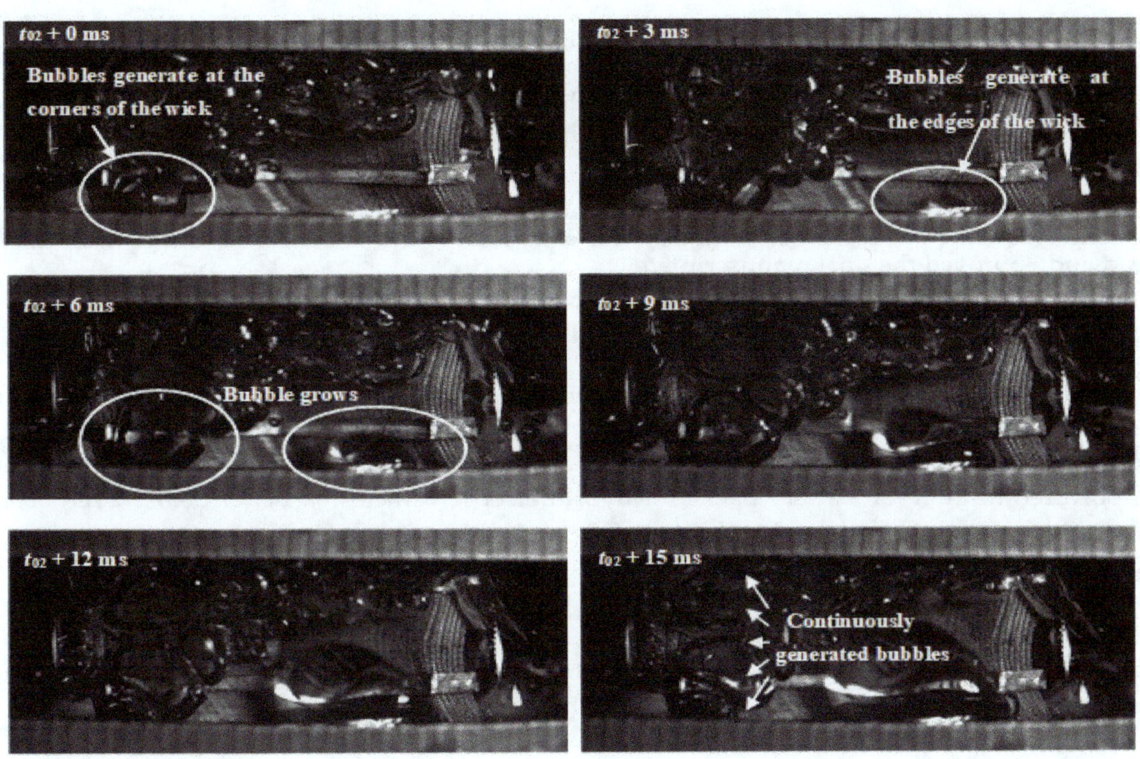

Figure 11. Evolution of vapor–liquid two-phase flow inside the wick thermosyphon (H = 15 mm, φ = 50%, q = 20 W/cm^2).

4. Conclusions

A flat two-phase thermosyphon with transparent wall was manufactured and assembled in this paper, and a visualization experiment system was developed for the visual experimental study of the boiling and condensation phase change heat transfer process inside the cavity. Considering the unfavorable temperature oscillation, a cross-shaped wick was introduced to eliminate the temperature oscillation. The effects of heat flux and wick structure on the two-phase flow behaviors and thermal performance were analyzed. The following main conclusions were obtained:

(1) Contributed by the evaporation of the liquid microlayer and the heat transferring from the superheated liquid around, bubbles grow to a certain size on the evaporator surface and then depart from the surface by the combined effect of surface tension, buoyancy, and inertial force. As a result, a strong liquid disturbance caused by the bubble motions such as growth, rise, coalescence, and rupture can be observed during the experiment, which is beneficial for enhancing the phase change heat transfer processes including boiling and condensation.

(2) More bubbles are generated as more nucleation sites are activated on the heated surface when the heat flux increases; therefore, the liquid disturbance is intensified and the thickness of the condensate film is thinned, leading to higher convective heat transfer and condensation heat transfer. In addition, the thermal resistance of the two-phase thermosyphon charged with water is smaller than that charged with ethanol because of lower viscosity and higher latent heat of vaporization.

(3) When a wick is embedded into the cavity, more bubbles are generated and the detachment frequency is significantly increased, causing stronger liquid disturbance. Therefore, the heat is more easily transferred from the evaporator heat spot to the liquid in the cavity, leading to a decrease in the degree of superheat of the liquid. As a result, the temperature oscillation is obviously reduced.

(4) As a shortcut path is constructed when a wick is used in the cavity, the backflow efficiency can be improved, resulting in the improvement of thermal performance for the flat two-phase thermosyphon.

Author Contributions: F.Y. provided guidance and supervision; C.W. implemented the main research, discussed the results, and wrote the paper; M.C.Z. collected the data; L.Y.W. and J.S. checked and revised the paper; all authors read and approved the final manuscript.

References

1. Liang, G.T.; Mudawar, I. Review of pool boiling enhancement with additives and nanofluids. *Int. J. Heat Mass Transf.* **2018**, *124*, 423–453. [CrossRef]
2. Al-Zareer, M.; Dincer, I.; Rosen, M.A. Novel thermal management system using boiling cooling for high-powered lithium-ion battery packs for hybrid electric vehicles. *J. Power Source* **2017**, *363*, 291–303. [CrossRef]
3. Zhang, C.; Chen, Y.; Wu, R.; Shi, M. Flow boiling in constructal tree-shaped minichannel network. *Int. J. Heat Mass Transf.* **2011**, *54*, 202–209. [CrossRef]
4. Amalfi, R.L.; Salamon, T.; Lamaison, N.; Marcinichen, J.B.; Thome, J.R. Two-phase liquid cooling system for electronics—Part 3: Ultracompact liquid-cooled condenser. In Proceedings of the 16th Intersociety Conference on Thermal and Thermomechanical Phenomena in Electronic Systems (ITherm), Orlando, FL, USA, 30 May–2 June 2017; pp. 687–695.
5. Ling, Z.; Cao, J.; Zhang, W.; Zhang, Z.; Fang, X.; Gao, X. Compact liquid cooling strategy with phase change materials for li-ion batteries optimized using response surface methodology. *Appl. Energy* **2018**, *228*, 777–788. [CrossRef]

6. Zhang, C.; Deng, Z.; Chen, Y. Temperature jump at rough gas–solid interface in couette flow with a rough surface described by cantor fractal. *Int. J. Heat Mass Transf.* **2014**, *70*, 322–329. [CrossRef]
7. Zhang, C.; Chen, Y.; Deng, Z.; Shi, M. Role of rough surface topography on gas slip flow in microchannels. *Phys. Rev. E* **2012**, *86*. [CrossRef] [PubMed]
8. Chen, Y.; Zhang, C.; Shi, M.; Yang, Y. Thermal and hydrodynamic characteristics of constructal tree-shaped minichannel heat sink. *AIChE J.* **2009**, *56*, 2018–2029. [CrossRef]
9. Najim, M.; Feddaoui, M.B.; Nait Alla, A.; Charef, A.; Kabeel, A.E. New cooling approach using successive evaporation and condensation of a liquid film inside a vertical mini-channel. *Int. J. Heat Mass Transf.* **2018**, *122*, 895–912. [CrossRef]
10. Chen, Y.; Shen, C.; Shi, M.; Peterson, G.P. Visualization study of flow condensation in hydrophobic microchannels. *AIChE J.* **2014**, *60*, 1182–1192. [CrossRef]
11. Chen, X.; Ye, H.; Fan, X.; Ren, T.; Zhang, G. A review of small heat pipes for electronics. *Appl. Therm. Eng.* **2016**, *96*, 1–17. [CrossRef]
12. Qu, J.; Li, X.J.; Cui, Y.Y.; Wang, Q. Design and experimental study on a hybrid flexible oscillating heat pipe. *Int. J. Heat Mass Transf.* **2017**, *107*, 640–645. [CrossRef]
13. Naphon, P.; Wongwises, S.; Wiriyasart, S. Application of two-phase vapor chamber technique for hard disk drive cooling of pcs. *Int. Commun. Heat Mass Transf.* **2013**, *40*, 32–35. [CrossRef]
14. Chen, Y.; Gao, W.; Zhang, C.; Zhao, Y. Three-dimensional splitting microfluidics. *Lab Chip* **2016**, *16*, 1332–1339. [CrossRef] [PubMed]
15. Chen, Y.P.; Deng, Z.L. Hydrodynamics of a droplet passing through a microfluidic t-junction. *J. Fluid Mech.* **2017**, *819*, 401–434. [CrossRef]
16. Deng, Z.; Liu, X.; Zhang, C.; Huang, Y.; Chen, Y. Melting behaviors of pcm in porous metal foam characterized by fractal geometry. *Int. J. Heat Mass Transf.* **2017**, *113*, 1031–1042. [CrossRef]
17. Chen, Y.; Zhang, C.; Shi, M.; Wu, J.; Peterson, G.P. Study on flow and heat transfer characteristics of heat pipe with axial "Ω"-shaped microgrooves. *Int. J. Heat Mass Transf.* **2009**, *52*, 636–643. [CrossRef]
18. Deng, Z.; Zheng, Y.; Liu, X.; Zhu, B.; Chen, Y. Experimental study on thermal performance of an anti-gravity pulsating heat pipe and its application on heat recovery utilization. *Appl. Therm. Eng.* **2017**, *125*, 1368–1378. [CrossRef]
19. Qu, J.; Wu, H.Y.; Cheng, P.; Wang, Q.; Sun, Q. Recent advances in mems-based micro heat pipes. *Int. J. Heat Mass Transf.* **2017**, *110*, 294–313. [CrossRef]
20. Lv, L.; Li, J. Managing high heat flux up to $500 w/cm^2$ through an ultra-thin flat heat pipe with superhydrophilic wick. *Appl. Therm. Eng.* **2017**, *122*, 593–600. [CrossRef]
21. Zhang, M.; Liu, Z.; Ma, G. The experimental investigation on thermal performance of a flat two-phase thermosyphon. *Int. J. Therm. Sci.* **2008**, *47*, 1195–1203. [CrossRef]
22. Chang, S.W.; Yu, K.C. Thermal performance of reciprocating two-phase thermosyphon with nozzle. *Int. J. Therm. Sci.* **2018**, *129*, 14–28. [CrossRef]
23. Fertahi, S.E.D.; Bouhal, T.; Agrouaz, Y.; Kousksou, T.; El Rhafiki, T.; Zeraouli, Y. Performance optimization of a two-phase closed thermosyphon through cfd numerical simulations. *Appl. Therm. Eng.* **2018**, *128*, 551–563. [CrossRef]
24. Narcy, M.; Lips, S.; Sartre, V. Experimental investigation of a confined flat two-phase thermosyphon for electronics cooling. *Exp. Therm. Fluid. Sci.* **2018**, *96*, 516–529. [CrossRef]
25. Wang, X.Y.; Wang, Y.F.; Chen, H.J.; Zhu, Y.Z. A combined cfd/visualization investigation of heat transfer behaviors during geyser boiling in two-phase closed thermosyphon. *Int. J. Heat Mass Transf.* **2018**, *121*, 703–714. [CrossRef]
26. Fadhl, B. Modelling of the Thermal Behaviour of a Two-Phase Closed Thermosyphon. Ph.D. Thesis, Brunel University London, London, UK, March 2016.
27. Zhang, G.; Liu, Z.; Wang, C. An experimental study of boiling and condensation co-existing phase change heat transfer in small confined space. *Int. J. Heat Mass Transf.* **2013**, *64*, 1082–1090. [CrossRef]
28. Chen, Y.; Yu, F.; Zhang, C.; Liu, X. Experimental study on thermo-hydrodynamic behaviors in miniaturized two-phase thermosyphons. *Int. J. Heat Mass Transf.* **2016**, *100*, 550–558. [CrossRef]
29. Zhang, G.; Liu, Z.; Wang, C. A visualization study of the influences of liquid levels on boiling and condensation co-existing phase change heat transfer phenomenon in small confined spaces. *Int. J. Heat Mass Transf.* **2014**, *73*, 415–423. [CrossRef]

30. Liu, Z.L.; Zheng, F.W.; Li, Y.X. Enhancing boiling and condensation co-existing heat transfer in a small and closed space by copper foam inserts. *Int. J. Heat Mass Transf.* **2017**, *108*, 961–971. [CrossRef]
31. Peng, H.; Li, J.; Ling, X. Study on heat transfer performance of an aluminum flat plate heat pipe with fins in vapor chamber. *Energy Convers. Manag.* **2013**, *74*, 44–50. [CrossRef]
32. Hassan, H.; Harmand, S. An experimental and numerical study on the effects of the flat heat pipe wick structure on its thermal performance. *Heat Transf. Eng.* **2015**, *36*, 278–289. [CrossRef]
33. Lips, S.; Lefevre, F.; Bonjour, J. Combined effects of the filling ratio and the vapour space thickness on the performance of a flat plate heat pipe. *Int. J. Heat Mass Transf.* **2010**, *53*, 694–702. [CrossRef]
34. Ju, Y.S.; Kaviany, M.; Nam, Y.; Sharratt, S.; Hwang, G.S.; Catton, I.; Fleming, E.; Dussinger, P. Planar vapor chamber with hybrid evaporator wicks for the thermal management of high-heat-flux and high-power optoelectronic devices. *Int. J. Heat Mass Transf.* **2013**, *60*, 163–169. [CrossRef]
35. Wu, S.; Yu, C.; Yu, F.; Chen, Y. Lattice boltzmann simulation of co-existing boiling and condensation phase changes in a confined micro-space. *Int. J. Heat Mass Transf.* **2018**, *126*, 773–782. [CrossRef]

4

Experimental Investigation of Flow-Induced Motion and Energy Conversion of a T-Section Prism

Nan Shao, Jijian Lian, Guobin Xu, Fang Liu, Heng Deng, Quanchao Ren and Xiang Yan *

State Key Laboratory of Hydraulic Engineering Simulation and Safety, Tianjin University, No. 92, Wei Jin Road, Nan Kai District, Tianjin 300072, China; shaonan@tju.edu.cn (N.S.); tju_luntan@126.com (J.L.); xuguob@sina.com (G.X.); fangliu@tju.edu.cn (F.L.); d_heng@yeah.net (H.D.); 2014205258@tju.edu.cn (Q.R.)

* Correspondence: xiangyan@tju.edu.cn

Abstract: Flow-induced motion (FIM) performs well in energy conversion but has been barely investigated, particularly for prisms with sharp sections. Previous studies have proven that T-section prisms that undergo galloping branches with high amplitude are beneficial to energy conversions. The FIM experimental setup designed by Tianjin University (TJU) was improved to conduct a series of FIM responses and energy conversion tests on a T-section prism. Experimental results are presented and discussed, to reveal the complete FIM responses and power generation characteristics of the T-section prism under different load resistances and section aspect ratios. The main findings are summarized as follows. (1) Hard galloping (HG), soft galloping (SG), and critical galloping (CG) can be observed by varying load resistances. When the load resistances are low, HG occurs; otherwise, SG occurs. (2) In the galloping branch, the highest amplitude and the most stable oscillation cause high-quality electrical energy production by the generator. Therefore, the galloping branch is the best branch for harvesting energy. (3) In the galloping branch, as the load resistances decrease, the active power continually increases until the prism is suppressed from galloping to a vortex-induced vibration (VIV) lower branch with a maximum active power P_{harn} of 21.23 W and a maximum η_{out} of 20.2%. (4) Different section aspect ratios (α) can significantly influence the FIM responses and energy conversions of the T-section prism. For small aspect ratios, galloping is hardly observed in the complete responses, but the power generation efficiency ($\eta_{out,0.8}$ = 27.44%) becomes larger in the galloping branch.

Keywords: flow-induced motion; sharp sections; T-section prism; load resistances; section aspect ratios; energy conversion

---

1. Introduction

The flow-induced motion (FIM) phenomenon [1] widely exists in the civil engineering field, and it can lead to the failure of oscillating structures such as solar receiver tubes [2–4], long-span bridges [5], parallel twin bridges [6], offshore risers [7], and aircraft [8]. Many early studies have devoted great efforts to the suppression [9,10] of FIM. With the study of FIM increasing significantly [11–14], researchers [15–17] are gradually paying more attention to the potential of FIM energy. Many creative structures [18–20] have been proposed for exploiting this energy, especially for the vortex-induced vibration (VIV) [21] and galloping [22,23] responses of FIM.

VIV occurs due to the alternating shedding of vortices from either side of the bluff cylinder [1]. Galloping occurs due to the forces acting on a prism as it is subjected to periodic variations in the angle of attack of the flow [24]. An isolated smooth circular cylinder can only undergo VIV, while galloping is rarely observed [25]. On the contrary, galloping is easily observed for non-circular-section prisms such as rectangular prisms, triangular prisms, and passive turbulence control (PTC) circular cylinders, etc.

There are some studies on rectangular prisms [26–29] to measure the aerodynamic forces during unsteady galloping oscillations, and some studies report that the incident angle has a significant influence on the performance of prisms. Research on triangular prisms has concluded that the stability of galloping strongly depends on the incident flow orientation and on the aspect ratio, as well as on the mass and damping [30–33]. In addition, two different types of galloping, soft galloping (SG) and hard galloping (HG), were first discovered during trials on PTC circular cylinders by Park [21]. SG is self-excited by VIV, and HG cannot be self-excited by VIV, but HG can be excited by external excitation at a high-flow velocity. Furthermore, Zhang, J. et al. [34,35] also discovered the phenomenon of SG and HG in a study on triangular prisms, and pointed out two HG types: I-type HG and II-type HG. Subsequently, Lian, J. et al. [36] found that the VIV and galloping of the triangular prism in complete FIM responses depended on the damping, stiffness, and mass of the system, as well as the section of the prism.

Research on VIV and galloping with energy conversion targets has developed rapidly in recent years. FIM power generation equipment such as VIVACE (Vortex-Induced Vibration for Aquatic Clean Energy) was successively proposed and gradually optimized. By inventing the virtual damper spring system [37,38] (V_{CK}) and the PTC [39], Bernitsas et al., in the University of Michigan, carried out many experiments to further understand FIM responses and power generation for circular cylinders at high damping [37], a high Reynolds number [21,39,40], variable stiffness [41,42], and variable mass [41], etc. It was found that a high Reynolds number, high damping, and low mass helped to improve power generation capacity and oscillation strength. In addition, non-cylindrical oscillators such as PTC cylinders performed with outstanding advantages in power generation due to the high amplitude of galloping.

At present, most researchers pay more attention to oscillators with regular cross-sections, such as cylinders, triangular prisms, square prisms, and rectangular prisms. However, less attention is paid to oscillators with irregular cross-sections, such as the T-section. Recently, FIM tests on T-section prisms were conducted in Tianjin University [43]; the published study reported that T-section prisms also presented galloping responses. Actually, the T-section prism can be regarded as a simplified triangular prism with axisymmetric structures. Until now, previous studies on the FIM of T-section prisms have not been systematic, and have only focused on oscillation responses. The complete FIM responses and the power generation characteristics of T-section prisms have not been investigated systematically. In order to better understand the complete FIM responses and their power generation characteristics, a series of tests were conducted, specifically including the following three aspects:

(1) Experimental research on FIM responses to external forces were conducted to determine the complete FIM responses of T-section prisms;

(2) FIM power generation tests with different load resistances were carried out to investigate the power generation of the T-section prism in complete FIM responses;

(3) In order to guide the optimization design of the prism, FIM responses and power generation tests at different section aspect ratios were conducted.

2. Experiment Setup

2.1. Water Channel and Calibration of Flow Velocity

All experiments were conducted in a recirculating water channel at the State Key Laboratory of Hydraulic Engineering Simulation and Safety (SKL-HESS) of Tianjin University (shown in Figure 1). The recirculating water channel consisted of a water tank, a variable frequency power pump, a 2-m wide flow channel, a bend flow channel, a contraction section, and a 1-m wide flow channel. The whole length was about 50 m. The water tank dimensions were 5 m (length) × 5 m (width) × 2 m (height). The channel was made out of transparent tempered glass and it was powered by a 90-kW variable frequency power pump with a maximum speed of 490 r/min. It could recirculate 200 cubic meters of fresh water at flow rates up to 2600 L/s. The frequency range of water pump was 0.0–50.0 Hz, controlled by a frequency conversion controller (FCC). In the 1-m wide flow channel, the velocity variation range was 0.0–1.8 m/s. The water depth was 1.34 m.

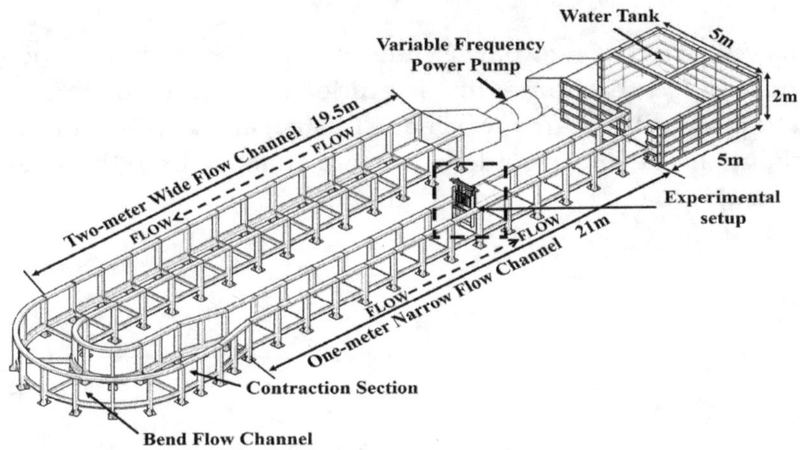

Figure 1. Recirculating water channel system.

In order to guarantee the authenticity of the results, two test instruments, a Pitot tube with a differential pressure transmitter and a propeller current meter [36], were employed to measure the flow velocity. The accuracy of the differential pressure transmitter was within ±0.1% of 6 KPa, which was the linear available range, and the resolution was within 0.01% FS (Full Scale). All of the data were recorded over a time interval of 60 s at a 40 Hz sampling rate. The probes of the Pitot tube and the propeller were placed 1 m in front of the T-section prism.

All experiments were conducted using the TrSL3 (20,000 < Re < 300,000) flow regime [13]. This study covered a range of Reynolds numbers of 45,133 ≤ Re ≤ 116,396 (0.516 m/s ≤ U ≤ 1.332 m/s). In order to describe the incoming flow, the flow velocity and the turbulence were both analyzed. We conducted three tests with different flow velocities. The velocity profile and the turbulence profile of the recirculating water channel were taken at a 15–100-cm deep area of the test section, as shown in Figure 2. In Figure 2a, the flow velocity near the bottom was smaller, due to viscous action. The average flow velocities in the vertical direction of three tests were 0.62 m/s, 0.84 m/s, and 1.22 m/s, respectively. The difference between the flow velocities in the oscillation range (40 cm–80 cm) was small. In Figure 2b, the turbulence grew as the flow velocity decreased and the depth increased. The average turbulences in the vertical direction of the three tests were 8%, 14%, and 24%, respectively. The difference in the turbulence in the oscillation range (40 cm–80 cm) was small as well. In a word, it could be ensured that the incoming flow in the oscillation range of the prism was a uniform flow.

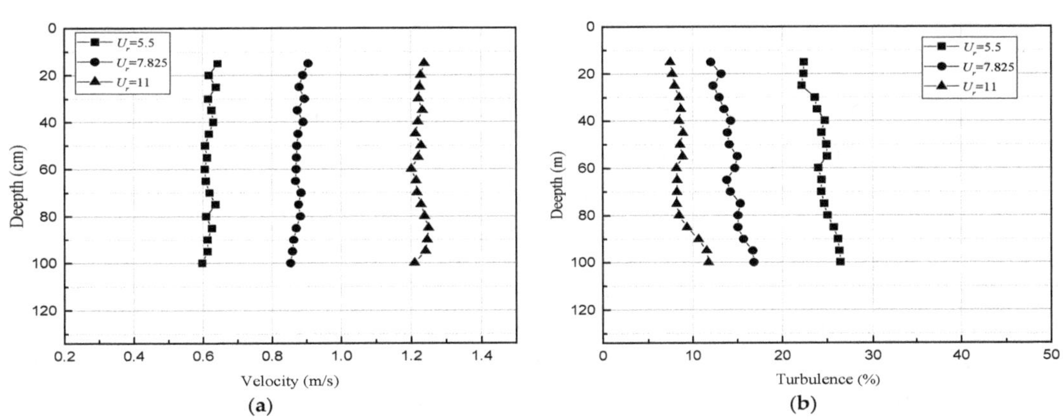

Figure 2. Incoming flow characteristics. (**a**) Velocity profile; (**b**) turbulence profile.

2.2. *Test Apparatus and Energy Conversion System*

2.2.1. Test Apparatus

The test apparatus consisted of two parts: the oscillation system and the transmission system. The oscillation system included the frame, linear guide ways, side struts, spring carrier structure, and springs (shown in Figure 3). The frame was made of steel and was fixed in a moving car at the top of the 1-m wide flow channel. The linear guideways were attached onto the steel frame, parallel to the side struts and perpendicular to the flow velocity direction.

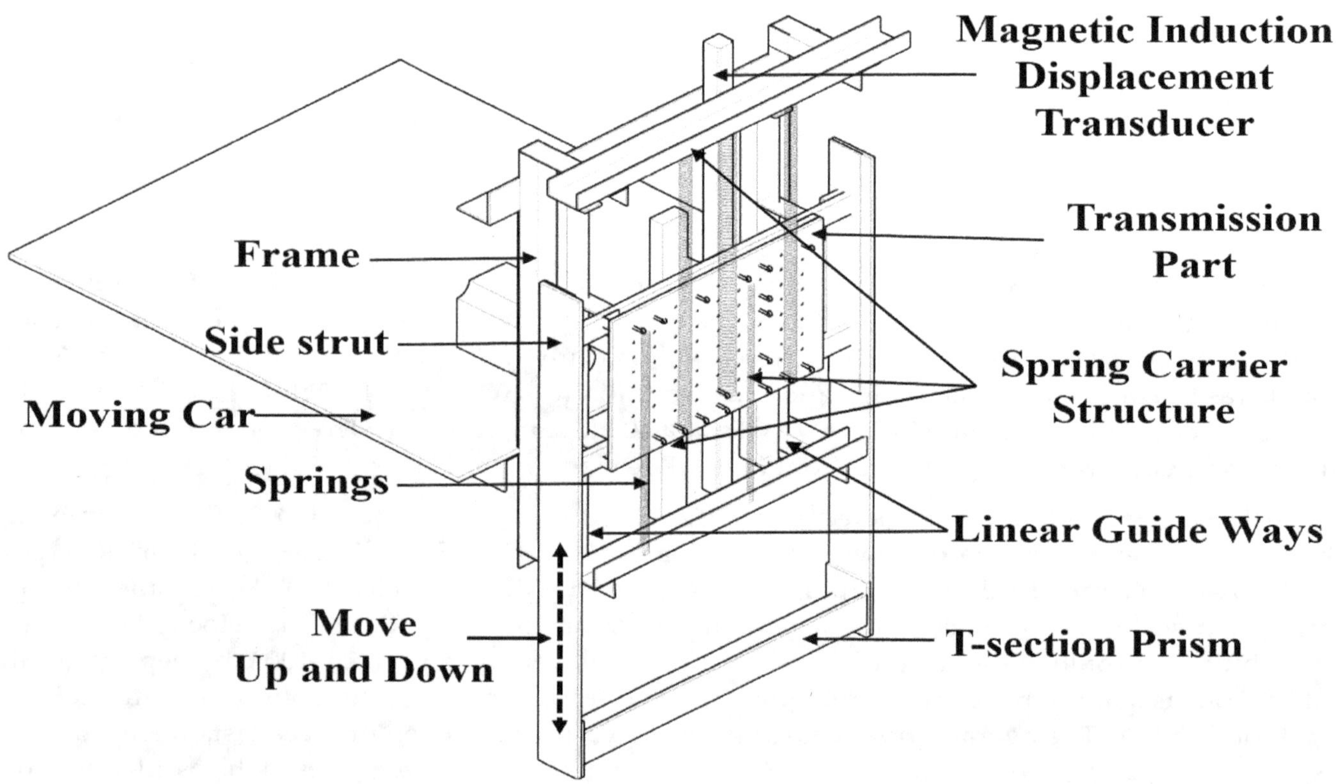

Figure 3. Test apparatus.

The transmission system included two side struts and a connective structure. The side struts, which were made of duralumin plate, were rigidly connected with the prism, which was immersed in the water. The connective structure was joined with the side struts and springs, and it was constrained to move on the linear guideways in a vertical direction by four linear bearings. Each side of the upper and the lower extensional springs was suspended vertically on the frame and the connection structure by the spring carrier structure.

2.2.2. Energy Conversion System

The rack was fixed into the transmission part that was connected to the rotor of the generator by the gears (shown in Figure 4). The linear motion of the prism was transferred to the rotational motion of the rotor. The generator was connected to the load resistances by the output wire, creating an electrical circuit. In this system, the mechanical energy of the prism was partly transformed to the electric energy of the generator and was dissipated by the load resistances.

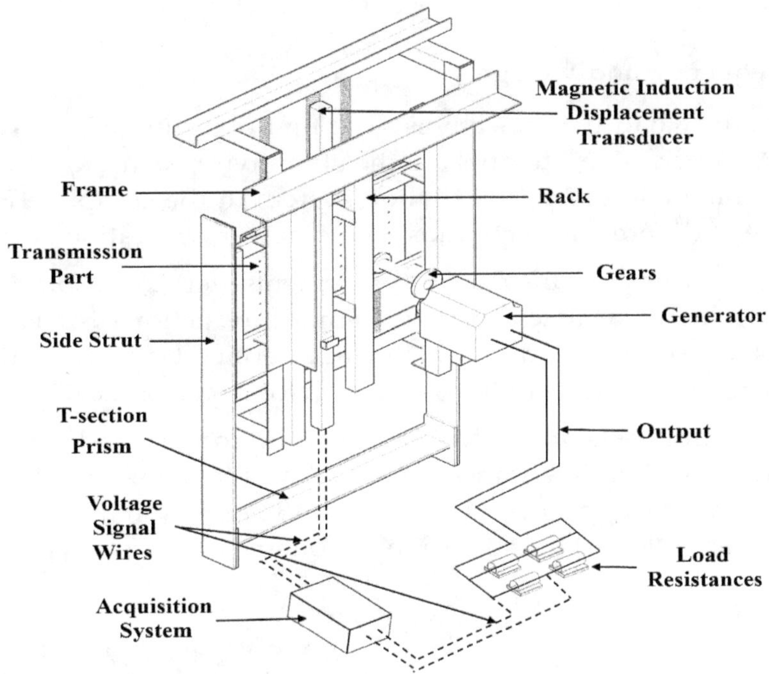

Figure 4. Energy conversion system.

2.2.3. T-Section Prisms

In the tests, the projection width of the T-section prism in the direction of the incoming flow (D) was 0.1 m, the prism length (L) was 0.9 m, the height of the prism cross-section (H) was 0.1 m, and the thickness (d) was 0.01 m. The prism was made of polymethyl methacrylate. A rectangular endplate was installed at both ends of the prism to reduce the effect of the boundary [30]. The thickness of the endplate was 0.01 m (shown in Figure 5).

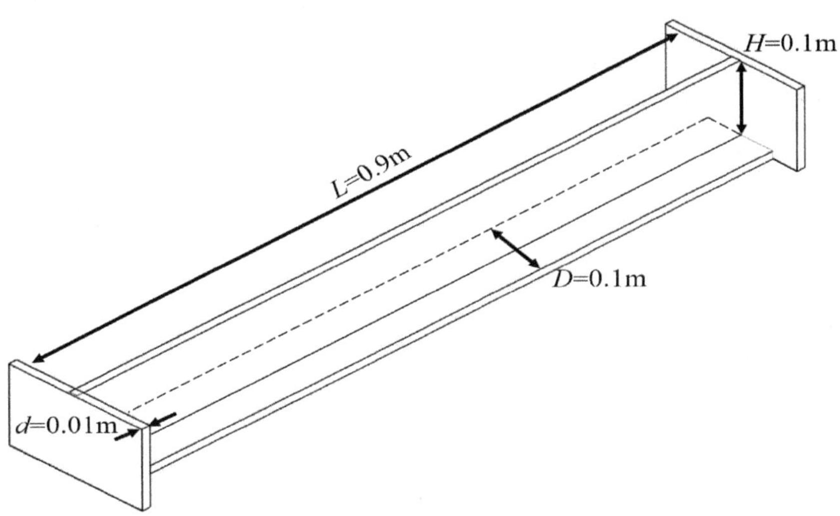

Figure 5. T-section prism size diagram.

In order to investigate the influence of the aspect ratio of the T-section prism on FIM responses and power generation characteristics, five T-section prisms with different H values (0.15 m, 0.12 m, 0.1 m, 0.09 m, and 0.08 m) were tested in the experiments. The corresponding section aspect ratios ($\alpha = H/D$) were 1.5, 1.2, 1.0, 0.9, and 0.8, respectively.

2.3. Test Methods and Sensors

2.3.1. Displacement, Frequency, and Voltage

The experimental study had two objectives. The first objective was to find the complete FIM responses of the T-section prism and to analyze the stability of each branch. The second aim was to estimate the energy conversion of the oscillation system in the complete FIM responses, and to investigate the influence of different aspect ratios on the power generation efficiency.

For oscillation characteristic tests, the main measurements were the displacements and frequencies of the prism. The displacements were tested by a magnetic induction displacement transducer with a direct current (DC) 24 V working voltage, which was supplied by external DC power. The testing range was 0–800 mm, with a sensitivity of 0.1% and an error range of $\pm$0.05%.

For the power generation tests, the connection method for the load resistances is presented in Figure 6. The minimum of the load resistances was 1 Ω, and the maximum was 50 Ω. In this study, 10 different load resistances were applied: 4 Ω, 8 Ω, 11 Ω, 13 Ω, 16 Ω, 18 Ω, 21 Ω, 31 Ω, 41 Ω, and 51 Ω. The experimental data was collected in the form of a voltage signal (shown in Figure 7) by the data acquisition system.

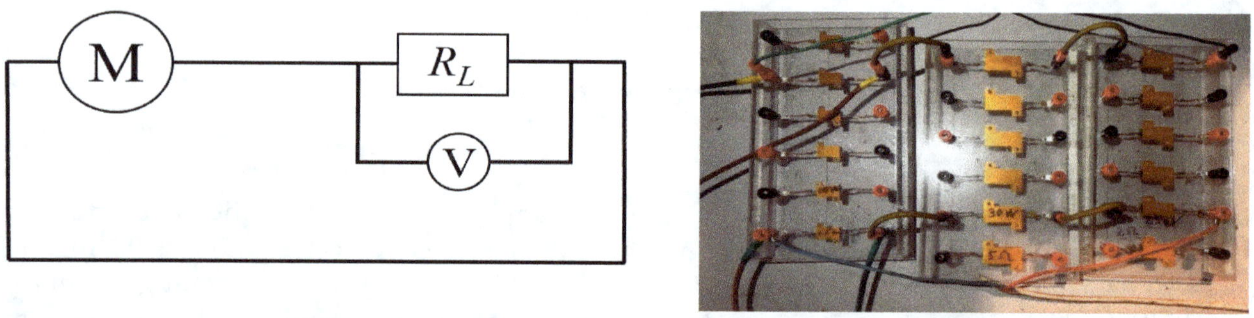

Figure 6. Voltage acquisition system: (**a**) contained circuit diagram; (**b**) connection resistances diagram.

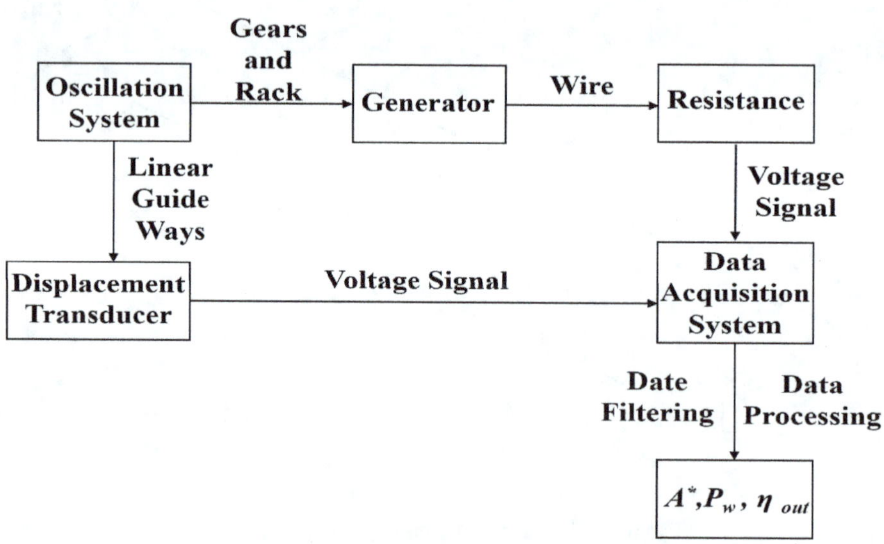

Figure 7. Test flow chart.

2.3.2. Active Power and Power Generation Efficiency

Active power (P_{harn}) and power generation efficiency (η_{out}) are the key parameters for evaluating the energy conversion capacity of the prism. In the tests, the system damping (ζ) was varied by changing the load resistances (R_L) [34]. The output voltage (u) of the generator was measured by the data acquisition system, while P_{harn} and η_{out} of the system were calculated by the following equations:

Instantaneous power expression: Equations (1)–(3)

$$P(t) = \frac{u^2(t)}{R_L} \tag{1}$$

where $P(t)$ is the instantaneous power, $u(t)$ is the instantaneous voltage, and R_L is the load resistance.

The active power is written as:

$$P_{harn} = \frac{1}{T}\int_0^T P(t)dt = \frac{1}{T}\int_0^T \frac{u^2(t)}{R_L}dt \tag{2}$$

where P_{harn} is the active power and T is a period of oscillation.

The power generation efficiency is derived as:

$$\eta_{out} = \frac{P_{harn}}{P_w} \tag{3}$$

where η_{out} is the power generation efficiency and P_w is the total power in the fluid, which is written as:

$$P_w = \frac{1}{2}\rho U^3 DL \tag{4}$$

where ρ is the water density, U is the incoming flow velocity, D is the projection width of the T-section prism in the direction of the incoming flow, and L is the prism length.

2.4. Calibration of Stiffness and Damping

A simple physical spring system designed by Xiang Yan et al. [36] was adopted to implement the stiffness of the system. First, free decay tests [35] with different spring stiffness values in the air were conducted to obtain the natural frequency (f_n) of the experimental system.

The stiffness (K) of the system varied from 800 to 1600 N/m. For each K, free decay tests were performed six times for the respective cases in air. f_n was then calculated using a simple averaging method. The test results are shown in Table 1. In these tests, the damping (C_{total}) of the experimental system was equal to the mechanical damping (C_m). The damping errors of different stiffnesses were within the allowable ranges of the tests (the average was ±5%).

Table 1. Free decay test results by varying stiffnesses.

K (N/m)	f_n (Hz)	m_{osc} (kg)	$C_{total} = C_m$ (N·s·m^{-1})	$\zeta_{total} = \zeta_m$
800	0.833	29.210	34.451	0.113
1000	0.930	29.302	34.032	0.094
1200	1.015	29.521	33.932	0.082
1400	1.087	30.046	34.903	0.085
1600	1.156	30.355	33.324	0.076

The damping ratio (ζ_{total}) of the test results could be determined by using the logarithmic decrement method, which is expressed as:

$$\zeta_{total} = \frac{1}{2\pi}\ln\left(\frac{A_i}{A_{i+1}}\right) \tag{5}$$

where A_i is the amplitude of the ith peak and ζ_{total} is the damping ratio of the experimental system.

The C_{total} of the oscillation system can be calculated as:

$$C_{total} = 2\zeta\sqrt{m_{osc}K} \quad (6)$$

where K is the system stiffness, C_{total} is the damping of the oscillation system, and m_{osc} is the system mass.

The FIM responses of the T-section prism with different ζ_{total} are described in this section. The ζ_{total} of the system is varied by changing R_L, and the parameters are shown in Table 2.

Table 2. Free decay test results by varying load resistances.

R_L (Ω)	f_n (Hz)	m_{osc} (kg)	C_{total} (N·s·m^{-1})	C_{harn} (N·s·m^{-1})	ζ_{total}
4.210	1.084	30.046	142.212	107.312	0.347
8.130	1.089	30.046	125.192	90.292	0.305
11.250	1.086	30.046	100.951	66.051	0.246
13.410	1.087	30.046	85.482	50.582	0.208
16.330	1.082	30.046	72.783	37.883	0.177
18.250	1.086	30.046	67.547	32.647	0.165
21.450	1.088	30.046	62.842	27.942	0.153
31.540	1.090	30.046	52.115	17.215	0.127
41.110	1.086	30.046	48.680	13.780	0.119
51.620	1.084	30.046	46.974	12.074	0.115
∞	1.087	30.046	34.903	0.000	0.085

3. Results and Discussion

3.1. Vibration Characteristic Analysis

3.1.1. Amplitudes and Frequencies

To explore the FIM performances of the T-section prism in water, a series of FIM tests with 10 different R_L were carried out at K = 1400 N/m (as seen in Section 2.4). In Figure 8a, A^* is the amplitude ratio ($A^* = A/D$), A denotes the average amplitude of continuous oscillation for 30 s, D is the projection width of the T-section prism in the direction of the incoming flow, U_r is the reduced velocity ($U_r = U/(D{\cdot}f_n)$), U is the incoming flow velocity, and f_n is the natural frequency in air. In Figure 8b, f^* is the frequency ratio ($f^* = f_{osc}/f_n$), and f_{osc} is the main frequency of oscillation obtained from the displacement time-history curves by the Fast Fourier Transformation (FFT) method. A^* and f^* vary with U_r; U and the Reynolds number Re for all R_L cases are plotted in Figures 8–11, respectively.

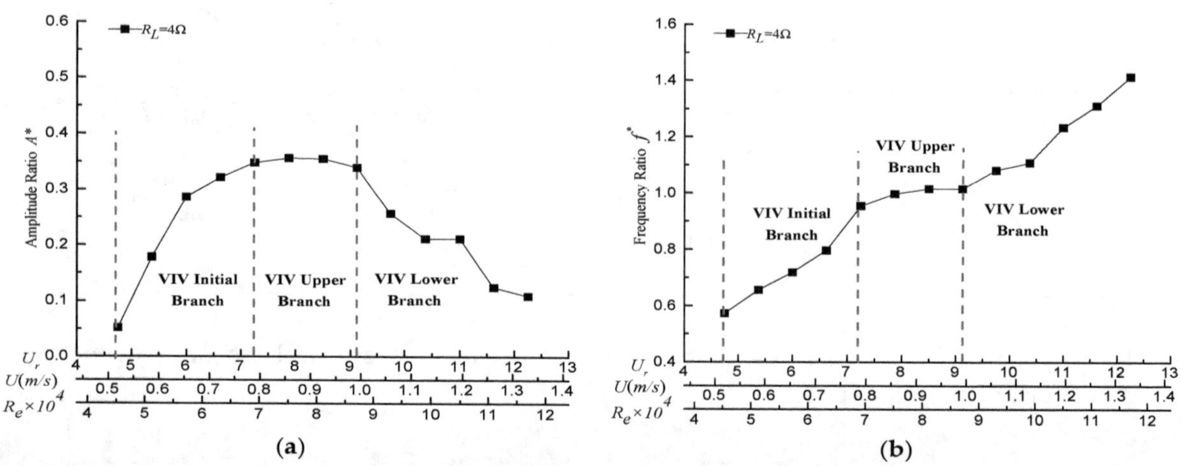

Figure 8. Vortex-induced vibration (VIV) oscillation characteristics (R_L = 4 Ω, ζ = 0.347): (**a**) response of the amplitude ratio; (**b**) response of the frequency ratio.

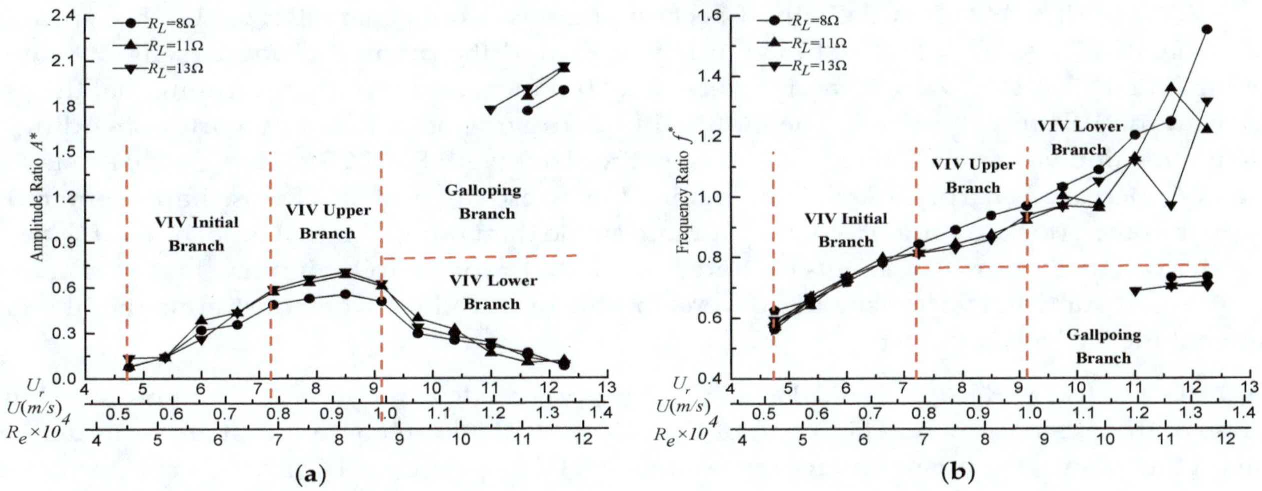

Figure 9. High galloping (HG) oscillation characteristics 8 Ω ≤ R_L ≤ 13 Ω (0.208 ≤ ζ ≤ 0.305): (**a**) response of the amplitude ratio; (**b**) response of the frequency ratio.

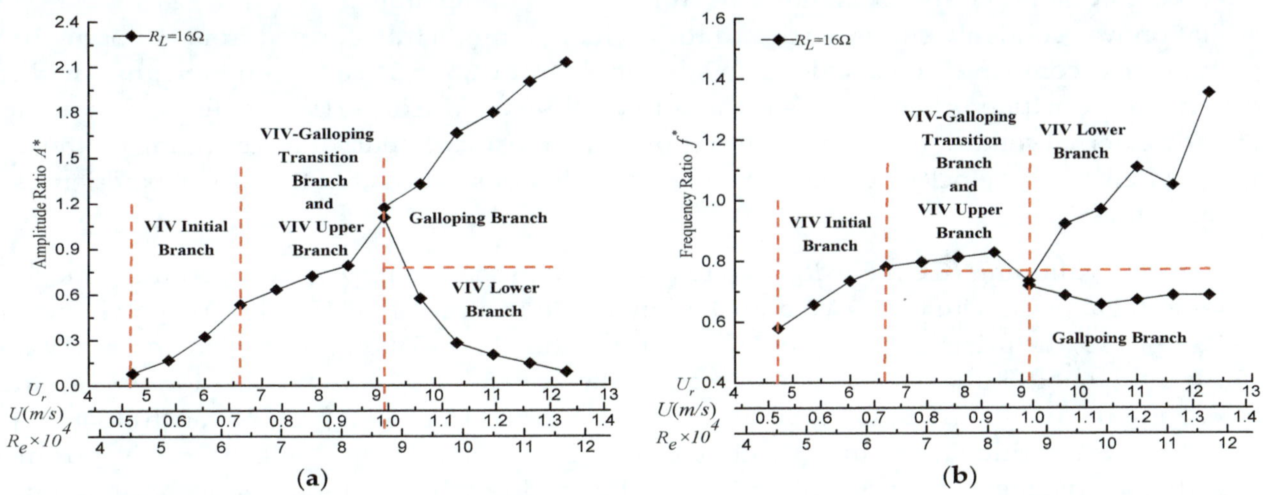

Figure 10. Critical galloping (CG) oscillation characteristics R_L = 16 Ω (ζ = 0.177): (**a**) response of the amplitude ratio; (**b**) response of the frequency ratio.

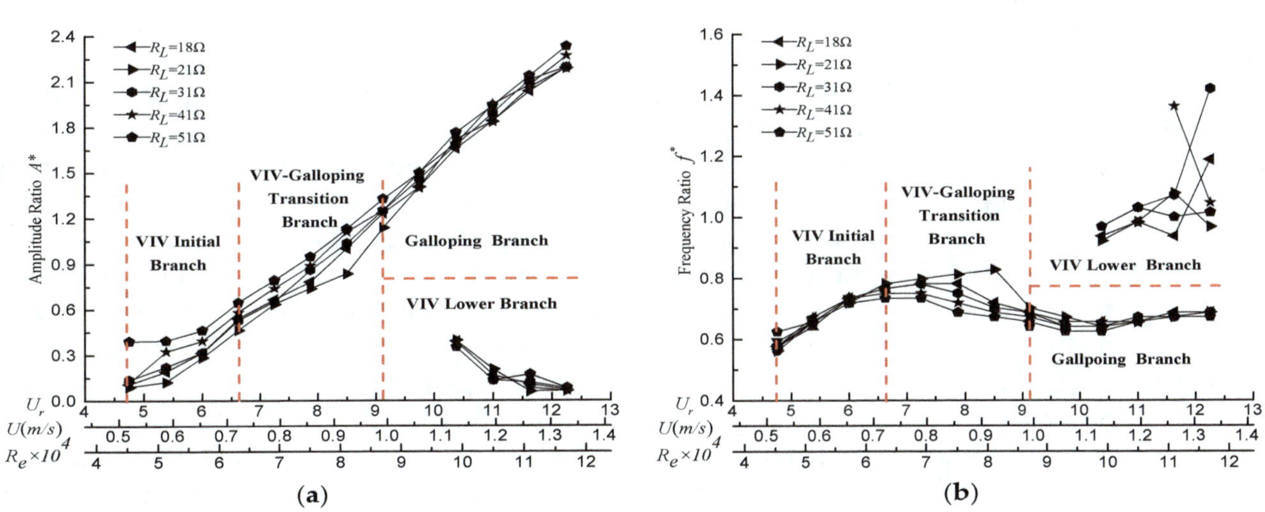

Figure 11. Soft galloping (SG) oscillation characteristics 18 Ω ≤ R_L ≤ 51 Ω (0.115 ≤ ζ ≤ 0.165): (**a**) response of the amplitude ratio; (**b**) response of the frequency ratio.

(1) VIV: For $R_L = 4\ \Omega$ ($\zeta = 0.347$), the T-section prism was only characterized by the typical VIV in the range of $4.75 \leq U_r \leq 12.25$ (shown in Figure 8), and the prism did not present any signs of galloping. For $4.75 \leq U_r \leq 7.25$, A^* and f^* were small but increased rapidly, indicating that the prism underwent the VIV initial branch. The motion of the prism was induced by vortex shedding, but the oscillation and vortex shedding were not synchronized well. For $7.25 \leq U_r \leq 9.125$, A^* stayed at 0.35, and f^* increased sharply and was maintained at approximately 1. The oscillation entered the VIV upper branch (lock-in range), which was more stable than the VIV initial branch. For $U_r \geq 9.125$, A^* rapidly collapsed down to about 0.1, and f^* increased rapidly to a high level ($f^* >> 1$). In this range, the oscillation and vortex shedding were not synchronized well, demonstrating that the prism underwent the VIV lower branch.

(2) HG: For $8\ \Omega \leq R_L \leq 13\ \Omega$ ($0.208 \leq \zeta \leq 0.305$), two modes of motion could be observed in the response of the prism (shown in Figure 9). If the prism oscillated freely (without any external forces acting on the prism), the prism underwent complete VIV responses as U_r increased or decreased in the range of $4.75 \leq U_r \leq 12.25$. However, when U_r decreased from 12.25 and the prism was manually pushed by a threshold initial displacement (exceeding $1 \times$ D), the prism first entered the galloping branch, accompanied by a large amplitude ($A^* \approx 2.3$) and a stable frequency ($f^* = 0.7$). As U_r decreased, A^* decreased gradually and the oscillation was still in the galloping branch. When U_r reached 11–11.625, the oscillation was suddenly suppressed, and the oscillation mode transformed from galloping to VIV (lower branch), accompanied by a sudden collapse in A^* from 1.8 to 0.2, and a sudden jump in f^* from 0.7 to 1. It was concluded that the prism cannot be self-excited from VIV to galloping in any case, without any external conditions (such as a threshold initial displacement) or forces acting on the prism, no matter whether the velocity decreases or increases, demonstrating that the prism experiences HG responses in these cases.

(3) Critical galloping (CG): For $R_L = 16\ \Omega$ ($\zeta = 0.177$), the system presented both complete VIV responses and galloping while the two branches intersected (shown in Figure 10). When U_r increased from 4.75 to 12.25, the prism experienced a VIV initial branch, a VIV upper branch, and a VIV lower branch. If a threshold initial displacement was applied at $U_r = 12.25$, the oscillation was suddenly enhanced, and the oscillation mode dramatically transformed from VIV (initial branch) to galloping, accompanied by a sudden jump in A^* from 0.08 to 2.12 as well as a sudden collapse in f^* from 1.35 to 0.68. Afterwards, the oscillation entered the galloping branch. A^* gradually decreased, with the decrease of U_r, and f^* was maintained at about 0.7. No oscillation suppression phenomena occurred until the oscillation underwent the VIV upper branch where the intersection point was $U_r = 9.125$. However, at $U_r \geq 9.125$, galloping could be induced by external excitation. This condition is actually a critical oscillation mode between HG and SG, which can be referred to as "critical galloping" (CG).

(4) SG. For $18\ \Omega \leq R_L \leq 51\ \Omega$ ($0.115 \leq \zeta \leq 0.165$), two modes of motion were observed in the responses of the prism. If the prism oscillated freely (without any external forces acting on the prism), the T-section prism experienced the VIV initial branch, followed by the VIV-galloping transition branch, and ended with the galloping branch (shown in Figure 11). In the range of $4.75 \leq U_r \leq 6.625$, A^* and f^* were small but increased rapidly, indicating that the prism underwent the VIV initial branch. In the range of $6.625 \leq U_r \leq 9.125$, A^* continued to grow with a strong uptrend to 1.5, while f^* decreased to about 0.7 with a slight downward trend. For $U_r \geq 9.75$, A^* continued to increase up to 2.36, while f^* almost remained at 0.7. The oscillation underwent the fully developed galloping branch. On the other hand, at $U_r = 12.25$, external suppression was applied to the prism; thus, the oscillation presented a low $A^* \approx 0.2$ and a high $f^* \approx 1.4$, the system experienced the VIV lower branch until $U_r \leq 10.375$, and the oscillation mode returned to galloping. For $6.625 \leq U_r \leq 10.375$, the oscillation was self-excited from VIV to galloping (regardless of whether the velocity increased or decreased). If U_r continued to increase, the oscillation would be maintained within the range of galloping. For $U_r \geq 10.375$, if there was an external suppression, the oscillation would be converted to the VIV lower branch and could not be self-excited to galloping.

It can be concluded that the oscillation responses of the T-section prism were similar to those of the triangular prism and the PTC circular cylinder. The occurrence conditions of different FIM branches of the T-section prism are listed in Table 3.

Table 3. VIV and galloping occurrence conditions with $K = 1400$ N/m.

Oscillation Form	R_L	U_r	External Excitation	External Suppression	Self-Excited by VIV
VIV	4 Ω	4.75 ~12.25			√
HG	8 ~13 Ω	11 ~12.25	√ (HG)		
CG	16 Ω	9.125 ~12.25		√ (VIV Low B)	√
SG	18 ~51 Ω	10.375 ~12.25		√ (VIV Low B)	√

3.1.2. Time-History and Frequency Spectrum of Each Branch

In order to explore the oscillation characteristics of each branch, four typical time-history curves and frequency spectra are depicted in this section (plotted in Figures 12 and 13). The physical parameters of the oscillation system are listed in Table 4.

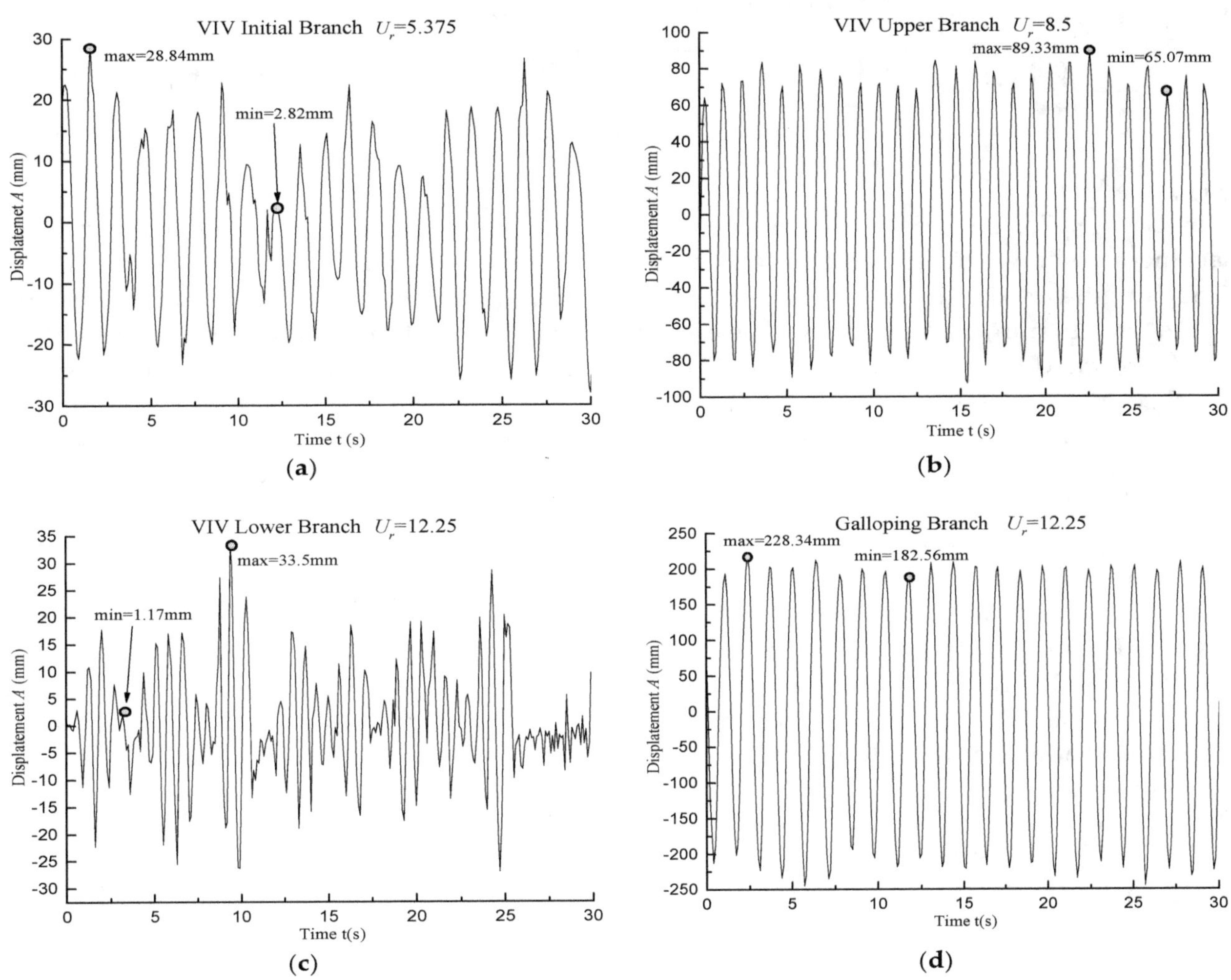

Figure 12. Typical flow-induced motion displacement time-history curves: (**a**) VIV initial branch; (**b**) VIV upper branch; (**c**) VIV lower branch; (**d**) galloping branch.

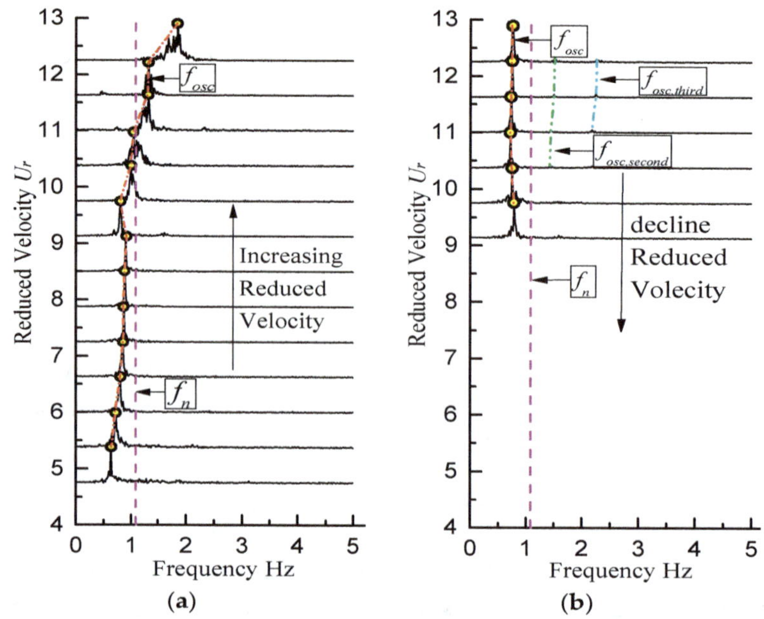

Figure 13. Typical oscillation frequency spectrum: (**a**) complete VIV branch; (**b**) galloping branch.

Table 4. Oscillation form and parameters list.

R_L (Ω)	C_{total} (N·s·m^{-1})	C_{harn} (N·s·m^{-1})	ζ	Oscillation Form
16	72.783	37.883	0.177	CG

(1) Time-History

For U_r = 5.375, the prism underwent the VIV initial branch. The time-history curve of the displacement is shown in Figure 12a. It was observed that the minimum of the positive displacement was 2.82 mm, the maximum was 28.84 mm, and the difference between the maximum and minimum was 1.25A (A = 16 mm), indicating that the amplitude was low and the oscillation was unstable.

As U_r reached 8.5, the oscillation mode entered the VIV upper branch, as shown in Figure 12b. It was observed that the minimum of the positive displacement was 65.07 mm, the maximum was 89.33 mm, and the difference between the maximum and minimum was 0.31A (A = 78 mm). Moreover, the prism underwent a more stable oscillation, although there was a certain fluctuation response amplitude.

For U_r = 12.25, it was observed that the minimum of the positive displacement was 1.17 mm, the maximum was 33.50 mm, and the difference between the maximum and minimum was 4.04A (A = 8 mm). The oscillation amplitude fluctuation was large and the performance was very unstable, as shown in Figure 12c, demonstrating that the prism underwent the VIV lower branch.

For U_r = 12.25 (galloping branch), the minimum of the positive displacement was 182.56 mm, the maximum was 228.34 mm, and the difference between the maximum and minimum was 0.21A (A = 212 mm), indicating that the oscillation amplitude was large and the fluctuation was small, as shown in Figure 12d.

Both stability and amplitude are key parameters for energy harvesting. Thus, the galloping branch is the most suitable branch for energy extraction and utilization.

(2) Frequency Spectrum

The frequency spectrum is important for describing the oscillation characteristics and mechanical energy. The oscillation frequency of the prism was extracted by using the FFT method from the displacement time-history curves (shown in Figure 13). Figure 13a plots the frequency spectrum of the complete VIV branch and Figure 13b plots the frequency spectrum of the galloping branch, while the trends of f_n and f_{osc} are marked by dashed lines.

For $4.75 \leq U_r \leq 6.625$, the prism underwent the VIV initial branch, and the frequency band was wider and exhibited double peaks in the frequency spectrum, revealing that periodicity of oscillation and mechanical energy was poor.

In the range of $6.625 \leq U_r \leq 9.125$, the prism entered the VIV upper branch, the frequency band became narrower, and the main frequency was clearly observed. The periodicity of oscillation was significantly enhanced, the prism oscillated more stably, and the main frequency was concentrated at 0.88 Hz.

For $U_r \geq 9.125$, two modes of oscillation were observed. When the prism oscillated freely (without external excitation), the prism went into the VIV lower branch, the main frequency was not obvious, the frequency band widened, and the energy was very dispersed.

On the other hand, if the prism was manually pushed by a threshold initial displacement (exceeding $1 \times D$), the prism entered the galloping branch with a narrowed frequency band and evident first dominant frequency, as shown in Figure 13b. It was noted that in the galloping branch, the oscillation was mainly caused by the instability of the lift force due to the T-section prism with sharp sections. Because of the lower energy and the higher frequency of the vortex shedding, the effect of vortex shedding on the prism was weaker, resulting in the second and third dominant frequencies being hardly observed in the frequency spectrum.

The conclusions of the time-history curves and frequency spectra are as follows: (1) different fluctuations are observed in each branch—the VIV upper branch and the galloping branch have higher amplitudes and more stable oscillations; (2) the VIV upper branch and the galloping branch have a better periodicity and frequency spectrum. In summary, the VIV upper branch and the galloping branch are more suitable for energy harvesting, but the galloping branch is better (shown in Table 5).

Table 5. Oscillation characteristics of different branches.

Branch	VIV Initial	VIV Upper	VIV Lower	Galloping
Amplitude	low	high	low	highest
Stability	bad	good	bad	best
Frequency	bad	good	bad	best

3.1.3. Summary

The specific findings are listed as follows:

(1) With an increase of damping (decrease in load resistances), the T-section prism oscillation mode gradually changes from SG ($0.115 \leq \zeta \leq 0.165$, $18\ \Omega \leq R_L \leq 51\ \Omega$) to CG ($\zeta = 0.177$, $R_L = 16\ \Omega$), and eventually to HG ($0.208 \leq \zeta \leq 0.305$, $8\ \Omega \leq R_L \leq 13\ \Omega$). The oscillation mode of SG, HG, and CG all contain VIV branches and galloping branches, the only difference being whether these can be self-excited from VIV to galloping.

(2) The analysis of the displacement time-history curves and the frequency spectra demonstrates that the stability and intensity of the VIV upper branch and the galloping branch are both better-performing, and that the maximum amplitude of the galloping branch is larger. It can be concluded that at a high velocity ($U_r \geq 10.375$, $U \geq 1.128$ m/s) the galloping branch is better for harvesting energy, and at a lower velocity ($6.625 \leq U_r \leq 9.125$, $0.720 \leq U \leq 0.992$ m/s) the VIV upper branch is better. Power generation with different load resistances will be presented in the next section.

3.2. Power Generation Analysis

3.2.1. Active Power Analysis

The variations of active power P_{harn} of the generator versus incoming flow velocity U, reduced flow U_r, and Reynolds number Re are plotted in Figure 14. P_{harn} is the average of the instantaneous power under continuous oscillation for 30 s. For $U_r \geq 6$, the oscillation system started to output electrical energy and P_{harn} was smaller, as the system experienced the complete VIV responses and the P_{harn} did not exceed 5 W. As galloping occurred, P_{harn} increased to over 20 W. The details are described as follows:

(1) For 18 $\Omega \leq R_L \leq$ 51 Ω, SG occurred.

As U_r increased, the prism experienced the VIV initial branch, the VIV-galloping transition branch, and the galloping branch. P_{harn} monotonically increased with U_r and the maximum P_{harn} appeared at the galloping branch ($P_{harn,\max}$ = 13 W, R_L = 18 Ω, U_r = 12.25). It was noted that the higher the U_r, the lower the R_L, and the more energy was converted.

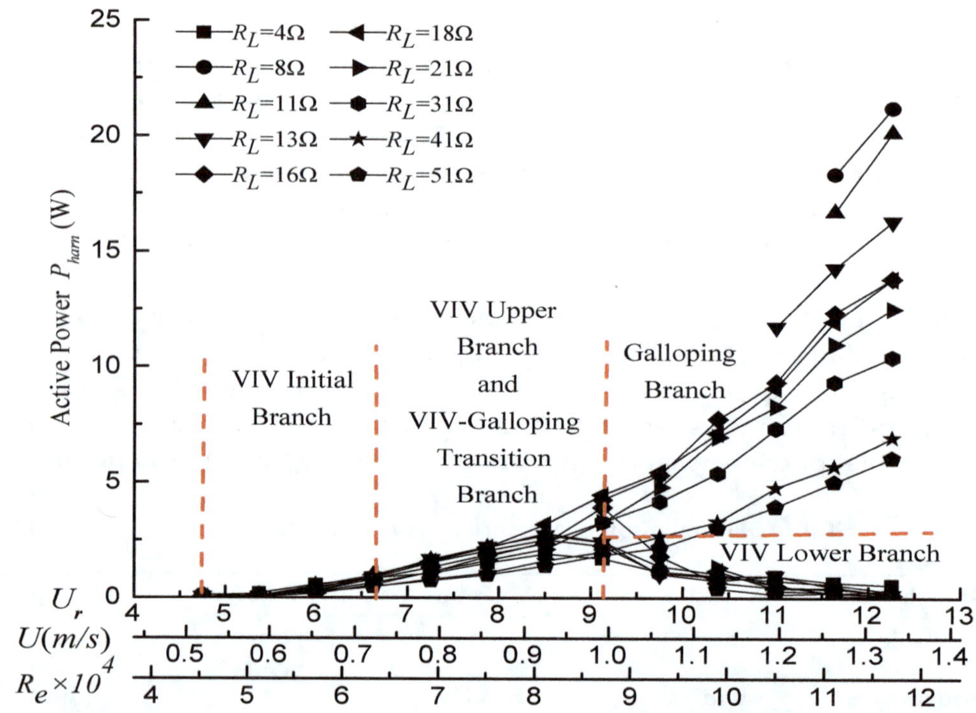

Figure 14. T-section prism power generation active power.

(2) For 8 $\Omega \leq R_L \leq$ 16 Ω, HG occurred, and two oscillation modes were observed.

(i) Without external excitation (such as forces, threshold initial displacement, etc.)

The prism only experienced VIV branches. P_{harn} rose with the increase of U_r at the VIV initial branch and at the VIV upper branch ($P_{harn,\max}$ = 3.92 W, R_L = 16 Ω, U_r = 9.125). Afterwards, P_{harn} decreased to lower than 1 W, as U_r rose up to 12.25 in the VIV lower branch.

(ii) An external excitation was applied to the prism.

At U_r = 12.25, if the threshold initial displacement was applied to the prism, the oscillation directly jumped into the galloping branch. As U_r decreased, P_{harn} dropped rapidly. In the present tests, the maximum P_{harn} was 21.23 W (U = 1.332 m/s, Re = 116396, R_L = 8 Ω). The maximum P_{harn} was close and it was slightly lower than that of the PTC circular cylinder ($P_{harn,\max}$ = 23.54 W), as reported by Lin, D [44].

(3) For R_L = 4 Ω, only VIV branches were observed.

The variation of P_{harn} versus U_r for R_L = 4 Ω was similar to those for 8 $\Omega \leq R_L \leq$ 16 Ω in VIV branches. Within the test flow velocity range, the prism did not experience any forms of galloping. The maximum P_{harn} was 2.5 W at U_r = 8.5 (VIV upper branch).

3.2.2. Efficiency Analysis

The variations of efficiency η_{out} versus incoming flow velocity U, reduced flow U_r, and Reynolds number Re are plotted in Figure 15. η_{out} was calculated using Equation (3) (shown in Section 2.3).

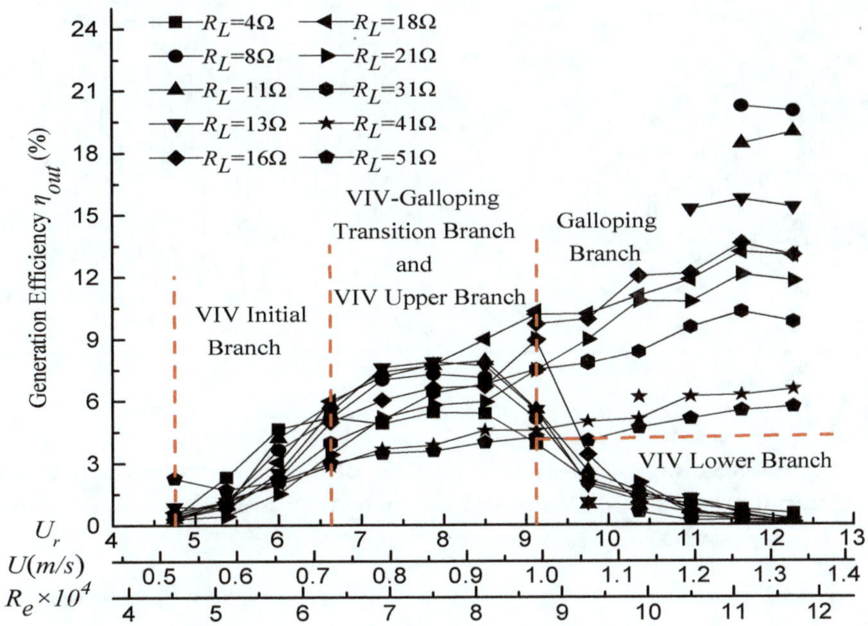

Figure 15. T-section prism power generation efficiency.

It was observed that η_{out} did not exceed 9% in VIV branches (whether in SG or HG responses). The corresponding load resistance was R_L = 16 Ω and the corresponding velocity was U_r = 9.125 (U = 0.992 m/s, Re = 86703). On the contrary, the maximum η_{out} reached 20.2% in the galloping branch, which was close to the test results of the smooth (22%) and the PTC (28%) circular cylinders. The corresponding load resistance was R_L = 8 Ω, and the corresponding velocity was U_r = 11.625 (U = 1.264 m/s, Re = 110,458). It is concluded that the energy conversion efficiency in the galloping branch is much higher than that of the VIV branches for the T-section prism.

In addition, in most of the galloping branches, especially for small load resistances (R_L < 31) or high damping ratios (ζ > 0.127), a slight declining trend was observed for U_r exceeding 11.625 in the present tests. It was indicated that U_r = 11.625 was the optimal velocity.

3.2.3. Stability Analysis

In order to describe the power generation stability of each oscillation branch, the instantaneous voltage time-history curves and instantaneous power time-history curves of four oscillation branches are discussed in this section (shown in Figures 16 and 17).

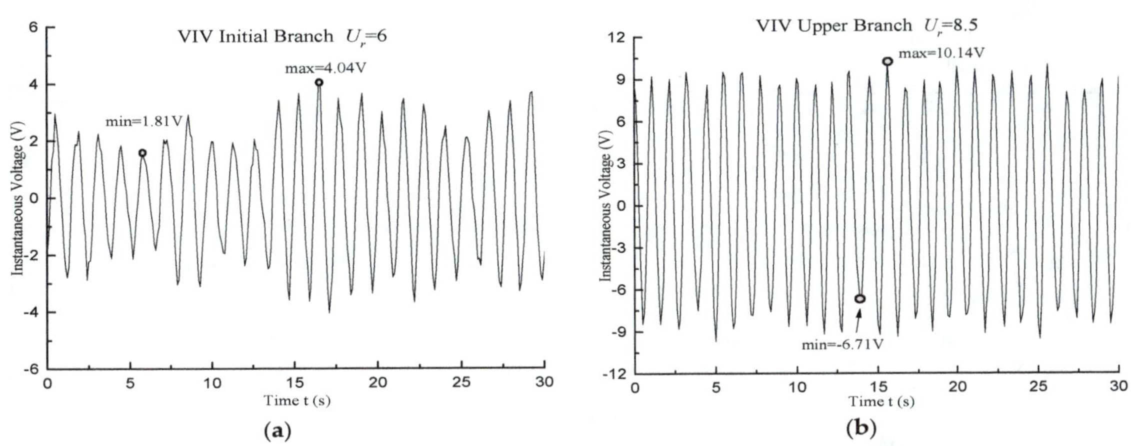

Figure 16. *Cont.*

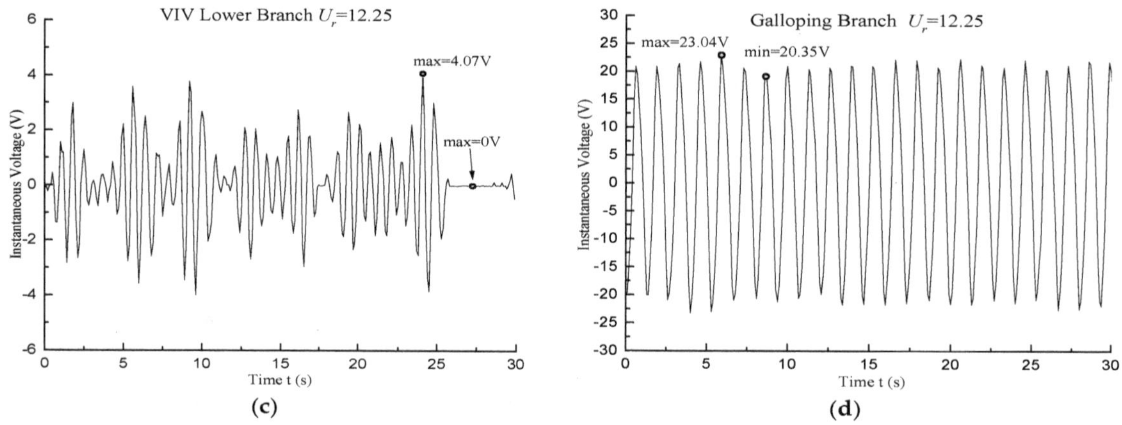

Figure 16. Instantaneous voltage time-history curves of T-section prism power generation with $R_L = 16\ \Omega$: (**a**) VIV initial branch; (**b**) VIV upper branch; (**c**) VIV lower branch; (**d**) galloping branch.

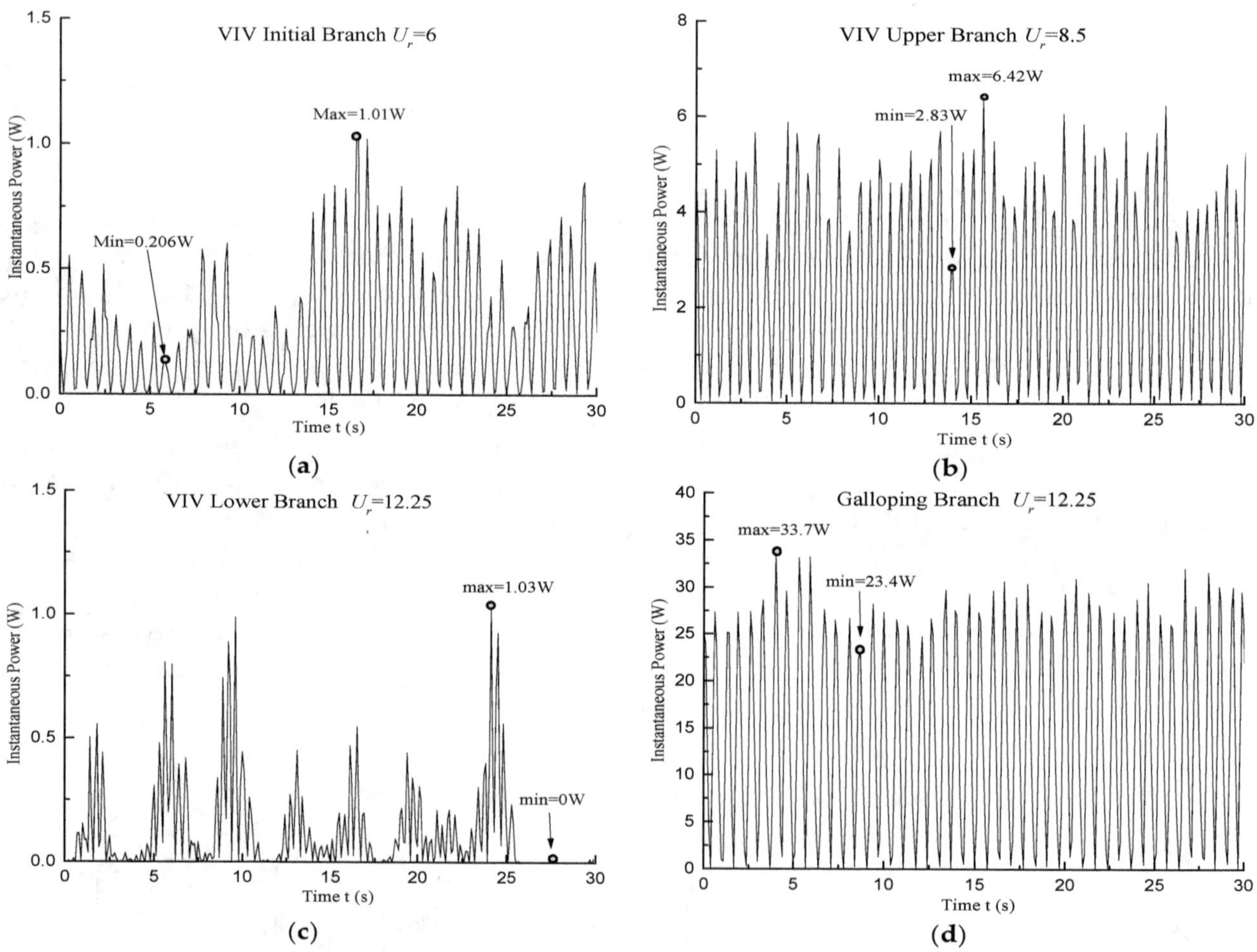

Figure 17. Power generation instantaneous power time-history curves of the T-section prism with $R_L = 16\ \Omega$: (**a**) VIV initial branch; (**b**) VIV upper branch; (**c**) VIV lower branch; (**d**) galloping branch.

Figure 16a,c show that the instantaneous voltage peaks fluctuated greatly when the prism underwent the VIV initial branch (V_{max} = 4.04 V, V_{min} = 1.81 V) and the VIV lower branch (V_{max} = 4.07 V, V_{min} = 0 V). This led directly to the instability of the instantaneous power of the VIV initial branch (P_{max} = 1.01 W, P_{min} = 0.206 W) and the VIV lower branch (P_{max} = 1.03 W, P_{min} = 0 W), as shown in Figure 17a,c. Additionally, the instantaneous voltages of the VIV initial branch and the

VIV lower branch were small, so that the powers were not as high as the instantaneous active power in the VIV upper branch (P_{max} = 6.42 W, P_{min} = 2.83 W) and in the galloping branch (P_{max} = 33.7 W, P_{min} = 23.4 W), as shown in Figure 17b,d. The peak instantaneous voltages of the VIV upper branch (V_{max} = 10.14 V, V_{min} = −6.71 V) and the galloping branch (V_{max} = 23.04 V, V_{min} = 20.35 V) had small fluctuations, as shown in Figure 16b,d. The reason for this was that the two branches had better oscillation stability. In addition, the instantaneous power of the galloping branch was much higher and the quality of the output energy was better.

In order to further discuss the stability of the harnessed power, the difference coefficient C_v was introduced as follows:

$$C_v = \frac{\sigma_p}{P_{harn,max}} \tag{7}$$

where σ_p is the square deviation of all peak powers, and $P_{harn,max}$ is the average of all peak powers. It was noted that a higher C_v corresponded to a more unstable power output; otherwise, it corresponded to a more stable power output. The power generation difference coefficient C_v was calculated based on Equation (6), and is shown in Figure 18. In different branches, the stability of the oscillation was quite different. The performances are described as follows:

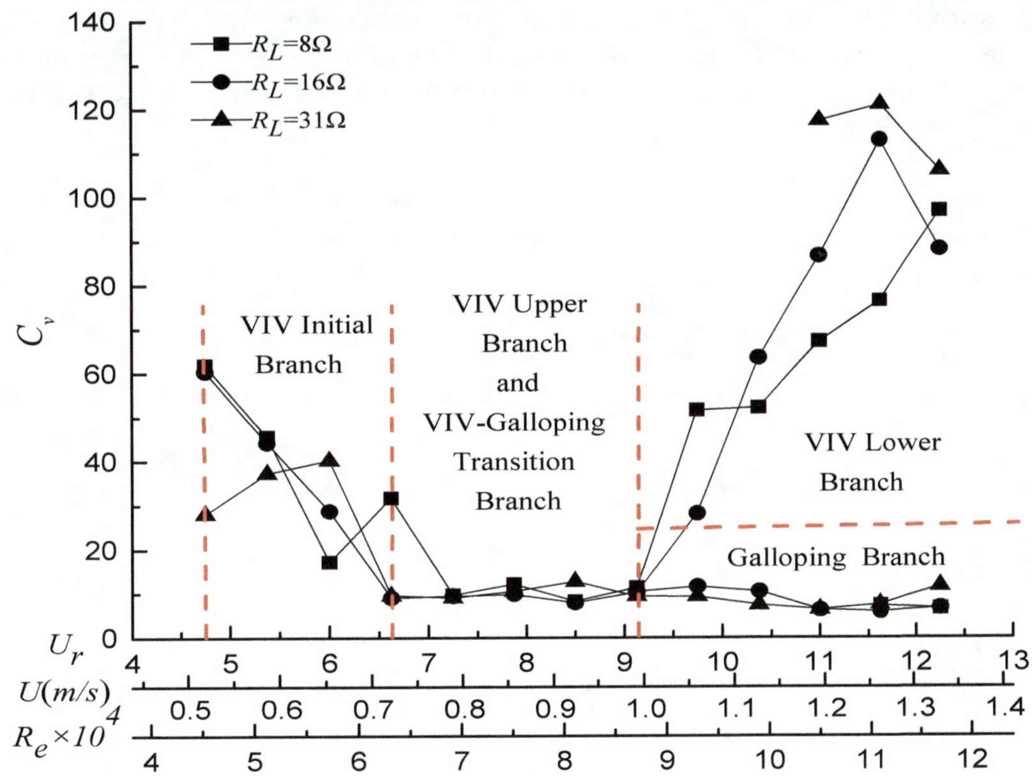

Figure 18. Power generation difference coefficient C_v value.

For 4.75 ≤ U_r ≤ 6.625, the value of C_v exceeded 20 due to the unstable oscillation in the VIV initial branch, indicating that the power generation stability was poor. For 6.625 ≤ U_r ≤ 9.125, as U_r increased, C_v dropped to lower than 10 and remained almost stable due to the stable oscillation of the prism. This indicated that the stability of the power output was good at the VIV upper branch. For 9.125 ≤ U_r ≤ 12.25, two performances of C_v could be observed. In the VIV lower branch, C_v increased rapidly and exceeded over 100 at U_r = 11.625. This indicated that the power output was more unstable than any branches, due to the poor oscillation of the prism. On the contrary, in the galloping branch, C_v remained at approximately 10, indicating that the power output was perfectly stable due to the perfectly stable oscillation of the prism.

It is concluded that the galloping branch was the best of all FIM branches to obtain a stable power output. The results of the FIM power generation are summarized in Table 6.

Table 6. Power generation characteristics of different branches.

Branch	VIV Initial	VIV Upper	VIV Lower	Galloping
Active Power (W)	0.02–1.67	1.53–4.46	0.12–3.92	2.18–21.22
Efficiency (%)	0.38–7.6	3.45–8.9	0.11–7.9	4.07–20.2
Stability	bad	good	bad	best

3.3. Effects of Aspect Ratios on Oscillation and Energy Conversion

3.3.1. Effects of Aspect Ratios on Oscillation

Different cross-section ratios (α) can significantly influence the angle of attack and lift force, resulting in different performances by the FIM responses and the energy conversion of the T-section prism [43].

In this section, five tests for different α values were conducted. The five α values were 0.8, 0.9, 1, 1.2, and 1.5. In addition, the physical parameters of the system were summarized as: m_{osc} = 29.521 kg, K = 1200 N/m, and R_L = 16 Ω. The variation of amplitude ratio A^* and the efficiency of η_{out} versus the reduced velocity U_r, the incoming flow velocity U, and the Reynolds number Re for the five α cases are plotted in Figures 19 and 20.

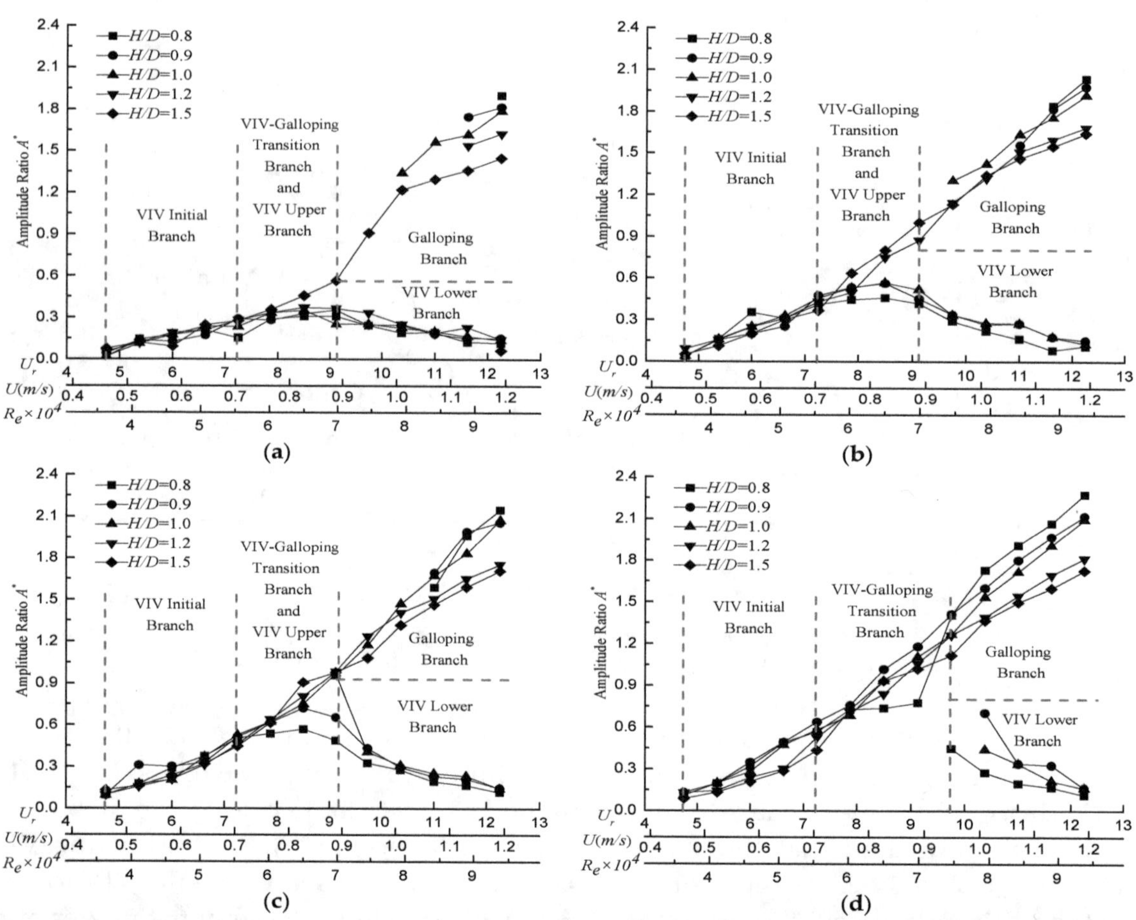

Figure 19. Oscillation characteristics with different section aspect ratios at K = 1200 N/m: (**a**) amplitude ratio (4 Ω, ζ = 0.347); (**b**) amplitude ratio (8 Ω, ζ = 0.305); (**c**) amplitude ratio (11 Ω, ζ = 0.246); (**d**) amplitude ratio (16 Ω, ζ = 0.177).

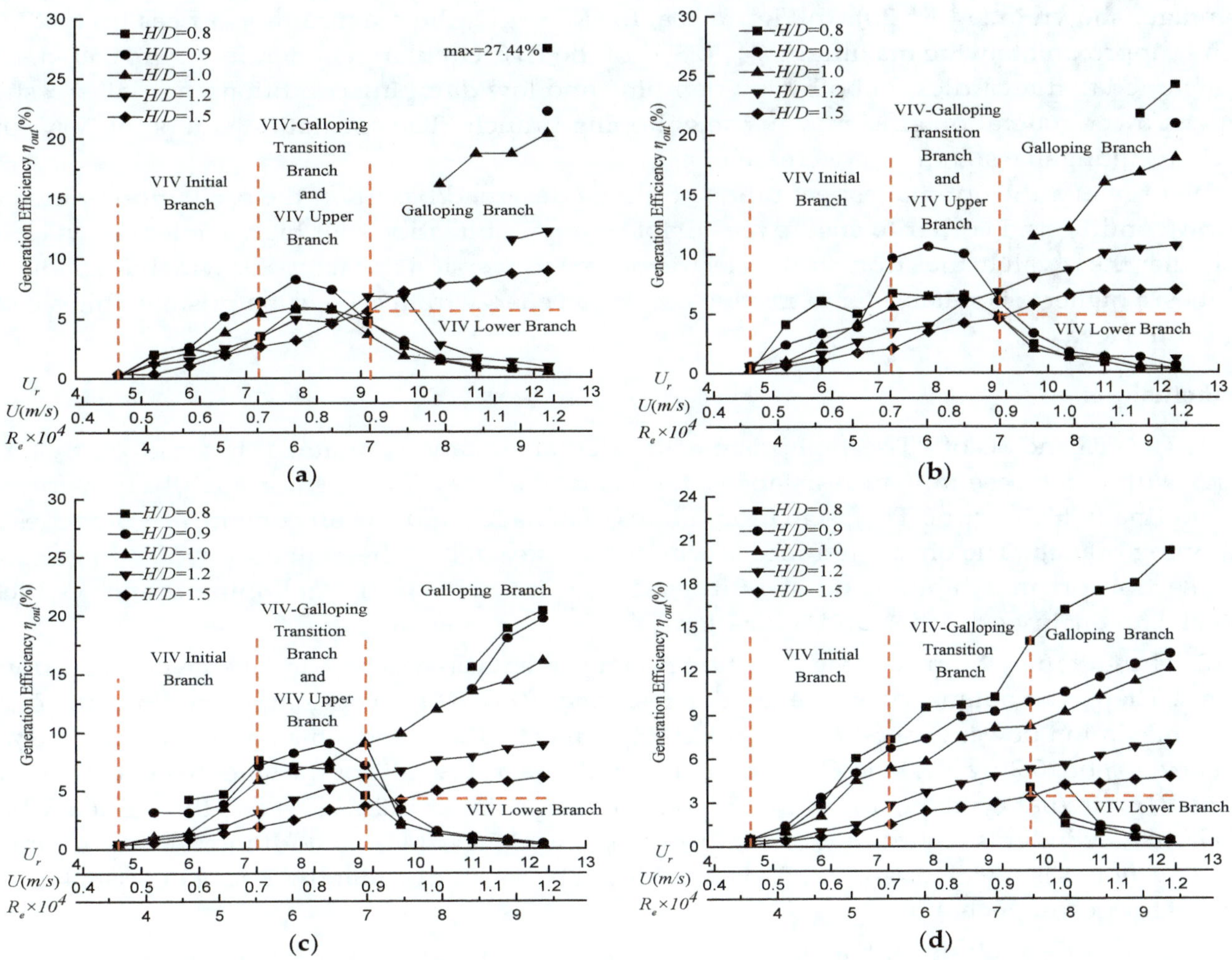

Figure 20. Power generation efficiency with different section aspect ratios at K = 1200 N/m: (**a**) amplitude ratio (4 Ω, ζ = 0.347); (**b**) amplitude ratio (8 Ω, ζ = 0.305); (**c**) amplitude ratio (11 Ω, ζ = 0.246); (**d**) amplitude ratio (16 Ω, ζ = 0.177).

From the previous results of this paper, it was found that the T-section prism could not be self-excited from VIV to galloping at high damping conditions with α = 1. In Figure 19a, for ζ = 0.347, only SG occurred at α = 1.5. For the prisms of other α values forced by initial displacement, HG occurred at a high velocity.

In Figure 19b,c, for ζ = 0.305 and ζ = 0.246 in the prisms of α = 1.5 and α = 1.2, SG was observed, while in the prisms of other α, HG was observed with external excitation. In Figure 19d, for ζ = 0.177, SG occurred under all conditions. For $1.2 \leq \alpha \leq 1.5$, the VIV lower branch could not be observed with external suppression. Based on these results, it can be concluded that a higher α value is more likely to cause SG under the same conditions. The rise of α is beneficial to the development of the oscillation mode from HG to SG. In addition, A^* decreases with the increase of α. Therefore, a reasonable α must be selected to ensure good oscillation characteristics, as well as to guarantee better electrical energy resources.

3.3.2. Effects of Aspect Ratios on Energy Conversion

Figure 20 presents the variation of the η_{out} with the five α values, which can be summarized as follows. In the present tests, T-section prisms with a higher α have a good advantage at high damping conditions in terms of oscillation, but their performances in power generation efficiency are not very prominent. On the contrary, the prisms with smaller α show a more outstanding power generation advantage (shown in Figure 20). In Figure 20a, for R_L = 4 Ω, the maximum η_{out} rises up to 27.44%, which is approximately the maximum η_{out} (28%) of the PTC circular cylinder. In Figure 20b–d, it can be concluded that regardless of both high damping and low damping conditions, a smaller α shows a better power generation efficiency in the galloping branch. The oscillation enters the VIV lower branch, resulting in a significant decrease in η_{out}.

In summary, the optimal aspect ratio should be designed based on the oscillation branch and the flow conditions, in order to ensure reasonable energy utilization. For high stable flow velocities, the smaller the α value, the stronger the electric power conversion capacity. For widely variable flow velocities, a higher α is suitable for extracting oscillation energy from a T-section prism within a certain range of the test.

4. Conclusions

(1) The HG and SG of a T-section prism with external excitation are similar to those of a triangular prism. With a decrease in load resistances, the SG of the T-section prism gradually transforms to CG, and finally HG occurs. The oscillation modes of SG, HG, and CG all contain VIV branches and a galloping branch. The only difference is whether these can be self-excited from VIV to galloping. The T-section prism exhibits good stability and high intensity in the galloping branch, which is beneficial for energy extraction and utilization.

(2) The galloping branch of the T-section prism presents an efficient and stable power output in the tests. There is an optimal power generation resistance R_L of 8 Ω, an optimum damping ratio ζ_{total} of 0.305, a maximum power generation P_{harn} of 21.23 W at U_r = 12.25, and a maximum power generation efficiency η_{out} of 20.2% at U_r = 11.625. Compared with the test results of the smooth circular cylinder and the PTC circular cylinder, η_{out} of the smooth circular cylinder is 22% [39] and the efficiency of the PTC circular cylinder is higher, at η_{out} = 28% and P_{harn} = 23.54 W [44]. In addition, the maximum η_{out} of the T-section prism with α = 0.8 is 27.44%, and it can be concluded that the T-section prism has great potential for power generation.

(3) In the present tests with aspect ratios from 0.8 to 1.5, the smaller section aspect ratio shows a stronger ability to convert electrical energy. The aspect ratio has a great influence on the energy collection for the T-section prism. Thus, it is necessary to optimize the section aspect ratio of the prism to enhance the power generation efficiency.

Author Contributions: J.L. and X.Y. designed the experiments; X.Y., N.S., Q.R., and H.D. performed the experiments; X.Y. and N.S. analyzed the data; X.Y. built and wrote the majority of the manuscript text; all authors reviewed the manuscript; F.L. and G.X. made contributions to the structure and language of the manuscript.

Acknowledgments: All workers from the State Key Laboratory of Hydraulic Engineering Simulation and Safety of Tianjin University are acknowledged. The authors are also grateful for the assistance of the editor and three anonymous reviewers for their professional and pertinent comments and suggestions, which were greatly helpful for further quality improvement of this manuscript.

Abbreviations

D	Projection width of the prism in the direction of the incoming flow
H	Height of the T-section prism cross-section
L	Prism length
α	Section aspect ratio H/D
K	System stiffness
C_{total}	System damping coefficient
C_{harn}	Electromagnetic damping coefficient
m_{osc}	Original mass, prism mass, transmission mass, and one-third of the spring mass [2]
U	Incoming flow velocity
U_r	Reduced velocity $U/(f_n \cdot D)$

R_L	Load resistance
ρ	Water density
A	Average of the amplitudes under continuous oscillation for 30 s
f_{osc}	Prime frequency of oscillation
$A^* = A/D$	Amplitude ratio A/D
$f^* = f_{osc}/f_n$	Frequency ratio
f_n	Natural frequency in air
ζ	System damping ratio
C_v	Critical velocity
η_{out}	Power generation efficiency
P_{harn}	Generation active power

References

1. Blevins, R.D. *Flow-Induced Vibration*, 3rd ed.; Van Nostrand Reinhold: New York, NY, USA, 1990.
2. Barriga, J.; Ruiz-de-Gopegui, U.; Goikoetxea, J.; Coto, B.; Cachafeiro, H. Selective coatings for new concepts of parabolic trough collectors. *Energy Procedia* **2014**, *49*, 30–39. [CrossRef]
3. Cachafeiroa, H.; de Arevaloa, L.F.; Vinuesaa, R.; Lopez-Vizcaino, R.; Luna, M. Analysis of vacuum evolution inside Solar Receiver Tubes. *Energy Procedia* **2015**, *69*, 289–298. [CrossRef]
4. Cachafeiroa, H.; de Arevaloa, L.F.; Vinuesaa, R.; Goikoetxea, J.; Barriga, J. Impact of solar selective coating ageing on energy cost. *Energy Procedia* **2015**, *69*, 299–309. [CrossRef]
5. Mannini, C.; Belloli, M.; Marra, A.M.; Bayati, I.; Giappino, S.; Robustelli, F.; Bartoli, G. Aeroelastic stability of two long-span arch structures: A collaborative experience in two wind tunnel facilities. *Eng. Struct.* **2016**, *119*, 252–263. [CrossRef]
6. Seo, J.-W.; Kim, H.-K.; Park, J.; Kim, K.T.; Kim, G.N. Interference effect on vortex-induced vibration in a parallel twin cable-stayed bridge. *J. Wind Eng. Ind. Aerodyn.* **2013**, *116*, 7–20. [CrossRef]
7. Assi, G.R.; Bearman, P.W.; Tognarelli, M.A. On the stability of a free-to-rotate short-tail fairing and a splitter plate as suppressors of vortex-induced vibration. *Ocean Eng.* **2014**, *92*, 234–244. [CrossRef]
8. Dowell, E.H.; Tang, D. Nonlinear aeroelasticity and unsteady aerodynamics. *AIAA J.* **2002**, *40*, 1697–1707. [CrossRef]
9. Assi, G.R.; Bearman, P.W.; Kitney, N. Low drag solutions for suppressing vortex induced vibration of circular cylinders. *J. Fluids Struct.* **2009**, *25*, 666–675. [CrossRef]
10. Park, H.; Bernitsas, M.M.; Kumar, R.A. Selective roughness in the boundary layer to suppress flow-induced motions of circular cylinder at 30,000 $\leq$ Re $\leq$ 120,000. *J. Offshore Mech. Arct. Eng.* **2012**, *134*, 041801. [CrossRef]
11. Sarpkaya, T. Fluid forces on oscillating cylinders. *NASA STI/Recon Tech. Rep. A* **1978**, *104*, 275–290.
12. Williamson, C.H.K.; Roshko, A. Vortex formation in the wake of an oscillating cylinder. *J. Fluids Struct.* **1988**, *2*, 355–381. [CrossRef]
13. Zdravkovich, M.M. *Flow around Circular Cylinders*; Oxford University Press: New York, NY, USA, 1997; Volume 1.
14. De Meij, A.; Vinuesa, J.F.; Maupas, V.; Waddle, J.; Price, I.; Yaseen, B.; Ismail, A. Wind energy resource mapping of Palestine. *Renew. Sustain. Energy Rev.* **2016**, *56*, 551–562. [CrossRef]
15. Sousa, V.C.; de M Anicézio, M.; De Marqui, C., Jr.; Erturk, A. Enhanced aeroelastic energy harvesting by exploiting combined nonlinearities: Theory and Experiment. *Smart Mater. Struct.* **2011**, *20*, 094007. [CrossRef]
16. Zhang, M.; Zhao, G.; Wang, J. Study on Fluid-Induced Vibration Power Harvesting of Square Columns under Different Attack Angles. *Geofluids* **2017**, *2017*, 6439401. [CrossRef]
17. Zhou, S.; Wang, J. Dual serial vortex-induced energy harvesting system for enhanced energy harvesting. *AIP Adv.* **2018**, *8*, 075221. [CrossRef]
18. Bernitsas, M.M.; Raghavan, K.; Ben-Simon, Y.; Garcia, E.M.H. VIVACE (vortex induced vibration aquatic clean energy): A new concept in generation of clean and renewable energy from fluid flow. *J. Offshore Mech. Arct. Eng.* **2008**, *130*, 041101–041115. [CrossRef]
19. Chamorro, L.P.; Hill, C.; Neary, V.S.; Gunawan, B.; Arndt, R.E.A.; Sotiropoulos, F. Effects of energetic coherent motions on the power and wake of an axial flow turbine. *Phys. Fluids* **2015**, *27*, 055104. [CrossRef]

20. Chamorro, L.P.; Lee, S.J.; Olsen, D.; Milliren, C.; Marr, J.; Arndt, R.E.A.; Sotiropoulos, F. Turbulence effects on a full-scale 2.5 MW horizontal-axis wind turbine under neutrally stratified conditions. *Wind Energy* **2015**, *18*, 339–349. [CrossRef]
21. Park, H.R.; Kumar, R.A.; Bernitsas, M.M. Enhancement of fluid induced vibration of rigid circular cylinder on springs by localized surface roughness at $3 \times 10^4 \leq \text{Re} \leq 1.2 \times 10^5$. *Ocean Eng.* **2013**, *72*, 403–415. [CrossRef]
22. Daqaq, M.F. Characterizing the response of galloping energy harvesters using actual wind statistics. *J. Sound Vib.* **2015**, *357*, 365–376. [CrossRef]
23. Hemon, P.; Amandolese, X.; Andrianne, T. Energy harvesting from galloping of prisms: A wind tunnel experiment. *J. Fluids Struct.* **2017**, *70*, 390–402. [CrossRef]
24. Alonso, G.; Meseguer, J. A parametric study of the galloping stability of two dimensional triangular cross-section bodies. *J. Wind Eng. Ind. Aerodyn.* **2006**, *94*, 241–253. [CrossRef]
25. Williamsona, C.H.K.; Govardhanb, R. A brief review of recent results in vortex-induced vibrations. *J. Wind Eng. Ind. Aerodyn.* **2008**, *96*, 713–735. [CrossRef]
26. Mannini, C.; Marra, A.M.; Massai, T.; Bartoli, G. Interference of vortex-induced vibration and transverse galloping for a rectangular cylinder. *J. Fluid Struct.* **2016**, *66*, 403–423. [CrossRef]
27. Liu, B.; Hamed, A.M.; Jin, Y.; Chamorro, L.P. Influence of vortical structure impingement on the oscillation and rotation of flat plates. *J. Fluid Struct.* **2017**, *70*, 417–427. [CrossRef]
28. Bokaian, A.R.; Geoola, F. Hydroelastic instabilities of square cylinders. *J. Sound Vib.* **1984**, *92*, 117–141. [CrossRef]
29. Vinuesa, R.; Prus, C.; Schlatter, P.; Nagib, H.M. Convergence of numerical simulations of turbulent wall-bounded flows and mean cross-flow structure of rectangular ducts. *Meccanica* **2016**, *51*, 3025–3042. [CrossRef]
30. Lian, J.; Yan, X.; Liu, F.; Zhang, J. Analysis on Flow Induced Motion of Cylinders with Different Cross Sections and the Potential Capacity of Energy Transference from the Flow. *Shock Vib.* **2017**, *2017*, 4356367. [CrossRef]
31. Camarri, S.; Salvetti, M.V.; Buresti, G. Large-eddy simulation of the flow around a triangular prism with moderate aspect-ratio. *J. Wind Eng. Ind. Aerodyn.* **2006**, *94*, 309–322. [CrossRef]
32. Iungo, G.V.; Buresti, G. Experimental investigation on the aerodynamic loads and wake flow features of lowaspect-ratio triangular prisms at different wind directions. *J. Fluid Struct.* **2009**, *25*, 1119–1135. [CrossRef]
33. Alonso, G.; Sanz-Lobera, A.; Meseguer, J. Hysteresis phenomena in transverse galloping of triangular cross-section bodies. *J. Fluid Mech.* **2012**, *33*, 243–251. [CrossRef]
34. Zhang, J.; Liu, F.; Lian, J.; Yan, X.; Ren, Q. Flow Induced Vibration and Energy Extraction of an Equilateral Triangle Prism at Different System Damping Ratios. *Energies* **2016**, *9*, 938. [CrossRef]
35. Zhang, J.; Xu, G.; Liu, F.; Lian, J.; Yan, X. Experimental investigation on the flow induced vibration of an equilateral triangle prism in water. *Appl. Ocean Res.* **2016**, *61*, 92–100. [CrossRef]
36. Lian, J.; Yan, X.; Liu, F.; Zhang, J.; Ren, Q.; Yan, X. Experimental Investigation on Soft Galloping and Hard Galloping of Triangular Prisms. *Appl. Sci.* **2017**, *7*, 198. [CrossRef]
37. Lee, J.H.; Bernitsas, M.M. High-damping High-Reynolds VIV tests for energy harnessing using the VIVACE converter. *Ocean Eng.* **2011**, *38*, 1697–1712. [CrossRef]
38. Hai, S.; Kim, E.S.; Bernitsas, M.P.; Bernitsas, M.M. Virtual Spring–Damping System for Flow-Induced Motion Experiments. *J. Offshore Mech. Arct. Eng.* **2015**, *137*, 061801.
39. Bernitsas, M.M.; Ben-Simon, Y.; Raghavan, K.; Garcia, E.M. The VIVACE converter: Model tests at high damping and Reynolds number around 10^5. *J. Offshore Mech. Arct. Eng.* **2009**, *131*, 011102. [CrossRef]
40. Chang, C.C.; Kumar, R.A.; Bernitsas, M.M. VIV and galloping of single circular cylinder with surface roughness at $3.0 \times 10^4 \leq \text{Re} \leq 1.2 \times 10^5$. *Ocean Eng.* **2011**, *38*, 1713–1732. [CrossRef]
41. Hai, S.; Kim, E.S.; Nowakowski, G.; Mauer, E.; Bernitsas, M.M. Effect of mass-ratio, damping, and stiffness on optimal hydrokinetic energy conversion of a single, rough cylinder in flow induced motions. *Renew. Energy* **2016**, *99*, 936–959.
42. Ma, C.; Sun, H.; Nowakowski, G.; Mauer, E.; Bernitsas, M.M. Nonlinear piecewise restoring force in hydrokinetic power conversion using flow induced motions of single cylinder. *Ocean Eng.* **2016**, *128*, 1–12. [CrossRef]

43. Lian, J.; Ren, Q.; Liu, F.; Yan, X.; Zhang, J. Experimental Study of Flow-Induced Vibration Characteristics of T-shape Cross-Section Oscillator. *J. Exp. Mech.* **2017**, *32*, 216–222. (In Chinese)
44. Lin, D. *Research on Flow Induced Motion of Multiple Circular Cylinders with Passive Turbulence Control*; Chong Qing University: Chongqing, China, 2013. (In Chinese)

5

Numerical Study of the Magnetic Field Effect on Ferromagnetic Fluid Flow and Heat Transfer in a Square Porous Cavity

Mohamed F. El-Amin [1,2,*], Usama Khaled [3,4] and Abderrahmane Beroual [5]

1 College of Engineering, Effat University, 21478 Jeddah, Saudi Arabia
2 Department of Mathematics, Faculty of Science, Aswan University, Aswan 81528, Egypt
3 Department of Electrical Engineering, College of Engineering, King Saud University, P.O. Box. 800, 11421 Riyadh, Saudi Arabia; ukhaled@ksu.edu.sa
4 Department of Electrical Engineering, Faculty of Energy Engineering, Aswan University, Aswan 81528, Egypt
5 Ecole Centrale de Lyon, University of Lyon, AMPERE CNRS UMR 5005, 36 Avenue Guy de Collongue, 69134 Ecully, France; Abderrahmane.Beroual@ec-lyon.fr
* Correspondence: mohamed.elamin.kaust@gamil.com

Abstract: A numerical study of ferromagnetic-fluid flow and heat transfer in a square porous cavity under the effect of a magnetic field is presented. The water-magnetic particle suspension is treated as a miscible mixture and, thus, the magnetization, density and viscosity of the ferrofluid are obtained. The governing partial-differential equations were solved numerically using the cell-centered finite-difference method for the spatial discretization, while the multiscale time-splitting implicit method was developed to treat the temporal discretization. The Courant–Friedrichs–Lewy stability condition (CFL < 1) was used to make the scheme adaptive by dividing time steps as needed. Two cases corresponding to Dirichlet and Neumann boundary conditions were considered. The efficiency of the developed algorithm as well as some physical results such as temperature, concentration, and pressure; and the local Nusselt and Sherwood numbers at the cavity walls are presented and discussed. It was noticed that the particle concentration and local heat/mass transfer rate are related to the magnetic field strength, and both pressure and velocity increase as the strength of the magnetic was increased.

Keywords: magnetic field; ferrofluid; porous cavity; heat transfer; mass transfer; numerical modeling

1. Introduction

Ferromagnetic fluids are smart fluids [1,2], composed of magnetized nanoparticles suspended in a liquid-based medium, such as oil and water. Ferromagnetic fluids have been used in different engineering and environmental applications, such as enhanced oil recovery (EOR). The idea of using ferrofluids, such as iron oxide, Fe_2O_3, and zinc oxide, ZnO, under an external magnetic field is to control the ferrofluids' movement in porous media [3–5]. Heat transfer in porous media involves a wide range of applications such as heat exchangers, oil/gas recovery, geophysical systems, nuclear waste disposal, chemical reactors, thermal insulation of buildings, and drying processes. Chegenizadeh et al. [6] provided a classification of the most popular nanoparticles under optimal operational conditions. El-Amin et al. [7,8] conducted some studies on modeling nanoparticle transport in porous media. Suleimanov et al. [9] and Ryoo et al. [10] conducted an experimental investigation of using nanoparticles in enhanced oil recovery application.

Flows in cavities have many technological applications, such as heat exchangers, cooling systems for electronic equipment, and environmental flows. Sheremet and Pop [11] investigated the steady laminar mixed convection inside a lid-driven square cavity filled with a water-based nanofluid. Chalambaz et al. [12] studied the heat and mass transfer in a square porous cavity with differential temperature and concentration at the sidewalls, and they found that the heat transfer of the mixture and the mass transfer of the other phase can be maximized for specific values of the Lewis number of one phase. Carvalho and de Lemos [13] presented the problem of laminar free convection within a square porous cavity filled with a saturated fluid. Javed et al. [14] presented a numerical study for free convection through a square enclosure filled with a ferrofluid-saturated porous medium under a uniform magnetic field.

Numerical simulation is an important tool enabling engineers and scientists to predict the transport phenomena, test, and optimize an appropriate intervention strategy for heat transfer [15]. The time-stepping method has been presented for the problems of fluid dynamics by using implicit-type time-marching procedures to resolve transients [16]. Martinez [17] solved the shallow water equations by using the time-splitting technique. The time viscosity-splitting method was used for the Boussinesq problem by Zhang and Qian [18]. A time-splitting Fourier spectral method has been developed for approximating singular solutions of the Gross–Pitaevskii equation [19]. El-Amin et al. [20–23] used a multiscale time-splitting strategy to manage different time-step sizes for different physics. In this research, a multiscale adaptive time-splitting scheme is introduced to simulate the problem of magnetic-field effects on ferrofluids and heat transfer with a single-phase flow in a porous cavity considering variation of the nanoparticle concentration.

2. Problem Definition

The problem of a single-phase flow with ferromagnetic fluid flow and heat transfer in a square porous cavity under the effect of an external magnetic field is considered in this study. The ferromagnetic fluids gain properties of both liquid and magnetized solid particles. If the ferromagnetic fluid is subjected to an external magnetic field, it flows toward the magnetic field, and the flow resistance increases. In the absence of the magnet, nearby ferrofluids act as normal liquid. Figure 1 shows the schematic diagram of the 2D square porous cavity of Cases 1 and 2. In Case 1, the cavity walls are kept at a low temperature/concentration, T_0, C_0, except the inlet at the center of the left wall is kept at a higher temperature/concentration T_h, C_h (Dirichlet boundary condition). Initially, the cavity was filled with a pure water, then the ferrofluid suspension injection begins. The flow boundary condition of the inlet is defined by the existing velocity, $u_{x,in}$. Above and below inlet, the no-flow boundary condition, $\frac{\partial p}{\partial x} = 0$, is assumed, as circulation may exist close to the inlet. However, far away from the inlet region, the circulation is expected to vanish, and a constant pressure boundary condition, p_0, is assumed. In Case 2, the walls are assumed adiabatic ($\frac{\partial C}{\partial x} = 0$, $\frac{\partial C}{\partial y} = 0$), ($\frac{\partial T}{\partial x} = 0, \frac{\partial T}{\partial y} = 0$) as indicated in Figure 1b. The constant pressure, p_0, is assumed to represent the flow boundary condition. The inlet is kept at high temperature, T_h, high concentration, C_h, velocity $u_{x,in}$. Above and below inlet are kept at low temperature T_c, concentration, C_c, and no-flow ($\frac{\partial p}{\partial x} = 0$) boundary condition.

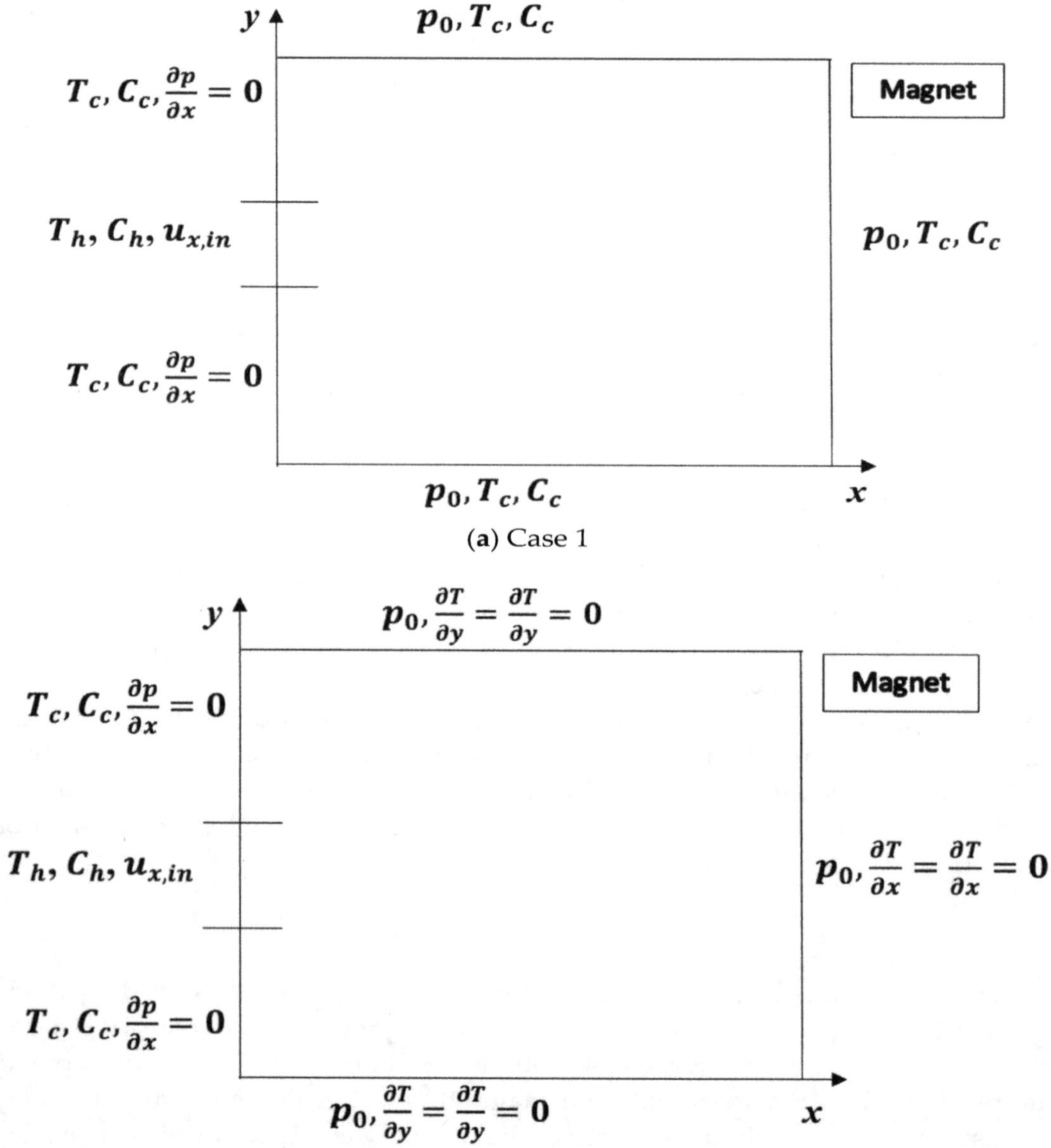

Figure 1. Schematic diagram of the 2D square porous cavity—(**a**) Case 1: cavity walls are kept at a low temperature/concentration except the inlet at center of the left wall is kept at a higher temperature/concentration (Dirichlet boundary condition) and inlet velocity (Neumann boundary condition); (**b**) Case 2: the cavity walls are adiabatic and isothermal with constant pressure (Dirichlet boundary condition), except that the inlet at center of the left wall is kept at a high temperature/concentration and the upper/lower of left wall is kept at a low temperature/concentration with no-flow Neumann boundary condition. The z-axis of the local magnet coordinate system is set perpendicular to the poles, the x-y plane.

3. Mathematical Modeling

The external permanent magnetic field produces attractive forces on the fluid. The external magnetic force with a variable concentration, and hence density, acts as a body force per unit volume, F_{mag} [4,24]:

$$F_{mag} = -\nabla\left[\mu_0 \int_0^H \left(\frac{\partial M}{\partial V}\right)_{H,T} dH\right] + \mu_0 M \nabla H \tag{1}$$

where μ_0 is the magnetic permeability. The specific volume, V, is defined by:

$$\partial V = \partial\left(\frac{1}{\rho}\right) = -\frac{1}{\rho^2}\partial\rho$$

Therefore, Equation (1) can be rewritten as:

$$F_{mag} = \mu_0\rho^2\nabla\left[\int_0^H\left(\frac{\partial M}{\partial\rho}\right)_{H,T}dH\right] + \mu_0\, M\nabla H \quad (2)$$

Such that:
$$\left(\frac{\partial M}{\partial\rho}\right)_{H,T} = \left(\frac{\partial M}{\partial a_1}\right)_{b_1,H,T}\frac{\partial a_1}{\partial\rho} + \left(\frac{\partial M}{\partial b_1}\right)_{a_1,H,T}\frac{\partial b_1}{\partial\rho}$$

The quantities $\frac{\partial a_1}{\partial\rho}$ and $\frac{\partial b_1}{\partial\rho}$ may be approximated around the points $(a_{1,0}, b_{1,0}) = (0,0)$, which represent the non-magnetized case, as:

$$\frac{\partial a_1}{\partial\rho} \approx \frac{a_1 - a_{1,0}}{\rho-\rho_0} = \frac{a_1}{\rho-\rho_0}$$

$$\frac{\partial b_1}{\partial\rho} \approx \frac{b_1 - b_{1,0}}{\rho-\rho_0} = \frac{b_1}{\rho-\rho_0}$$

The magnetization M is a function of the magnetic field strength H:

$$M = a_1\, tan^{-1}\,(b_1 H) \quad (3)$$

The two parameters a_1 and b_1 depend on the type of the ferromagnetic material and controlled both the initial susceptibility and the saturation magnetization. Therefore, Equation (1) can be written as:

$$F_{mag} = \frac{\mu_0\rho^2}{\rho-\rho_0}M\nabla H + \frac{\mu_0\rho^2}{\rho-\rho_0}\frac{a_1 b_1 H}{1+b_1^2H^2}\nabla H + \mu_0\, M\nabla H \quad (4)$$

The resulting magnetic induction (B) is a sum of the magnetic field strength (H) and the magnetization (M), i.e., they are related by the following linear relationship:

$$B = \mu_0\,(H+M) \quad (5)$$

Equation (5) is expressed in a scalar formulation, such that B, H, and M are scalars representing vector magnitudes. In this study, there is no free electric charge flow, so the Maxwell equations are reduced to the second Gauss law:

$$\nabla\cdot B = 0 \quad (6)$$

which means that the magnetic field induction is divergence-free. Ampere's law is reduced to the form:

$$\nabla\times H = 0 \quad (7)$$

which means that the magnetic field is vorticity-free.

Assuming that the medium is free of ferromagnetic materials and a permanent magnet is the source of the magnetic field, we can use the simple Equations (8) and (9), to calculate the components of H directly [4]. This assumption decouples the calculation of the external magnetic field from the ferrofluid distribution. At a given location the gradient of H can be calculated by simple first-order differencing in each of the coordinate directions. The z-axis of the local magnet coordinate system is set perpendicular to the poles, and the magnet is located at the upper right corner as indicated in Figure 1.

The flow simulation is in the x-y plane of the coordinate system (Equations (8) and (9) represent H_x and H_y in the x-axis and y-axis, respectively). The 2D magnetic field strength is given by [4]:

$$H_x = \frac{Br}{4\pi\mu_0}\left(\ln\left[\frac{y+b+\sqrt{(y+b)^2+(x-a)^2}}{y-b+\sqrt{(y-b)^2+(x-a)^2}}\,\frac{y-b+\sqrt{(y-b)^2+(x+a)^2}}{y+b+\sqrt{(y+b)^2+(x+a)^2}}\right] - \ln\left[\frac{y+b+\sqrt{(y+b)^2+(x-a)^2+L_p^2}}{y-b+\sqrt{(y-b)^2+(x-a)^2+L_p^2}}\,\frac{y-b+\sqrt{(y-b)^2+(x+a)^2+L_p^2}}{y+b+\sqrt{(y+b)^2+(x+a)^2+L_p^2}}\right]\right) \quad (8)$$

$$H_y = \frac{Br}{4\pi\mu_0}\left(\ln\left[\frac{x+a+\sqrt{(y-b)^2+(x+a)^2}}{x-a+\sqrt{(y-b)^2+(x-a)^2}}\,\frac{x-a+\sqrt{(y+b)^2+(x-a)^2}}{x+a+\sqrt{(y+b)^2+(x+a)^2}}\right] - \ln\left[\frac{x+a+\sqrt{(y-b)^2+(x+a)^2+L_p^2}}{x-a+\sqrt{(y-b)^2+(x-a)^2+L_p^2}}\,\frac{x-a+\sqrt{(y+b)^2+(x-a)^2+L_p^2}}{x+a+\sqrt{(y+b)^2+(x+a)^2+L_p^2}}\right]\right) \quad (9)$$

where L_p is the distance between the poles of the magnet, and B_r is the residual magnetization. If the ferromagnetic materials are heated to temperatures above the critical point (Curie temperature T_C), they become paramagnetic, whereas, below T_C, spontaneous magnetization occurs. The Curie temperature T_C for ferrofluids is high (e.g., for Fe_2O_3 is 948 K). The temperature dependence of the spontaneous magnetization at low temperatures, which is the case in this study, is given by Bloch's law [25]:

$$M(T) = M_0\left(1 - \left(\frac{T}{T_C}\right)^{\frac{3}{2}}\right) \quad (10)$$

where $M_0 = M(T = 0)$ is the spontaneous magnetization at absolute zero. In fact, there are several other mechanisms of temperature influence on the resulting magnetization of the magnetic fluid. One of the most significant models is the Langevin model which takes into consideration the role of the thermal fluctuations of the magnetic moments of the particles. The Langevin model is considered one the simplest cases, which can be written as:

$$M(T) = M_0\phi\left(\mathrm{cth}\,\xi - \frac{1}{\xi}\right) \quad (11)$$

such that:

$$\xi = \frac{\mu_0 \rho H}{kT} \quad (12)$$

The thermal fluctuations of the magnetic moments are related to the Brownian diffusion by the second fluctuation-dissipation theorem, which will be ignored in this study for simplicity.

The magnetization increases linearly with the ferrofluid mass fraction, C [4]:

$$M\,(C) = M(C = 1)\,C \quad (13)$$

Combining Equations (10) and (13), one may write:

$$M(T,C) = M(T = 0, C = 1)\left(1 - \left(\frac{T}{T_C}\right)^{\frac{3}{2}}\right)C \quad (14)$$

It is assumed that the volumes of pure water and ferrofluid are additive; therefore, the mixture density is defined as [4,24,26]:

$$\frac{1}{\rho} = \frac{1-C}{\rho_w} + \frac{C}{\rho_f} \quad (15)$$

where ρ_w is the density of the pure water component, and ρ_f is the density of ferromagnetic-particle component. The viscosity of the water–magnetic-particle mixture is calculated by the following linear relationship [4]:

$$\mu = \mu_w(1 + 1.35C) \tag{16}$$

where μ_w is the viscosity of the pure water. At a Reynolds number greater than 10, the inertial effect becomes significant. The inertial term, which is known as the Forchheimer term, can account for the nonlinear behavior of the pressure difference versus flow. Momentum conservation in porous media is represented by the extended non-Darcy's law [27]:

$$\mathbf{u} + \frac{F\sqrt{K}}{\nu}|\mathbf{u}|\mathbf{u} = -\frac{K}{\mu}(\nabla p - F_{mag} + g\rho_0(1 - \beta(T - T_r) - \beta^*(C - C_0))\nabla z) \tag{17}$$

where F is the inertia parameter, and ν is kinematic viscosity. In the above equation, the second term on the left-hand side represents the inertial term in Darcy's equation. The magnetic field effect on the fluid flow acts as a body force in the extended Darcy's law. Although this is an unsteady problem, however, the momentum (Darcy's law) has no time derivative term which is experimentally based. Actual fluid velocity varies throughout the pore space, due to the connectivity and geometric complexity of that space. This variable velocity can be characterized by its mean or average value. The convective flows are caused by the temperature difference in terms of the Boussinesq approximation, in which the density of the body force term is calculated by:

$$\rho = \rho_0(1 - \beta(T - T_0) - \beta^*(C - C_0)),\ \beta = -\frac{1}{\rho_0}\frac{\partial \rho}{\partial T},\ \beta^* = -\frac{1}{\rho_0}\frac{\partial \rho}{\partial C}$$

The mass conservation equation is represented by:

$$\nabla \cdot \mathbf{u} = 0 \tag{18}$$

The energy conservation equation in porous media can be represented as [23]:

$$\frac{\partial[(1-\varphi)\rho_s c_{p,s} + \varphi\rho c_p]T}{\partial t} + \rho\mathbf{u}\cdot\nabla T = \nabla\cdot[(1-\varphi)h_s + \varphi h]\nabla T \tag{19}$$

The concentration (solute transport) equation may be represented as [26]:

$$\varphi\frac{\partial C}{\partial t} + \nabla\cdot(\mathbf{u}C - D\nabla C) = 0 \tag{20}$$

In the above equations, C is the concentration, C_0 is the initial concentration, c_p is the heat capacity, D is the diffusion-dispersion tensor, g is the gravitation acceleration, h is the thermal conductivity of the fluid, h_s is the thermal conductivity of the solid, K is the permeability, T is the temperature, T_0 is the initial temperature, p is the fluid pressure, $\mathbf{u}$ is the fluid velocity vector, β is the thermal expansion coefficient, β^* is the solute expansion coefficient, φ represents the porosity of the porous media, ρ is the density of the fluid mixture, ρ_s is the density of the solid phase, and μ is the fluid viscosity. Substituting Equation (16) into Equation (18), one may obtain:

$$\nabla\cdot\mathbf{u} = -\nabla\cdot\frac{K}{\mu\left(1 + \frac{F\sqrt{K}}{\nu}|\mathbf{u}|\right)}(\nabla p - F_{mag} + g\rho_0(1 - \beta(T - T_r) - \beta^*(C - C_0))) = 0 \tag{21}$$

Calculating the heat transfer rate at boundaries/surfaces within fluids is of great interest in industrial and technological applications. The Nusselt number (Nu), a dimensionless number, is the ratio of convective to conductive heat transfer across the boundary. Unlike in other studies in the literature, the local Nusselt number was calculated on the wall instead of the average Nusselt number, because it gives the direct heat transfer rate between the wall and the adjacent fluid. The local Nusselt number measures the competition between convection and conduction heat flows and is defined by:

$$Nu_x = \frac{q_x L}{(T_{in} - T_0)k_T} = -\frac{L}{(T_{in} - T_0)} \frac{\partial T}{\partial x}\bigg|_{x=0} \tag{22}$$

$$Nu_y = \frac{q_y L}{(T_{in} - T_0)k_T} = -\frac{L}{(T_{in} - T_0)} \frac{\partial T}{\partial y}\bigg|_{y=0} \tag{23}$$

where L is the characteristic length; q_x, q_y are the local surface heat fluxes and defined as:

$$q_x = -k_T \frac{\partial T}{\partial x}\bigg|_{x=0} \tag{24}$$

$$q_y = -k_T \frac{\partial T}{\partial y}\bigg|_{y=0} \tag{25}$$

where k_T is the thermal conductivity of the fluid. Similarly, the local surface mass fluxes, j_x, j_y maybe defined as:

$$j_x = -D \frac{\partial C}{\partial x}\bigg|_{x=0} \tag{26}$$

$$j_y = -D \frac{\partial C}{\partial y}\bigg|_{y=0} \tag{27}$$

Therefore, the local Sherwood number which represents the ratio of the convective mass transfer to the rate of diffusive mass transport and given by:

$$Sh_x = \frac{j_x L}{(C_w - C_0)D} = -\frac{L}{(C_w - C_0)} \frac{\partial C}{\partial x}\bigg|_{x=0} \tag{28}$$

$$Sh_y = \frac{j_y L}{(C_w - C_0)D} = -\frac{L}{(C_w - C_0)} \frac{\partial C}{\partial y}\bigg|_{y=0} \tag{29}$$

Initially, the cavity is filled with pure water and, then, the ferrofluid suspension injection begins. In the following, two cases based on the types of boundary condition are described:

Case 1. All the walls of the square cavity are kept cold at temperature T_c and low concentration C_c except that the center of the left wall is kept at a high temperature T_h and a high concentration C_h, as shown in Figure 1a. The boundary conditions are described mathematically as follows:

at $x = 0, 0 \le y \le L, T = T_c, C = C_c, \frac{\partial p}{\partial x} = 0,$
at inlet $T_{in} = T_h, C_{in} = C_h, u = u_{x,in};$
at $0 \le x \le L, y = 0, T = T_c, C = C_c, p = p_0;$
at $x = 1, 0 \le y \le L, T = T_c, C = C_c, p = p_0;$
at $0 \le x \le L, y = L, T = T_c, C = C_c, p = p_0;$

where *in* refers to the injection location (inlet), and the velocity is taken to be zero on all cells (no-flow boundary conditions, i.e., impermeable wall) except for the central cells. The fluid exits from constant pressure boundary, p_0, which is a kind of exit boundary, as it is smaller than the entire pressure.

Case 2. All the walls of the square cavity are adiabatic and isothermal with constant Dirichlet pressure, except that the center of the left wall is kept hot at temperature T_h and high concentration C_h. The upper and lower left walls are kept at a cold temperature and a low concentration with no-flow Neumann boundary condition (impermeable wall), as shown in Figure 1b. The boundary conditions are described mathematically as follows.

at $x = 0, 0 \leq y \leq L, T = T_c, C = C_c, \frac{\partial p}{\partial x} = 0,$
at inlet $T_{in} = T_h, C_{in} = C_h, u = u_{x,in}.$
at $0 \leq x \leq L, y = 0, \frac{\partial T}{\partial y} = 0, \frac{\partial C}{\partial y} = 0, p = p_0.$
at $x = L, 0 \leq y \leq L, \frac{\partial T}{\partial x} = 0, \frac{\partial C}{\partial x} = 0, p = p_0.$
at $0 \leq x \leq L, y = L, \frac{\partial T}{\partial y} = 0, \frac{\partial C}{\partial y} = 0, p = p_0.$

4. Numerical Method

The governing equations were discretized numerically using the CCFD method for the spatial discretization. The CCFD method is considered one of the most commonly used finite-difference methods for flow and transport in porous media, because it is locally conservative. In the CCFD method, one uses velocity on edges of the cell, and pressure on the center of the cell (cell-centered). The permeability is calculated using the harmonic averaging on the center as well.

The backward Euler time discretization was used for the time derivative of equations of temperature and concentrations, with a time-splitting implicit technique [17,20]. The outer time discretization, which had a relatively large time step, was used for pressure. The subscript $k+1$ represents the current time step, while the subscript k represents the previous time step. The temperature and concentration had the same level of time discretization. The following procedures have been considered,

- The total time interval, $[0, T]$, is divided into a number of N_p time steps, namely, $0 = t_0 < t_1 < \ldots < t_{N_p=T}$, with a time-step of length $\Delta t^k = t^{k+1} - t^k$.
- For temperature and concentration, each interval, $\left(t^k, t^{k+1}\right)$, was divided into a number of $N_{p,C}$ subintervals, i.e., $\left(t^k, t^{k+1}\right) = \cup_{l=0}^{N_{p,c}-1}\left(t^{k,m}, t^{k,m+1}\right)$.
- The Courant–Friedrichs–Lewy stability condition (CFL < 1) is used to accomplish the time step-size adaptation.
- The time-step size of the pressure was taken to be larger than the time-step sizes of temperature and concentration.

The algorithm can be stated as:

- Calculating the pressure implicitly by coupling the continuity and momentum equations.
- Compute the velocity explicitly.
- Solve energy and concentration equations implicitly.
- Update porosity, permeability, and density.

Now, let us express the pressure equation as:

$$-\nabla\cdot\frac{K}{\mu\left(1+\frac{F\sqrt{K}}{\nu}\left|\mathbf{u}^k\right|\right)}\left(\nabla\cdot p^{k+1} - F_{mag}^k + g\rho_0\left(1-\beta\left(T^k - T_r\right) - \beta^*\left(C^k - C_0\right)\right)\nabla z\right) = q^{k+1} \quad (30)$$

The energy equation is computed implicitly as:

$$\frac{\left[(1-\varphi)\rho_s c_{p,s}+\varphi\rho\left(T^k,C^k\right)c_p\right]T^{k+1,m+1}-\left[(1-\varphi)\rho_s c_{p,s}+\varphi\rho\left(T^k\right)c_p\right]T^{k+1,m}}{\Delta t} + \rho\mathbf{u}^{k+1}.\nabla T^{k+1,m+1} = \nabla\cdot\left[(1-\varphi)h_s + \varphi h\right]\nabla T^{k+1,m+1} \quad (31)$$

Similarly, the concentration equation is computed implicitly as:

$$\varphi\frac{C^{k+1,m+1} - C^{k+1,m}}{\Delta t^m} + \nabla\cdot\left[\mathbf{u}^{k+1}C^{k+1,m+1} - D\nabla C^{k+1,m+1}\right] = 0 \quad (32)$$

In order to control the time-step size, the CFL condition (CFL < 1) is used. The CFL condition of the temperature and concentration equations is expressed as:

$$\text{CFL}_x = \frac{u_x \Delta t^{k,m}}{x}, \ \text{CFL}_y = \frac{u_y \Delta t^{k,m}}{\Delta y} \tag{33}$$

The numerical stability procedures can be summarized as follows:

- The pressure time-step, $\Delta t^{k,0} = \Delta t^k$ is taken as an initial time step for both temperature and concentration equations.
- After that, the conditions $\text{CFL}_x > 1$ and $\text{CFL}_y > 1$ are examined. If one of them is satisfied, the temperature/concentration time-step is divided into smaller steps.
- Therefore, the conditions CFL_x and CFL_y are recalculated based on the new time steps and so on, until satisfying the conditions $\text{CFL}_x < 1$ and $\text{CFL}_y < 1$.

5. Results and Discussion

To obtain physical insights, the system of the governing partial differential equations was solved numerically using the above numerical method. Two different cases were considered based on different boundary conditions, as in Figure 1a,b. In Case 1, all boundaries were of a Dirichlet type except that the west boundary was a no-flow boundary condition, which was a Neumann type. The physical and computational parameters are presented in Table 1. A value for the outer time loop k was selected, and the step size Δt^k was determined; then, the time-step size Δt^m was calculated based on the CFL conditions. In our calculation the velocity inlet was taken as 0.4756×10^{-8} m/s and the corresponding Reynolds number is 0.00143, which is suitable for flow in porous media.

Table 1. Physical parameter values.

Parameter	Description	Value	Units
a_1	Constant	10^4–10^5	m A^{-1}
b_1	Constant	10^{-6}–10^{-5}	m A^{-1}
a, b	Half of width/height of the magnet	0.02	m
Br	Residual magnetization	0:0.2	Tesla
C_0	Initial concentration	0	-
C_{in}	Inlet concentration	1	-
c_p	Heat capacity	800	J/Kg K
D	Diffusion coefficient	5×10^{-3}	m^2/S
h_s	Thermal conductivity of the solid	0.718	W/(m/K)
h	Thermal conductivity of the ferrofluid	0.6	W/(m/K)
g	Gravity acceleration	9.81	m/s^2
K	Permeability	100	md
L_p	Distance between poles	2.4	m
L_{in}	Inlet width	0.3	m
L	Cavity side length	5	m
md	Millidarcy	9.86923×10^{-16}	m^2
p_0	Initial pressure	1×10^6	N/m^2
T_0	Initial temperature	300	K
T_{in}	Inlet temperature	360	K
$u_{x,in}$	Inlet velocity	0.4756×10^{-8}	m/s
β	Thermal expansion coefficient	0.005	K^{-1}
β^*	Solute expansion coefficient	0.001	K^{-1}
φ	Porosity	0.3	-
μ	Water viscosity	0.001	Pa.s
μ_0	Magnetic permeability	1×10^{-7}	$N \cdot A^{-2}$
ρ_s	Solid media density	2500	kg/m^3
ρ_w	Pure water density	1000	kg/m^3
ρ_f	Particles density	8933	kg/m^3

Some error estimates are listed in Table 2 to examine the accuracy of the proposed scheme. The errors were calculated for temperature and concentration at various values of the time-step iteration, k. The results in Table 2 indicate that the error decreases as the number of time steps increases. In addition, the temperature error is very small and approaches zero; however, the concentration error needs bigger time step numbers to reach the desired approximation. Moreover, the calculated time step size (Δt^m) is plotted against its number m in Figure 2. Moreover, concentration profiles for different spatial grid size n_x and n_y of Case 1 are plotted in Figure 3 to show the grid independence test. The spatial discretization, $n_x \times n_y$, was chosen, respectively, as 50×50, 100×100, and 150×150. It is clear from Figure 3 that the differences between the three curves are very small, therefore, we choose the course grid, 50×50, for the computations in this study.

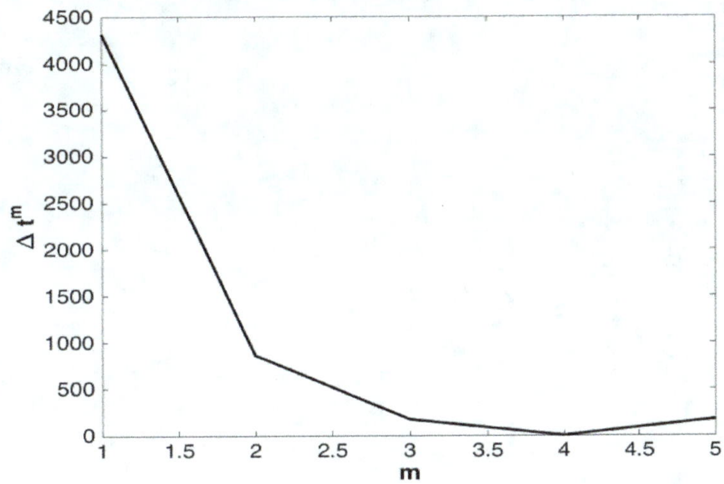

Figure 2. Variation of the adaptive time step size (Δt^m) against its number m of Case 1.

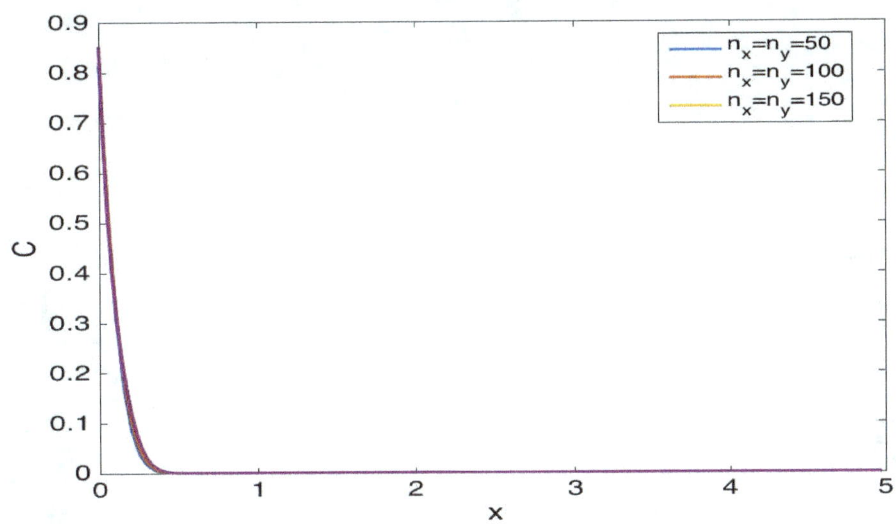

Figure 3. Concentration profiles for different spatial grid size n_x and n_y of Case 1.

Table 2. Error estimates for various values of k, Case 2.

k	$\lVert T^{k+1,l+1} - T^{k+1,l} \rVert$	$C^{k+1,m+1} - C^{k+1,m}$
50	5.0317×10^{-11}	0.0198
100	6.9070×10^{-11}	0.0099
300	3.0481×10^{-12}	0.0033

5.1. Results for Case 1

Figure 4a,b illustrates the temperature difference, $\Delta T = T - T_0$, contours of Case 1 without the magnetic-field effect (Br = 0.0) and with the magnetic-field effect (Br = 1.2), respectively. It seems that the magnetic field slightly enhances the temperature. The differences between of the two cases are similar, and it seems to be hard to notice the small differences (around 1 K); however, the differences are shown clearly in Figure 5, which shows the temperature profiles against the x-axis at y = 1 m, for various values of Br of Case 1. In Figure 6, the concentration profiles are plotted against the x-axis at y = 1 m, for various values of Br of Case 1. It is clear from Figure 6 that as the parameter Br increases, the ferrofluid concentration increases.

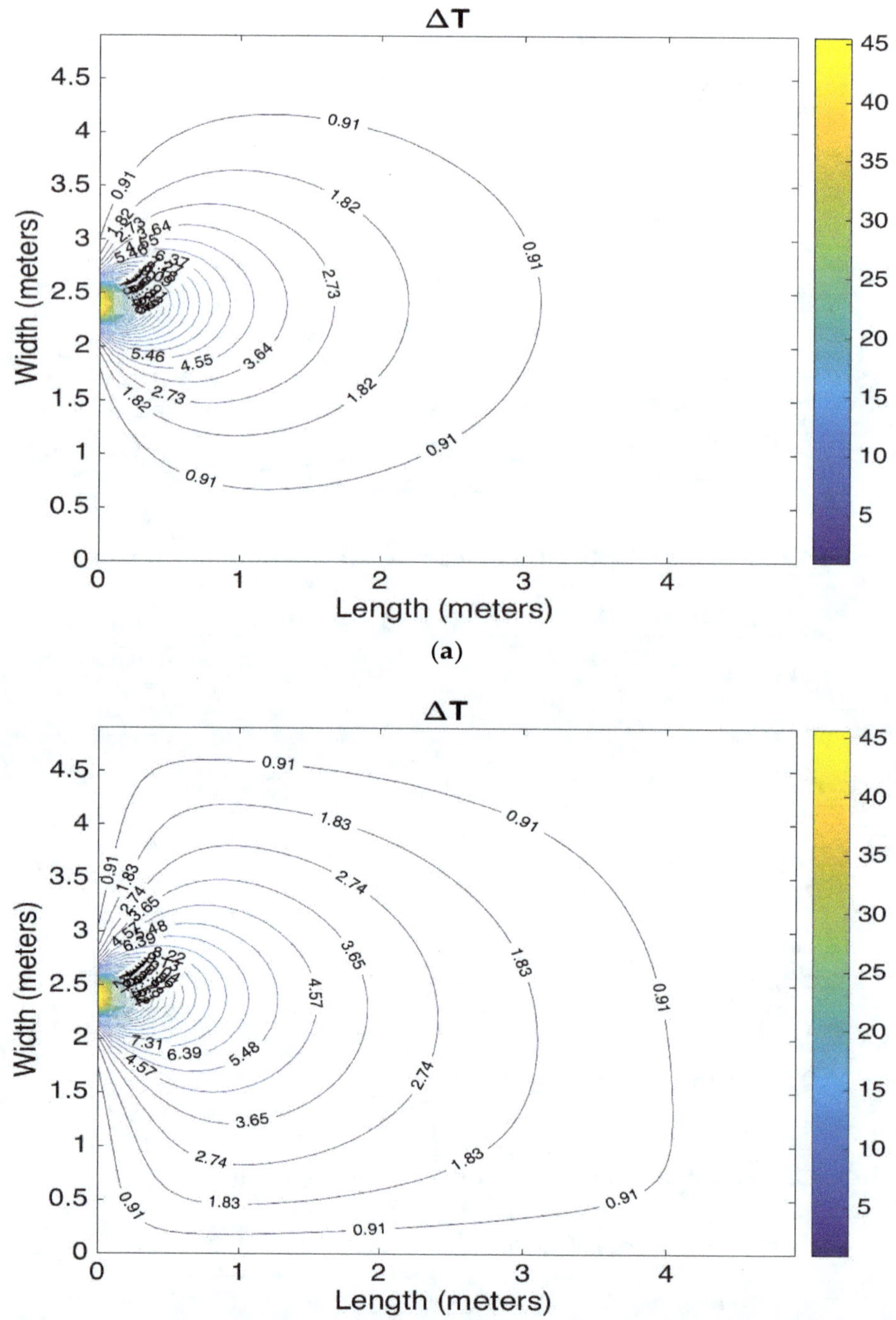

Figure 4. Temperature difference, $\Delta T = T - T_0$, contours of Case 1 with (**a**) Br = 0.0 and (**b**) Br = 1.2.

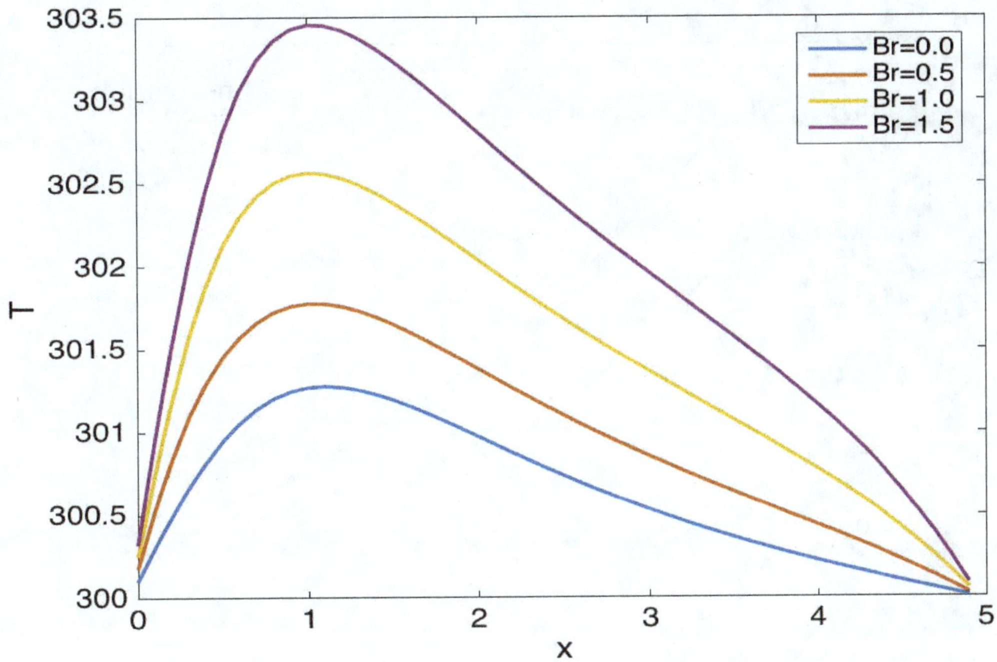

Figure 5. Temperature profiles against x-axis at $y = 1$ m, for various values of Br of Case 1.

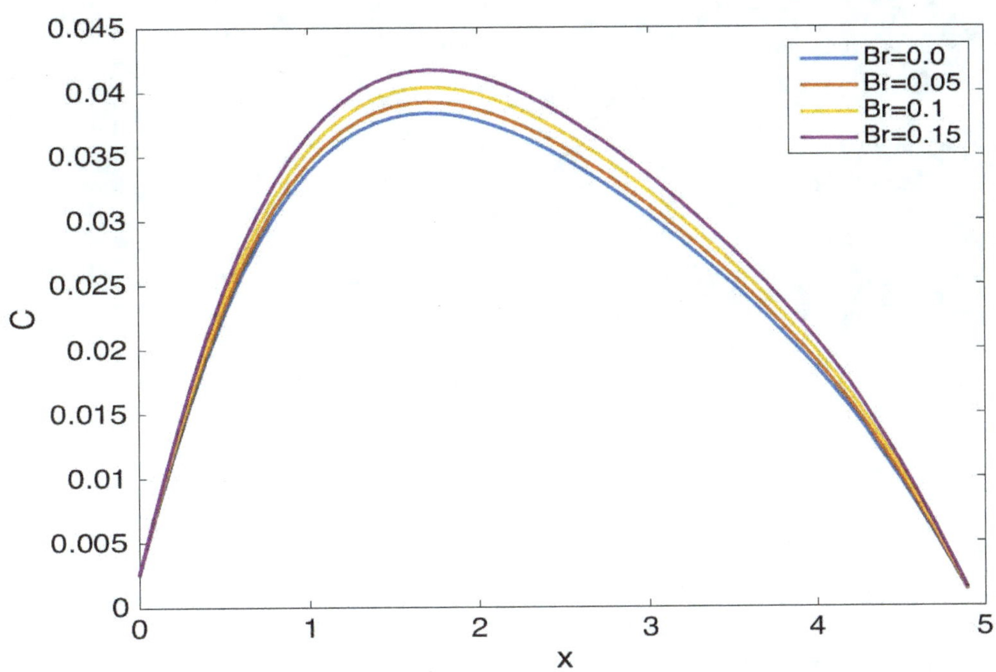

Figure 6. Concentration profiles against the x-axis at $y = 1$ m, for various values of Br of Case 1.

The pressure and velocity magnitude ($u_w = \sqrt{u_x^2 + u_y^2}$) profiles are plotted in Figure 7a,b, respectively, against the x-axis at $y = 1$ m, for various values of Br of Case 1. These figures show that, as Br increases, pressure and velocity increase. The maximum differences are located in the inlet vicinity and gradually decrease far away from it.

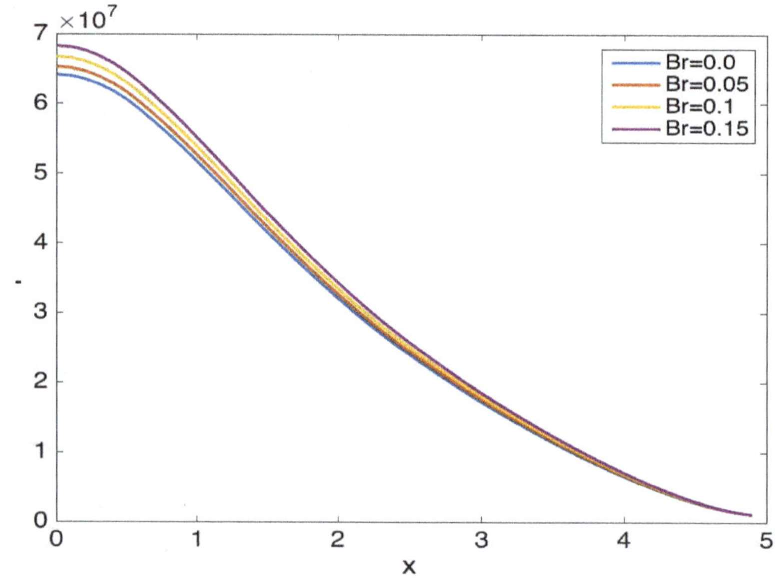

(a)

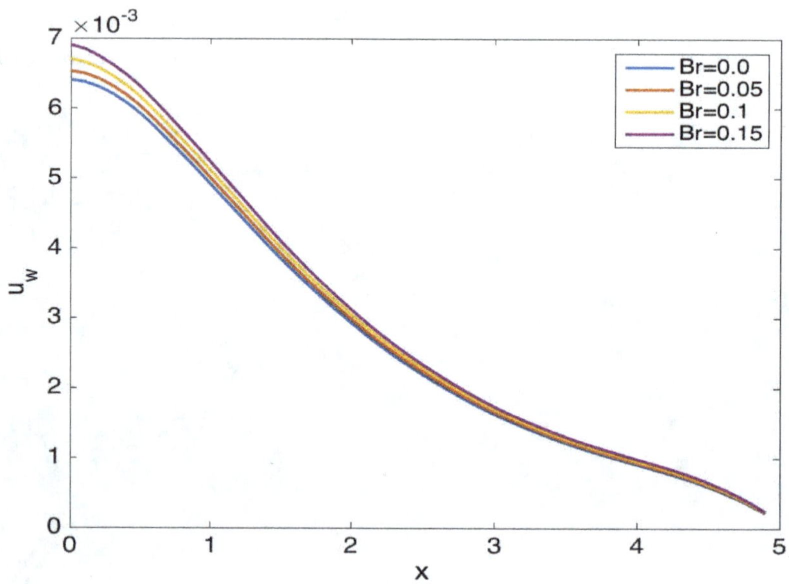

(b)

Figure 7. (**a**) Pressure and (**b**) average velocity profiles against x-axis at $y = 1$ m, for various values of Br of Case 1.

In Figure 8, the local Nusselt numbers are plotted along the left wall, right wall, upper wall, and lower wall of the porous cavity, for various values of the parameter Br of Case 1. The magnetic field has no effect on the left wall because the advection effect is much larger than the magnetic effect. The variation in magnetic-field strength has a small effect on the other walls; however, the differences are clear. The higher local Nusselt number is located in the center of the left wall, which indicates that the heat transfer rate is high in that region. Moreover, the heat transfer rate on the right wall has high values around the center of the wall and decreases gradually in the edge closure. The negative sign indicates that the heat transfer direction is from the fluid to the solid wall. The behavior of the heat transfer rate on the upper wall is similar to that on the right wall, but the maximum heat transfer rate is shifted to the bottom edge. Finally, the local Nusselt number on the lower wall has maximum values close to the left portion and then decreases gradually near the edges.

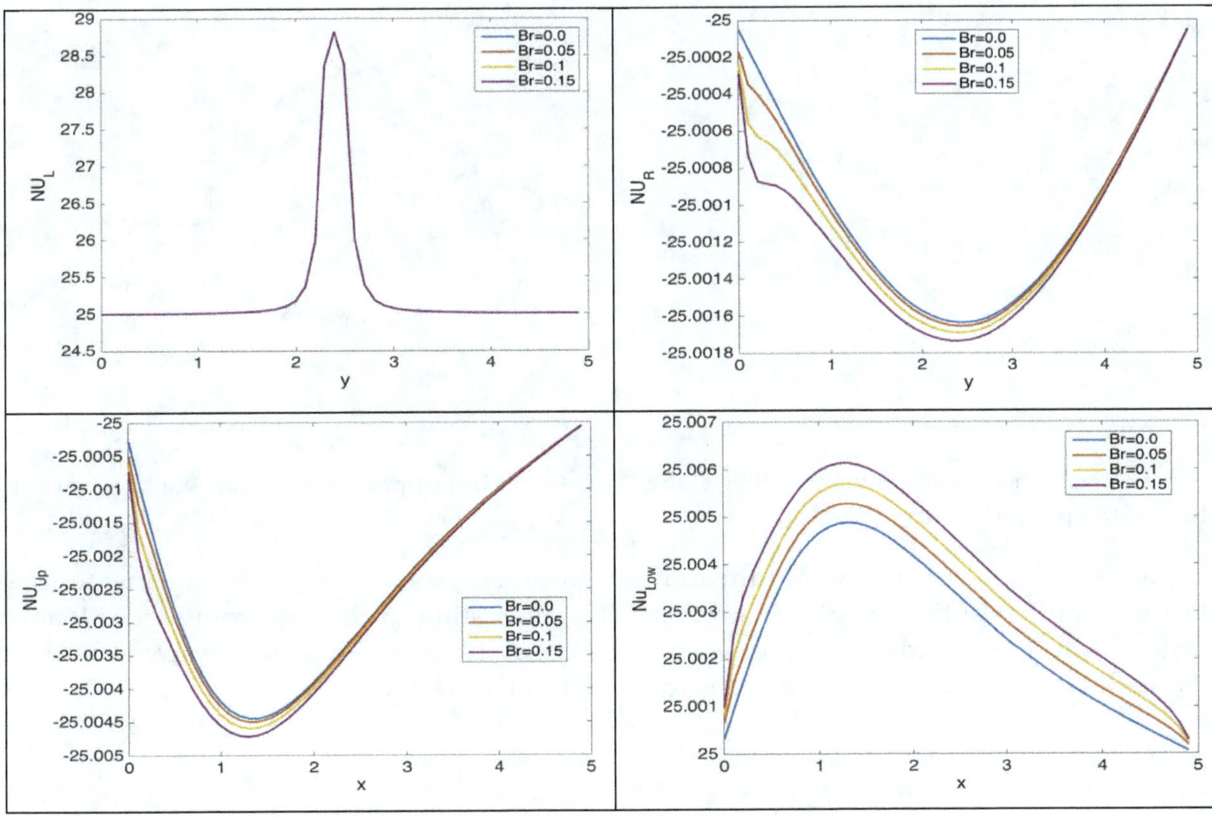

Figure 8. Local Nusselt number distributions along the left wall, right wall, upper wall, and lower wall for various values of the parameter Br of Case 1.

Figure 9 illustrates the local Sherwood number profiles along the left-, right-, upper-, and lower-walls. It can be seen from this figure that the local Sherwood number profiles along the left and right walls are symmetric around the center of y-axis and they decrease gradually far away from the center toward the upper to the lower corners. The mass transfer rate on the left-wall is high close to the inlet region and decreases from bottom to top. The negative sign indicates that mass transfer direction is from the fluid to the solid wall. It also may be noticed that the mass transfer rate on the right-wall is very small. The local Sherwood number on the upper- and left-walls has a similar behavior and same order of magnitude with different signs. The local mass transfer rate has maximum values at $x \approx 0.1$.

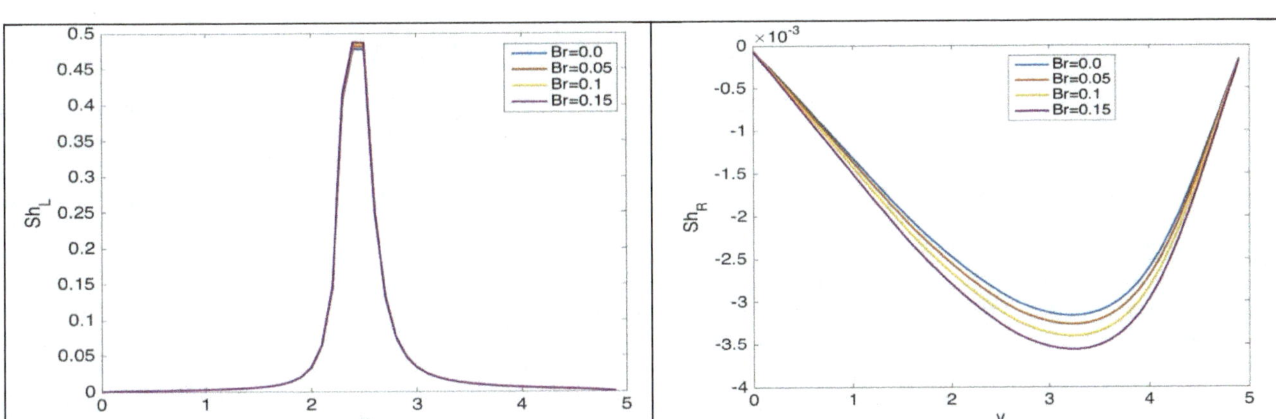

Figure 9. *Cont.*

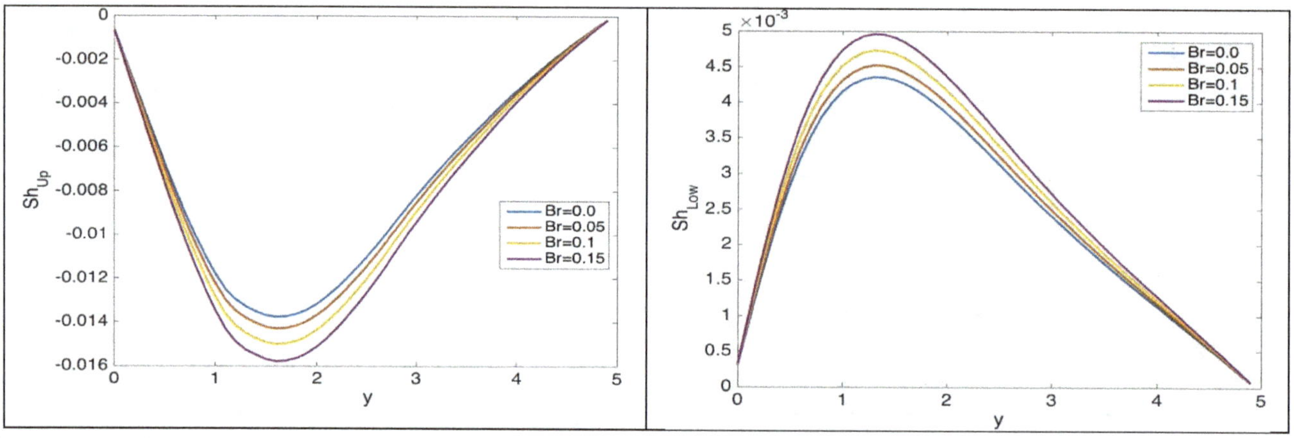

Figure 9. Local Sherwood number profiles along the **left**-, **right**-, **upper**-, and **lower**-walls for various values of the parameter Br; Case 1.

In Figure 10, the transient temperature and concentration profiles are plotted against the x-axis at $y = 4$, at different times of the ferrofluid injections. It is interesting to observe from Figure 10a that the temperature reaches the steady-state after around 2 days of injection, while the concentration reaches its steady-state after 2000 days of injection as indicated in Figure 10b.

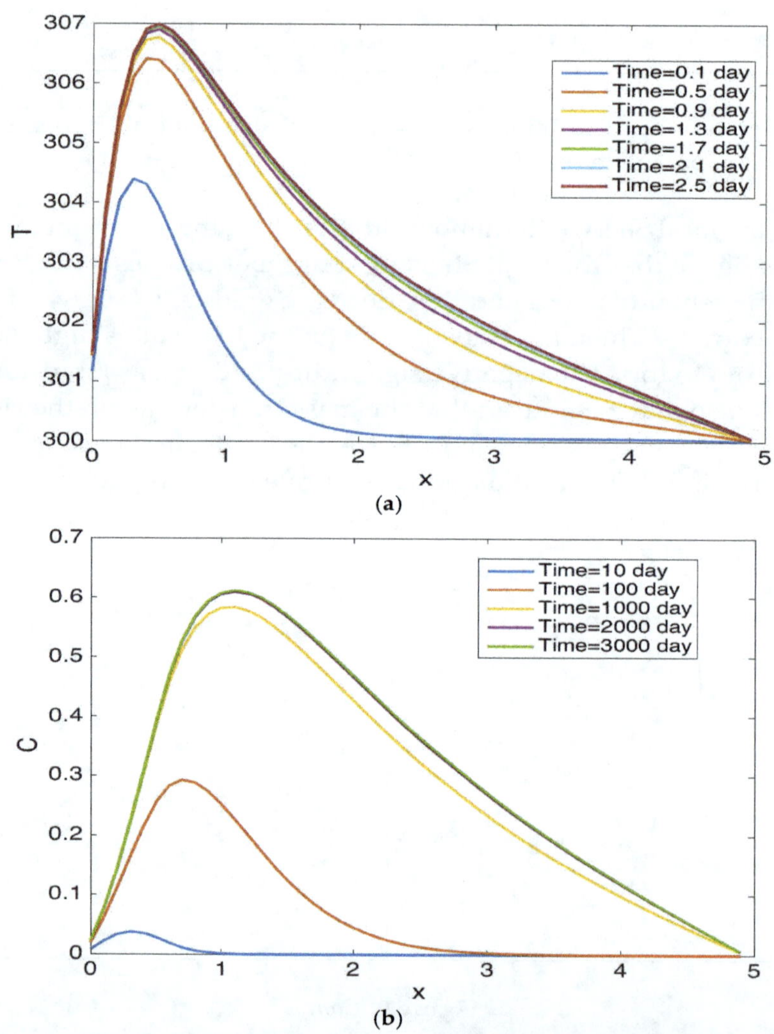

Figure 10. Transient (**a**) temperature, and (**b**) concentration profiles against x-axis at $y = 4$, of Case 1.

5.2. Results for Case 2

In Figure 11, the temperature profiles are plotted against the x-axis, for various values of Br of Case 2. This figure shows that the existence of the external magnetic field increases the temperature. The lowest temperature values are in the vicinity of the left wall, while the opposite happens at $y = 2.5$ m. The magnetic field has a small effect at the centerline as the advection dominates, whereas it has a clear effect far away from the high advection region (e.g., $y = 1$ m). As the parameter Br increases, the temperature increases, which may be interpreted as the magnetic field enhancing the kinetic energy, which, in turn, enhances the heat energy.

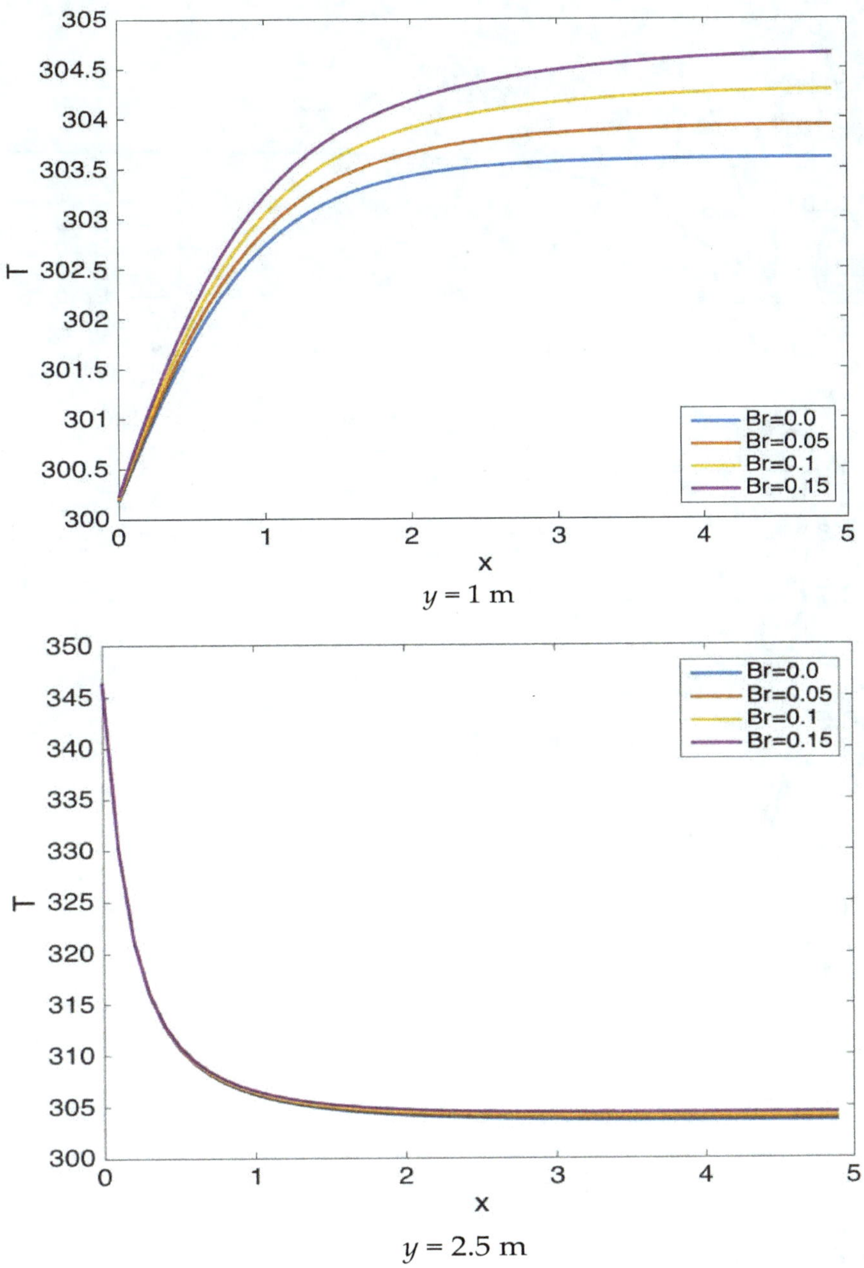

Figure 11. Temperature profiles against x-axis at $y = 1$ m, 2.5 m for various values of Br of Case 2.

The concentration profiles are plotted in Figure 12 against the x-axis at $y = 4$ m, 2.5 m, for various values of Br of Case 2. At the centerline ($y = 2.5$ m), the concentration has maximum values close to the inlet and decreases gradually far away from the inlet, whereas then opposite is true at $y = 4$ m. The concentration has minimum values close to the left wall and increases gradually far away from it.

As the magnetic field pulls the particles toward it based on its strength, the parameter Br increases the concentration of the ferrofluid.

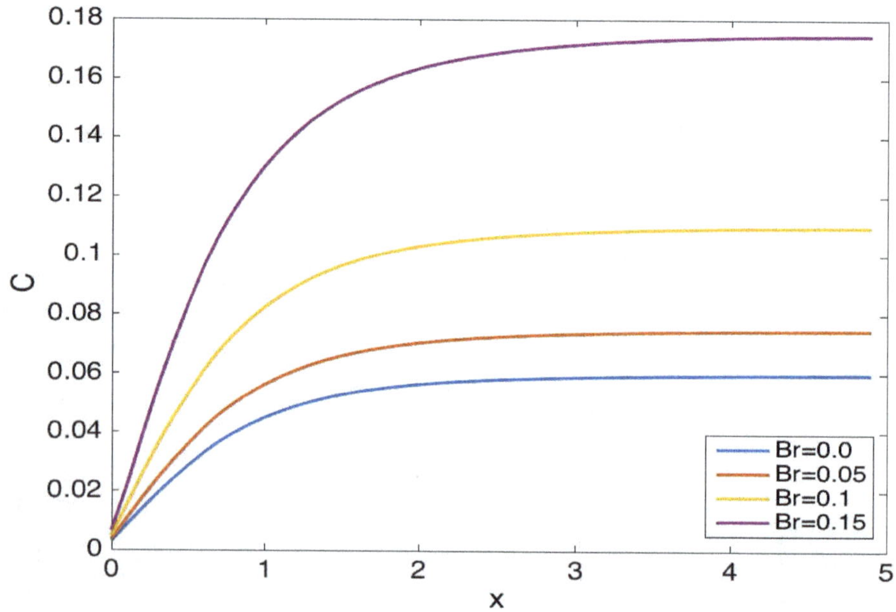

$y = 4$ m

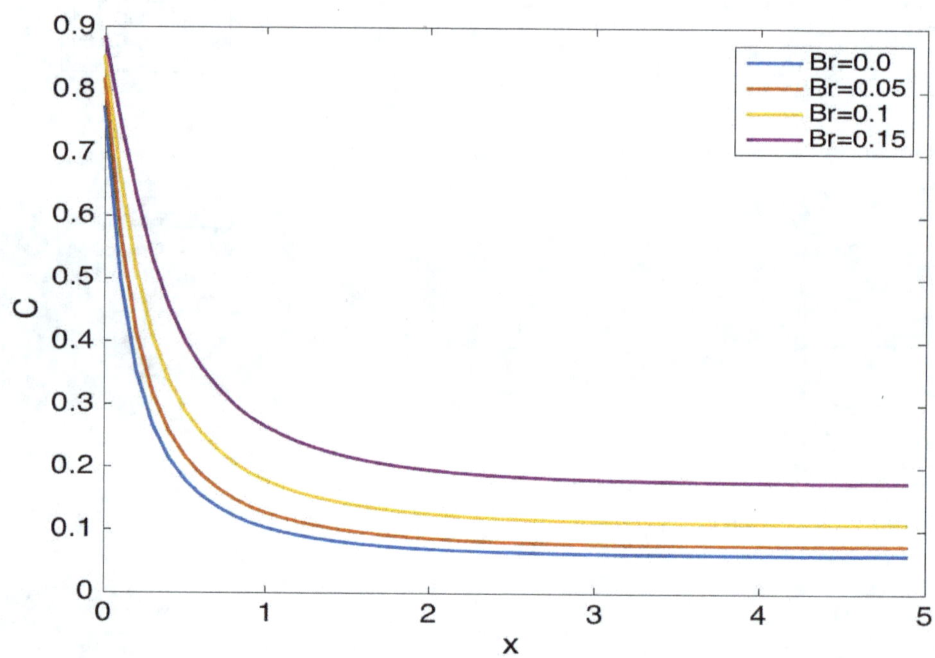

$y = 2.5$ m

Figure 12. Concentration profiles against x-axis at $y = 4$ m, 2.5 m, for various values of Br of Case 2.

The temperature distribution of Case 2 is shown in Figure 13a. It can be seen from this figure that the temperature is high around the inlet location. Then it decreases gradually far away from the inlet because the other walls are adiabatic with a constant pressure. This may be interpreted as the heat transfer mode changes from the mixed convection region close to inlet to the purely conduction region far away. Figure 13b shows the concentration distribution of Case 2. The behavior of the concentration is similar to the temperature distribution which can be interpreted similarly but with respect to mass transfer and solute convection.

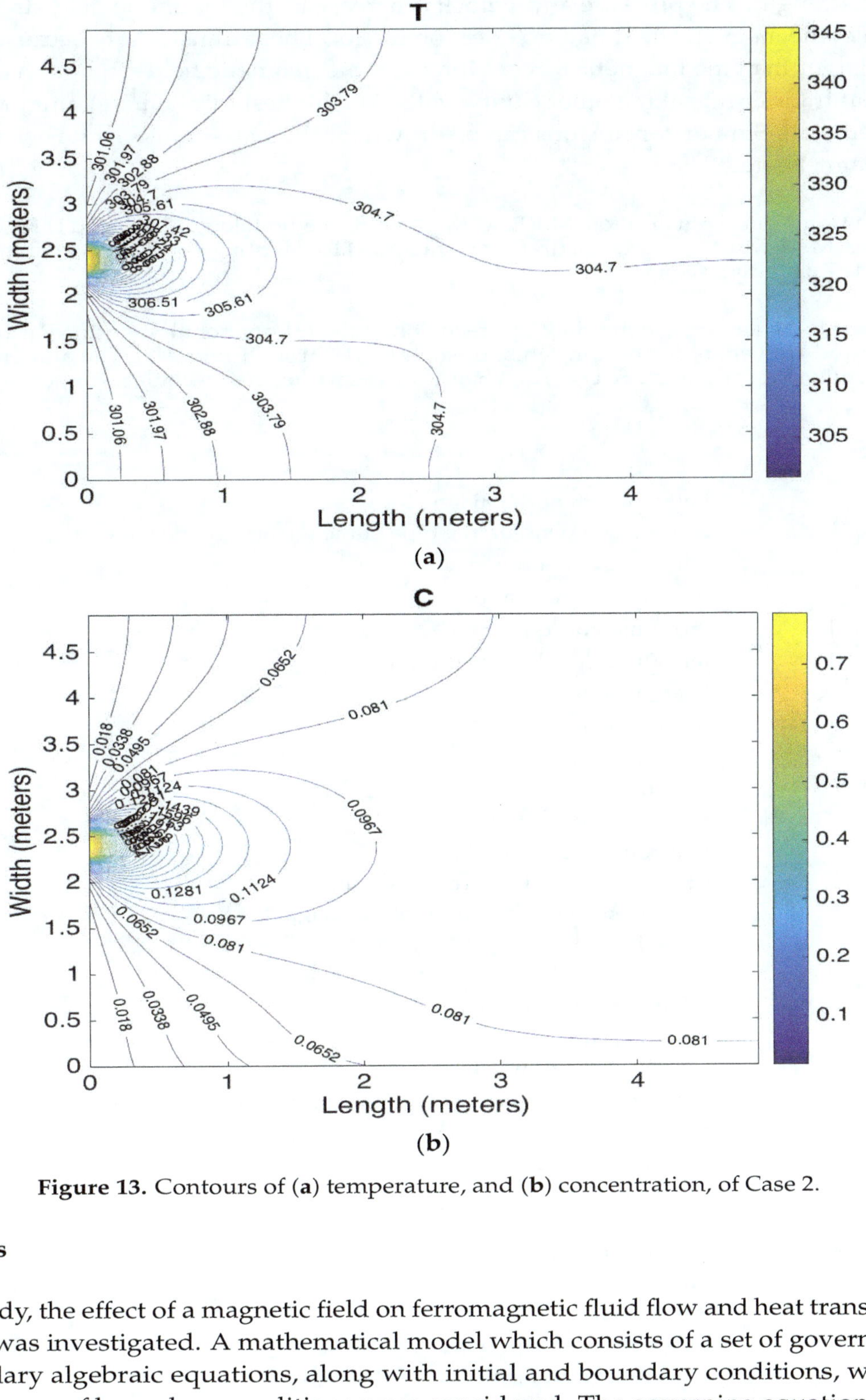

Figure 13. Contours of (**a**) temperature, and (**b**) concentration, of Case 2.

6. Conclusions

In this study, the effect of a magnetic field on ferromagnetic fluid flow and heat transfer in a square porous cavity was investigated. A mathematical model which consists of a set of governing equations and set of axillary algebraic equations, along with initial and boundary conditions, was developed. Two different cases of boundary conditions were considered. The governing equations were solved numerically using the multiscale time-splitting implicit method. An adaptive algorithm was built based on calculation of the Courant–Friedrichs–Lewy stability condition number. Error estimates for temperature and concentration at various values of iteration number were presented. Also, the grid independence of spatial discretization was examined. Numerical experiments were carried out to simulate some physical scenarios. It was found that the magnetic field increases the temperature and enhances the heat transfer rate. Also, the distribution of the particle concentration is related to the magnetic field strength. The pressure and velocity increase as the magnetic field strength increases. Close to the inlet the magnetic field had no effect on the local heat transfer rate because the advection force is much larger than the magnetic force. Moreover, the magnetic field strength had a small effect on the local heat transfer rate at the other cavity walls. The highest local Nusselt number occurs in the center of the left wall. Similar conclusions can be drawn for the mass transfer rate behavior under the effect of the magnetic field.

Author Contributions: Conceptualization, M.F.E., U.K. and A.B.; methodology, M.F.E. and U.K.; validation, M.F.E. and A.B.; analysis, M.F.E.; writing—original draft preparation, M.F.E.; writing—review and editing, M.F.E.; project administration, U.K.; funding acquisition, U.K.

Acknowledgments: The authors extend their appreciation to the International Scientific Partnership Program (ISPP) at King Saud University for funding this research work through ISPP#47. The authors also thank the Deanship of Scientific Research and RSSU at King Saud University for their technical support.

Nomenclature

A	[m]	half of width of the magnet,
a_1, b_1	[A m^{-1}]	constants depend on the type of the ferromagnetic material,
b	[m]	half of height of the magnet,
B_r	[Tesla]	residual magnetization,
C	[m m^{-3}]	ferrofluid concentration,
C_0	[m m^{-3}]	ferrofluid initial concentration,
c_p	[J Kg^{-1} K^{-1}]	heat capacity,
D	[m^2 s^{-1}]	diffusion coefficient,
F_{mag}	[N]	external magnetic force,
g	[m s^{-2}]	gravitation acceleration,
h	[J $K^{-1}m^{-1}s^{-1}$]	thermal conductivity,
H	[A m^{-1}]	magnetic field strength,
K	[m^2]	permeability of the porous medium,
L	[m]	distance between the poles of the magnet,
Nu	[-]	Nusselt number,
T	[K]	temperature,
T_r	[K]	reference temperature,
p	[Pa]	fluid pressure,
q	[kg s^{-1}]	external mass flow rate,
Q_c	[m^3s^{-1}]	rate of change of particle volume of a source/sink term,
Q_t	[m^3s^{-1}]	heat source term,
$\mathbf{u}$	[m s^{-1}]	fluid velocity vector,
x, y	[m]	Cartesian coordinates,
β	[K^{-1}]	thermal expansion coefficient,
β^*		solutal expansion coefficient,
Δt^k	[s]	time step for the loop k,
Δt^l	[s]	time step for the loop l,
Δt^m	[s]	time step for the loop m,
φ	[-]	porosity of porous media,
ρ	[Kg m^{-3}]	density of fluid mixture,
ρ_w	[Kg m^{-3}]	density of pure water component,
ρ_f	[Kg m^{-3}]	density of ferro-particles component,
ρ_s	[Kg m^{-3}]	density of solid phase,
μ	[m s^{-2}]	viscosity of water-magnetic-particles mixture,
μ_w	[m s^{-2}]	viscosity of water,
μ_0	[N A^{-2}]	magnetic permeability.

References

1. Scherer, C.; Figueiredo Neto, A.M. Ferrofluids: Properties and applications. *Braz. J. Phys.* **2005**, *35*, 718–727. [CrossRef]
2. Borglin, S.E.; Moridis, G.J.; Oldenburg, C.M. Experimental studies of the flow of ferrofluid in porous media. *Transp. Porous Media* **2000**, *41*, 61–80. [CrossRef]
3. Huh, C.; Nizamidin, N.; Pope, G.A.; Milner, T.E.; Wang, B. Hydrophobic Paramagnetic Nanoparticles as Intelligent Crude Oil Tracers. U.S. Patent Application No. 14/765,426, 15 January 2014.
4. Oldenburg, C.M.; Borglin, S.E.; Moridis, G.J. Numerical simulation of ferrofluid flow for subsurface environmental engineering applications. *Transp. Porous Media* **2000**, *38*, 319–344. [CrossRef]
5. Zahn, M. Magnetic fluid and nanoparticle applications to nanotechnology. *J. Nanopart. Res.* **2001**, *3*, 73–78. [CrossRef]
6. Chegenizadeh, N.; Saeedi, A.; Quan, X. Application of nanotechnology for enhancing oil recovery—A review. *Petroleum* **2016**, *2*, 324–333.
7. El-Amin, M.F.; Salama, A.; Sun, S. Numerical and dimensional analysis of nanoparticles transport with two-phase flow in porous media. *J. Pet. Sci. Eng.* **2015**, *128*, 53–64. [CrossRef]
8. El-Amin, M.F.; Salama, A.; Sun, S. Modeling and simulation of nanoparticles transport in a two-phase flow in porous media. In Proceedings of the SPE International Oilfield Nanotechnology Conference, Noordwijk, The Netherlands, 12–14 June 2012; SPE-154972-MS.
9. Suleimanov, B.A.; Ismailov, F.S.; Veliyev, E.F. Nanofluid for enhanced oil recovery. *J. Pet. Sci. Eng.* **2011**, *78*, 431–437. [CrossRef]
10. Ryoo, S.; Rahmani, A.; Yoon, K.Y.; Prodanovi, M.; Kots-mar, C.; Milner, T.E.; Johnston, K.P.; Bryant, S.L.; Huh, C. Theoretical and experimental investigation of the motion of multiphase fluids containing paramagnetic nanoparticles in porous media. *J. Pet. Sci. Eng.* **2012**, *81*, 129–144. [CrossRef]
11. Sheremet, M.A.; Pop, I. Mixed convection in a lid-driven square cavity filled by a nanofluid: Buongiorno's mathematical model. *App. Math. Comput.* **2015**, *266*, 792–808. [CrossRef]
12. Ghalambaz, M.M.; Sheremet, F.; Mikhail, A.; Pop, I. Triple-diffusive natural convection in a square porous cavity. *Transp. Porous Media* **2016**, *111*, 59–79. [CrossRef]
13. Carvalho, P.H.S.; de Lemos, M.J.S. Passive laminar heat transfer across porous cavities using thermal non-equilibrium model. *Numer. Heat Transf.* **2014**, *66*, 1173–1194. [CrossRef]
14. Javed, T.; Mehmood, Z.; Abbas, Z. Natural convection in square cavity filled with ferrofluid saturated porous medium in the presence of uniform magnetic field. *Phys. B Condens. Matter* **2017**, *506*, 122–132. [CrossRef]
15. Kleinstreuer, C.; Xu, Z. Mathematical modeling and computer simulations of nanofluid flow with applications to cooling and lubrication. *Fluids* **2016**, *1*, 16. [CrossRef]
16. Bhallamudi, S.M.; Panday, S.; Huyakorn, P.S. Sub-timing in fluid flow and transport simulations. *Adv. Water Res.* **2003**, *26*, 477–489.
17. Martinez, V. A numerical technique for applying time splitting methods in shallow water equations. *Comput. Fluids* **2018**, *169*, 285–295. [CrossRef]
18. Zhang, T.; Qian, Y. The time viscosity-splitting method for the Boussinesq problem. *J. Math. Anal. Appl.* **2017**, *445*, 186–211. [CrossRef]
19. Caliari, M.; Zuccher, S. Reliability of the time-splitting Fourier method for singular-solutions in quantum fluids. *Comput. Phys. Commun.* **2018**, *222*, 46–58. [CrossRef]
20. El-Amin, M.F.; Kou, J.; Sun, S. Discrete-fracture-model of multi-scale time-splitting two-phase flow including nanoparticles transport in fractured porous media. *J. Comput. Appl. Math.* **2018**, *333*, 327–349. [CrossRef]
21. El-Amin, M.F.; Kou, J.; Salama, A.; Sun, S. An iterative implicit scheme for nanoparticles transport with two-phase flow in porous media. *Procedia Comput. Sci.* **2016**, *80*, 1344–1353. [CrossRef]
22. El-Amin, M.F.; Kou, J.; Sun, S.; Salama, A. Adaptive time-splitting scheme for two-phase flow in heterogeneous porous media. *Adv. Geo-Energy Res.* **2017**, *1*, 182–189. [CrossRef]
23. El-Amin, M.F.; Kou, J.; Salama, A.; Sun, S. Multiscale adaptive time-splitting technique for nonisothermal two-phase flow and nanoparticles transport in heterogeneous porous media. In Proceedings of the SPE Reservoir Characterisation and Simulation Conference and Exhibition, Abu Dhabi, UAE, 8–10 May 2017; SPE-186047-MS.
24. Rosensweig, R.E. *Ferrohydrodynamics*; Cambridge University Press: Cambridge, UK, 1985.

25. Ashcroft, N.W.; Mermin, N.D. *Solid State Physics*; Holt, Rinehart and Winston: New York, NY, USA, 1976; ISBN 0-03-083993-9.
26. Herbert, A.W.; Jackson, C.P.; Lever, D.A. Coupled groundwater flow and solute transport with fluid density strongly dependent upon concentration. *Water Resour. Res.* **1988**, *24*, 1781–1795. [CrossRef]
27. El-Amin, M.F. Double dispersion effects on natural convection heat and mass transfer in non-Darcy porous medium. *Appl. Math Comput.* **2004**, *156*, 1–17. [CrossRef]

6

Experimental Study on the Fire-Spreading Characteristics and Heat Release Rates of Burning Vehicles using a Large-Scale Calorimeter

Younggi Park, Jaiyoung Ryu * and Hong Sun Ryou *

School of Mechanical Engineering, Chung-Ang University, 84, Heukseok-ro, Dongjak-gu, Seoul 06974, Korea; pyg0511@cau.ac.kr

* Correspondence: jairyu@cau.ac.kr (J.R.); cfdmec@cau.ac.kr (H.S.R.)

Abstract: In this article, large-scale experimental studies were conducted to figure out the fire characteristics, such as fire-spreading, toxic gases, and heat release rates, using large-scale calorimeter for one- and two-vehicle fires. The initial ignition position was the passenger seat, and thermocouples were attached to each compartment in the vehicles to determine the temperature distribution as a function of time. For the analysis, the time was divided into sections for the various fire-spreading periods and major changes, e.g., the fire spreading from the first vehicle to the second vehicle. The maximum temperature of 1400 °C occurred in the seats because they contained combustible materials. The maximum heat release rates were 3.5 MW and 6 MW for one and two vehicles, respectively. Since the time to reach 1 MW was about 240 s (4 min) before and after, the beginning of the car fire appears to be a medium-fast growth type. It shows the effect on the human body depending on the concentration of toxic substances such as carbon monoxide or carbon dioxide.

Keywords: fire-spreading characteristics; real vehicle experiments; toxic gases; temperature distributions; unsteady heat release rate

---

1. Introduction

Many vehicle fire studies are being carried out since vehicle accidents result in catastrophe due to various reasons such as drivers' negligence, electrical faults, deliberately lit and arson in parking lots [1], buildings, and tunnels, and so forth. To prevent vehicle disasters and suggest a guideline for evacuees in case of emergency, figuring out the fire-spreading mechanism and measuring other key parameters such as heat release rates and temperatures are necessary to install sprinklers, smoke control systems, and setting up fire extinguishers.

In fire researches, calorimeters, which can estimate a heat release rate, smoke generation, carbon dioxide, carbon monoxide, and so on, have been widely used for experimental study [2–5]. In the case of heat release rate, it has been regarded as one of the most important factors during experimental studied since it can be used to calculate the size of the fire and be more easily used in performance-based design (PBD). Large-scale tests are considered the most accurate way to secure fire-spreading effect and heat release rate, however, there is limited research, as it cost too much [6–8]. Katsuhiro et al. [6] found that the temperature distribution and maximum HRR reached 3 MW by changing the initial ignition location of the fire. In addition, they found that the fire spread radically after the windows were broken. Throughout related large-scale experiments, a single vehicle represents the heat release rate of 2.5–5 MW. However, most of the experimental studies were conducted only on one vehicle. Vehicle fires normally occur between two vehicles because of accidents. Li et al. [9] conducted large-scale experiments using two sedan vehicles to determine the change of the temperature distribution; however, in this study, the heat release rate was not considered, although it is an imperative part not only for analyzing the fire phenomena, but also suggesting guidelines for rescue as a function of time. Therefore, we investigated fire-spreading phenomena considering the effect of two vehicles fire scenario to figure out an important parameter such as heat release rates, fire-spreading time, and so forth.

A secondary factor that causes catastrophe in vehicle fires is the influence of toxic gases, e.g., carbon monoxide (CO) and carbon dioxide (CO_2), on human breathing and other functions [10–12]. Truchot et al. [13] investigated toxic-gas emissions from vehicle fires in tunnels. They did experimental studies using large-scale facilities, where the air flow in the tunnel could be controlled. They found and measured the toxic substances through a flame ionization detector/analyzer. It showed the toxic substances emitted in the tunnel shaft and analyzed the effects of various heat release rates. Smoke and toxic gases that are produced in the open are released into the air, due to their buoyancy. However, when a vehicle burns in an enclosed space, e.g., a tunnel or an indoor parking lot, the gas cannot escape to the atmosphere. Therefore, it is one of the main factors that should be considered in fire research to understand the concentration of the toxic substances generated when a vehicle fire occurs. However, there is limited information regarding the toxic gases that appear in the event of the fire.

Furthermore, many experimental studies are being actively carried out with pool-fire tests to represent fire accidents in tunnels or high-rise buildings [14–23]. Beak et al. [23] conducted tunnel-fire experiments using small-scale tunnels with pool fires, because it is challenging to experiment in real tunnels with real vehicles. He provided detailed information on the fire's heat release rate and its hole-plugging effect on tunnels; however, the results were obtained from pool-fire experiments, not real vehicles. It is clear that the results obtained from real products and pool-fire experiments will be different.

Therefore, it is necessary to analyze the fire-spreading characteristics of an actual vehicle and analyze the heat release rate and temperature distribution in order to understand the fire phenomena and its applications. Figure 1 represents a schematic diagram of the basic concept for this study. In this study, the fire-spreading characteristics, unsteady heat release rate, and the toxic substances in vehicle fires are measured using a large-scale calorimeter. All information, such as the change of heat release rate, temperature distributions and toxic gases as a function of time, gained throughout this research can be applied to other fire-related researches regarding vehicles and tunnels, parking lots, and so forth.

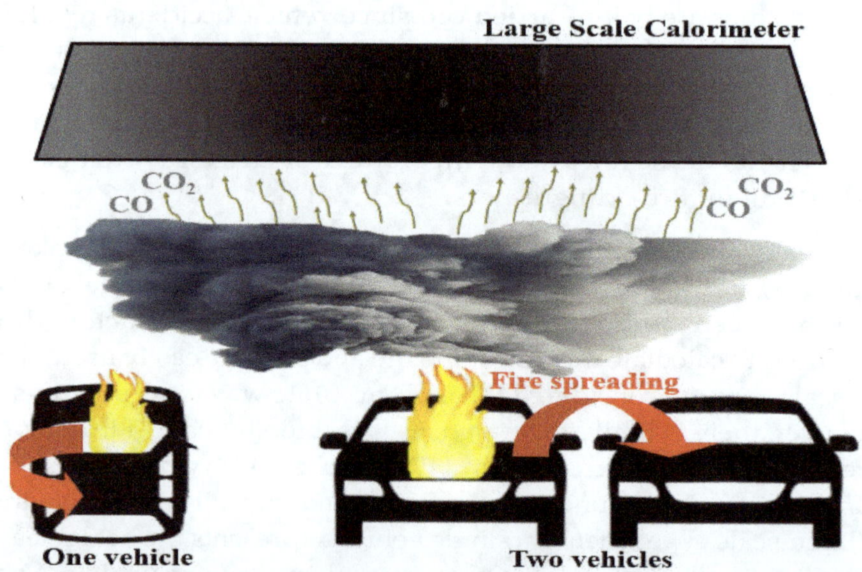

Figure 1. Schematic diagram of the basic concept for this study.

2. Experimental Setup and Conditions

Experiments were conducted using a large-scale calorimeter (LSC) from the Korea Institute of Construction Technology (KICT), applicable up to 10 MW, to determine the unsteady heat release rate and toxic gases, e.g., CO and CO_2, generated by the fire. The schematic diagram of LSC is represented in Figure 2 to figure out the values regarding change of the heat release rate and toxic substances as a function of time, and specifications of the experimental apparatus are represented in Table 1.

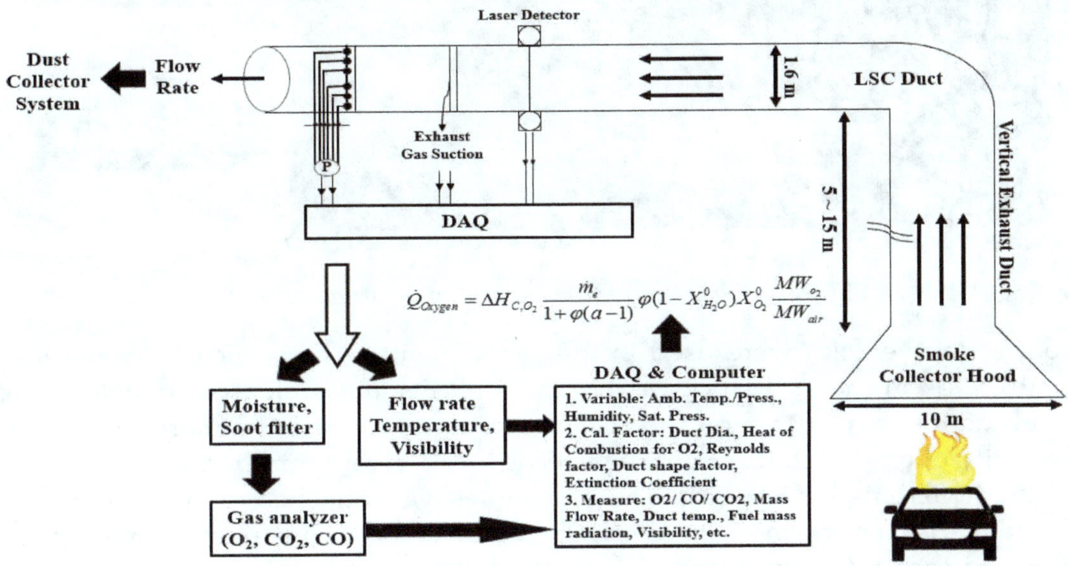

Figure 2. Schematic diagram of large-scale calorimeter (LSC) used to detect the changes in the heat release rate and toxic substances as a function of time.

Table 1. Specification of experimental apparatus.

Measurement	Specification
Duct pressure difference	Output: 4–20 mA, Range: 0–1245 Pa Model: PADT-D1000 Pa
Duct Temperature	K-type wire, Range: −200–1000 °C
Gas Analyzer	Output: 4–20 mA, Range: O_2 20.95%, CO_2 8%, CO 0.8% Model: Servomax 4100
Laser	Output: 0–8.4 mV, Range: 0–100% Model: 25-LHP-213-249
Load Cell	Output: 4–20 mA, Range: 0–3000 kg Manufacturer: Sartorious
Heat Flux	Plate Type, Range: −200–1000 °C Model: GTW-10-32-485A
Mass Flow	Output: 4–20 mA, Range: 0–2500 L/min Model: DPE-S

To investigate the fire-spreading characteristics and temperature distributions for one and two vehicles, K-type thermocouples (OMEGA, measuring range: −200–1260 °C) were attached to the engine room, bumpers, seats, and fuel tank. Temperature data were transmitted to a data-acquisition unit (DAQ) on a PC every second, and the experiments were recorded by a video recorder. The initial ignition location was assumed to be the passenger seat. Figure 3 represents the experimental setup and conditions. In the two-vehicle experiment, the distance between vehicles was set to 50 cm, and the thermocouples were attached in the same positions. The fuel was almost eliminated to prevent explosions during the experimental study.

Figure 3. Experimental setup and measurement tools.

Four-door sedan vehicles were used in the experiments. The vehicle size was 4.7 m (length) × 1.8 m (width) × 1.4 m (height), and was Hyundai's EF sonata released in 1998. These vehicles have not been used for about 10 years, and the experiment was carried out after the pressure of tires were removed with opening full front windows. The location and number of thermocouples in each compartment are represented in Figure 4. Thermocouples were attached to the engine room, bumpers, seats, and fuel tank. Sixteen thermocouples were attached to the engine room, four each at the top and bottom, and two on each side. Ten thermocouples were attached to the bumpers; five each on the front and rear. In addition, 14 thermocouples were attached to each seat in the vehicle interior, representing the head, waist, legs, and feet. Twelve thermocouples were attached to the fuel tank, one on each side, and four each on the top and bottom. Thus, 42 thermocouples were used to figure out the change of the temperature distribution as a function of time.

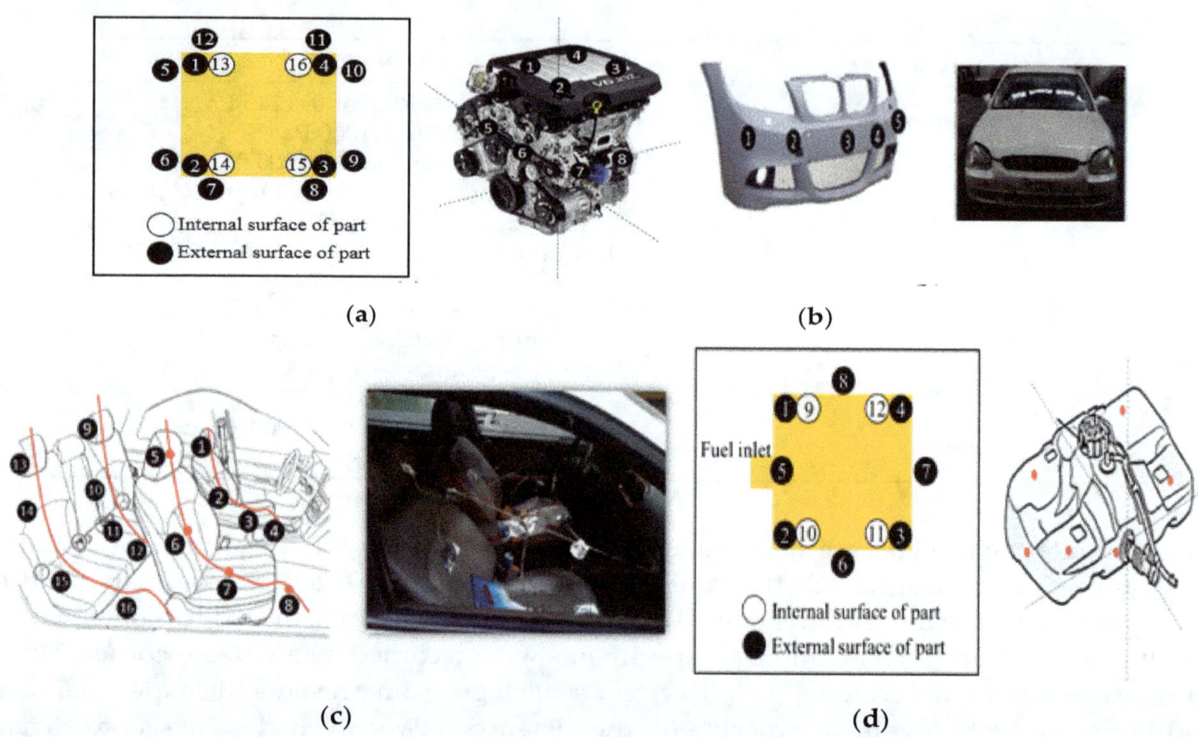

Figure 4. Location and number of thermocouples: (**a**) Engine room, (**b**) bumpers, (**c**) vehicle interior, and (**d**) fuel tank.

3. Results and Discussion

3.1. *Fire-Spreading Characteristics*

Fire-spreading over one and two vehicles as a function of time while on experiments are presented in Figure 5. Also, a flow-chart about fire-spreading inside a vehicle and two vehicles as a function of time are represented in Figure 6. These represent the major fire-spreading points as a function of time. As represented in the Figure 5 and the flow-chart, in the one-vehicle experiments, the fire spread to the driver seat and rear seat almost simultaneously, followed by the fuel tank, engine room, and finally the front and back bumpers. In the two-vehicle experiments, the fire followed the same sequence until it spread to the next vehicle, due to radiation. It then spread simultaneously to the rear seat and driver seat of the second vehicle, then the engine room, fuel tank, and bumpers in a regular sequence. Detailed information on the fire-spreading characteristics and special events as a function of time are represented in Table 2.

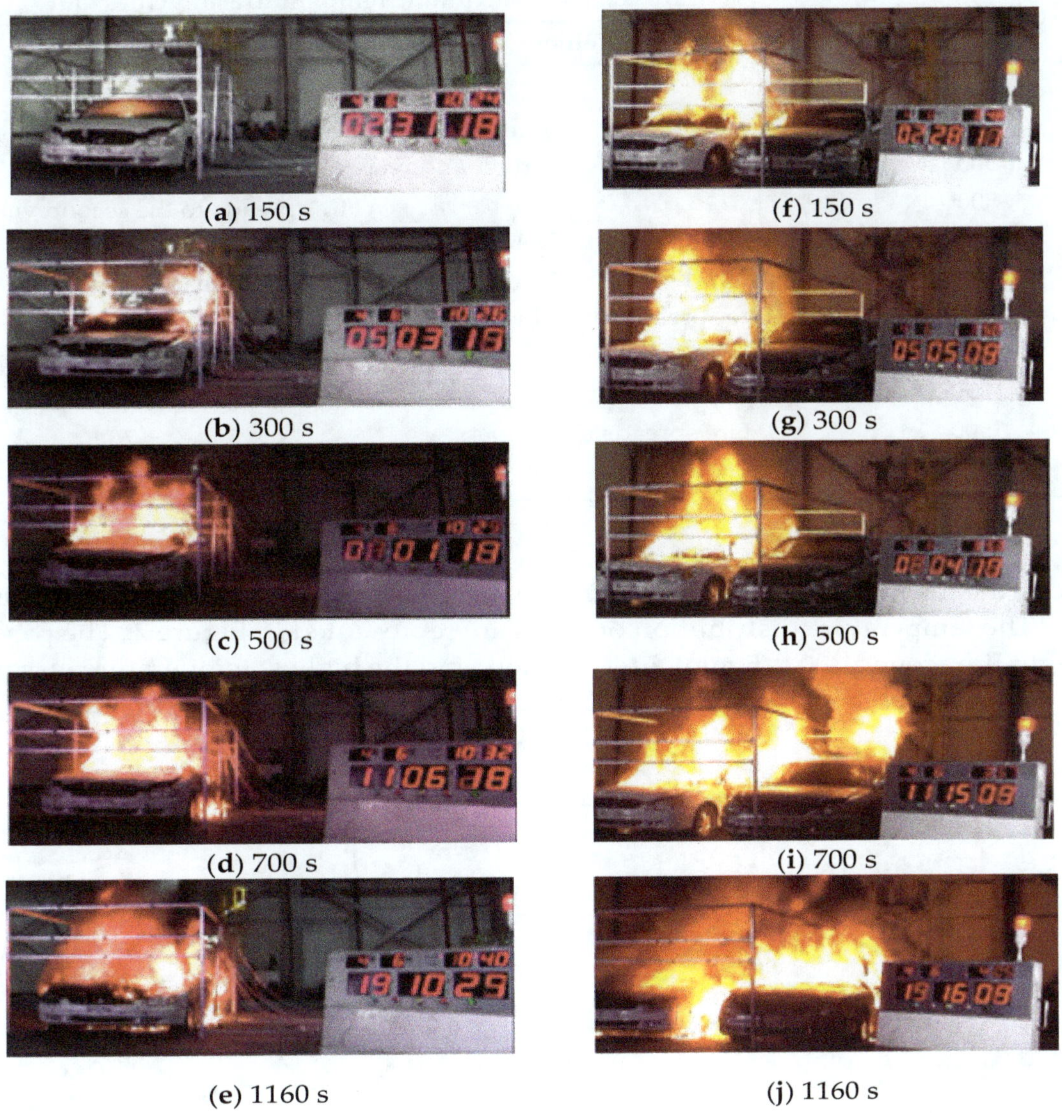

Figure 5. Fire spreading as a function of time for one and two vehicles; (**a**) 150 s after fires in one vehicle, (**b**) 300 s after fires in one vehicle, (**c**) 500 s after fires in one vehicle, (**d**) 700 s after fires in one vehicle, (**e**) 1160 s after fires in one vehicle, (**f**) 150 s after fires in two vehicles, (**g**) 300 s after fires in two vehicles, (**h**) 500 s after fires in two vehicles, (**i**) 700 s after fires in two vehicles, and (**j**) 1160 s after fires in two vehicles.

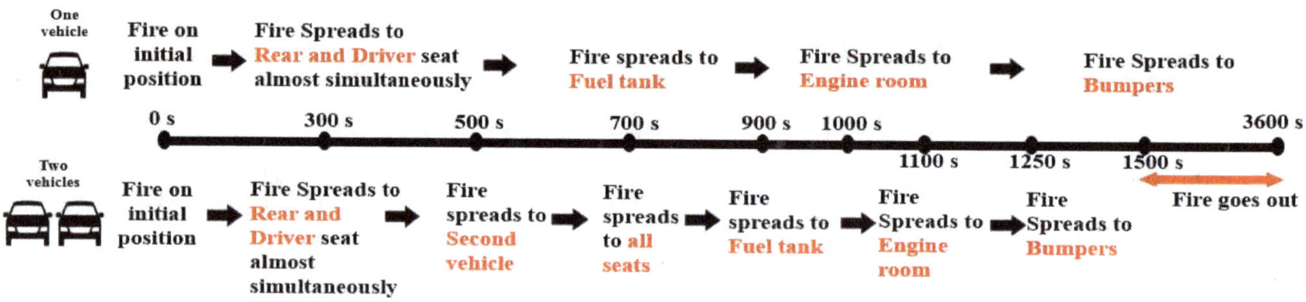

Figure 6. Major fire-spreading flow-chart.

Table 2. Detailed information on the fire-spreading characteristics and special events as a function of time.

Time (s) After Ignition	Fire Reaches Maximum Temperature and Goes Out	
	First Vehicle	Second Vehicle
0	Ignition in passenger seat	
150	Fires active in passenger seat	
300	Fire spreads to driver and rear seats almost simultaneously	
500	-	Fire spreads to the second vehicle
700	Fire spreads to fuel tank	Fire dramatically spreads to seats
900	-	Fire spreads to fuel tank
1000	Fire spreads to engine room	-
1100	-	Fire spreads to engine room
1250	-	Fire spreads to bumpers
1500	Fire spreads to bumpers	-
1500–3600	Fire goes out	Fire goes out

Therefore, the main temperature changes in the seats, engine room, fuel tank, and bumpers were represented. The temperature distribution on seats are represented in Figure 7. The passenger seat actively caught fire around 300 s (5 min) after ignition, and the highest temperature was observed in the interior, due to its combustible materials compared with the other compartments.

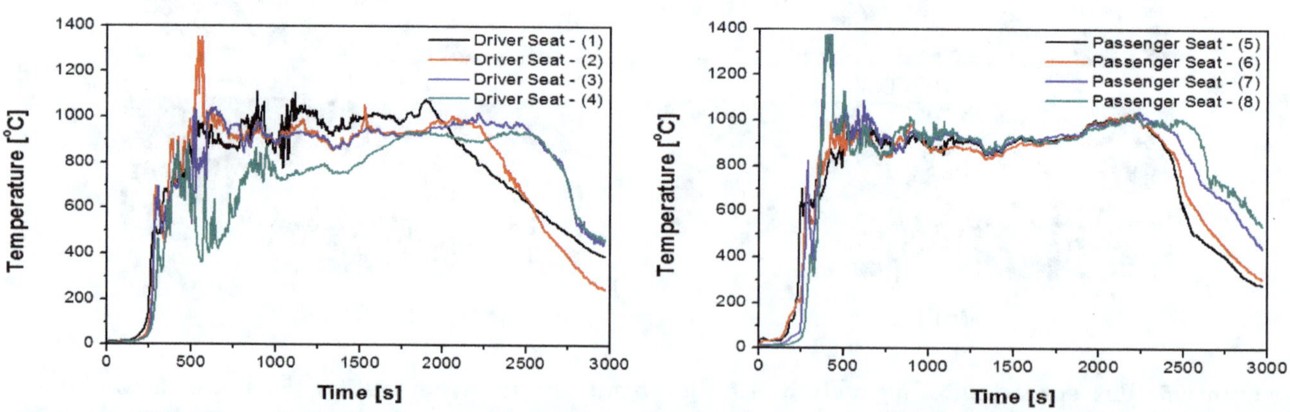

(a) (b)

Figure 7. Temperature distribution on seats of the first vehicle; (**a**) temperature of driver seat in first vehicle and (**b**) temperature of passenger seat in first vehicle.

The fire spread to the driver and rear seats simultaneously; however, the temperature differed slightly, as can be seen in Figure 8, which represents the temperature distribution of the driver and rear seats for the two sets of vehicles. The temperature increase in the driver seat and the rear seat can be observed almost simultaneously. The beginning of the temperature increase and the highest temperature are represented as the green and red lines, respectively. This means that the fire in the driver seat was larger than the rear seat, even though the fire-spreading time was similar because the driver seat was located closer to the fire than the rear seat.

To investigate the effect of temperature propagation on the number of vehicles, the driver seat and the rear seat inside the first vehicle are represented in Figure 8a,b. In the case of Figure 8c,d, it showed the driver seat and the rear seat inside the first vehicle when the two-vehicle experiment. After 500 s (8 min 20 s), the fire spread to the next vehicle, and the interior's temperature rapidly increased around 700 s (11 min 40 s), as shown in Figure 8c,d. Since the second passenger seat was located next to the first vehicle, the fire reached the thermocouple's temperature limit of 1370 °C.

The analysis was begun by dividing the time into sections for the various fire-spreading periods and major changes. Section 1 represented the period of the fire spreading from the passenger seat to the driver and rear seats after 300 s (5 min). Section 2 was 500 s (8 min 20 s) after the ignition, in which the fire spread to the next vehicle. Section 3 was when both the passenger seat and driver seat of the second vehicle caught fire. Section 4 was the period when the HRR reached maximum in the case of two-vehicle fire. These same sections were used for representing the heat release rates and the toxic substances in Sections 3.2 and 3.3, respectively.

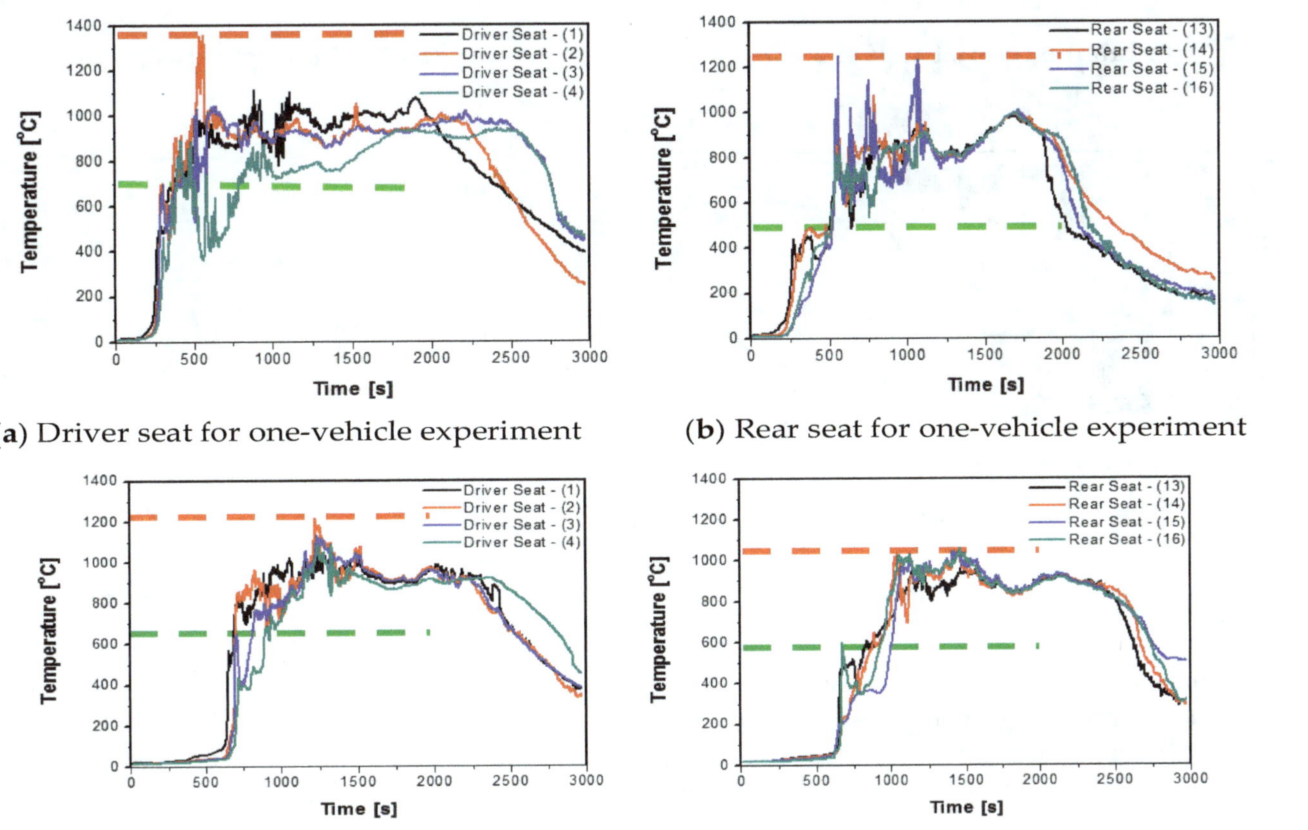

(**a**) Driver seat for one-vehicle experiment

(**b**) Rear seat for one-vehicle experiment

(**c**) Driver seat for two-vehicle experiment

(**d**) Rear seat for two-vehicle experiment

Figure 8. Temperature distribution results inside the driver and rear seats for one and two vehicles: (**a**) Temperature of driver seat inside the first vehicle for the one-vehicle experiment, (**b**) temperature of rear seat inside the first vehicle for the one-vehicle experiment, (**c**) temperature of driver seat inside the first vehicle for the two-vehicle experiment, and (**d**) temperature of rear seat inside the first vehicle for the two-vehicle experiment.

The temperature distribution in engine room and fuel tank were represented in Figure 9. The engine room was comprehensively burning within about 1200 s (20 min) of the fire ignition. In addition, the fire in the initial vehicle did not propagate directly to the engine room, but first spread to the seats. After about 1300 s (21 min 40 s), the fire spread to the adjacent vehicle. The fire went out 2500 s (41 min 20 s) after reaching the maximum temperature.

In the fuel tank, the temperature change was relatively small compared to the other combustibles. This was because most of the fuel was removed; however, in a real vehicle fire, it is the most vulnerable and dangerous compartment for the fire to reach.

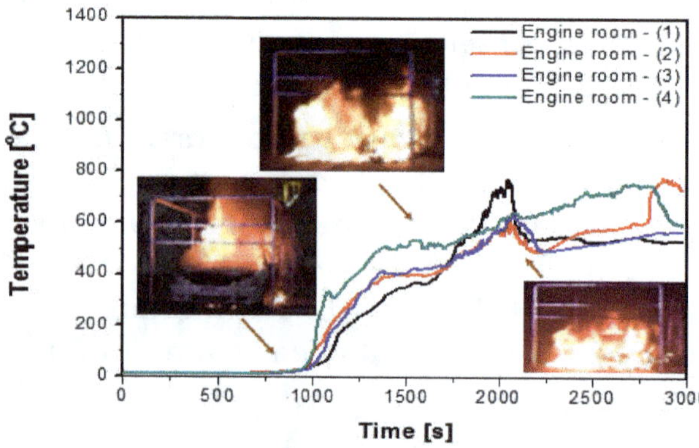

(**a**) Engine room; first vehicle

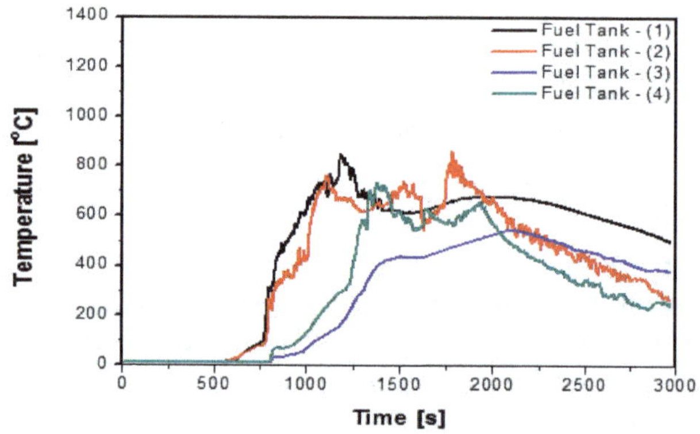

(**b**) Fuel tank; first vehicle

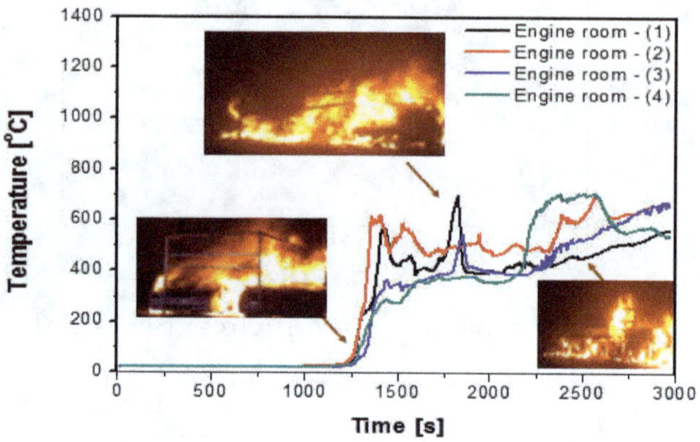

(**c**) Engine room; first vehicle

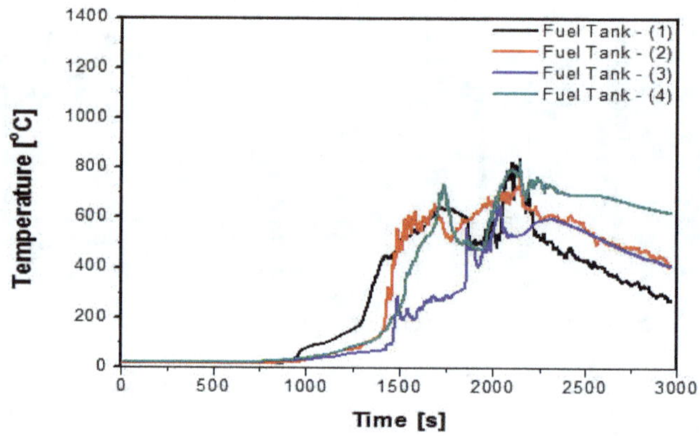

(**d**) Fuel tank; first vehicle

Figure 9. Temperature distribution in the engine room and fuel tank for one and two vehicles: (**a**) Temperature of engine room in first vehicle in the one-vehicle experiment, (**b**) temperature of fuel tank in first vehicle in the one-vehicle experiment, (**c**) temperature of engine room in first vehicle in the two-vehicle experiment, and (**d**) temperature of fuel tank in first vehicle in the two-vehicle experiment.

The temperature distributions at the bumpers are represented in Figure 10. For the front bumper, the temperature rose from the place nearest the initial fire vehicle. After about 1400 s (23 min 30 s), the temperature rose in the next vehicle. The temperature rose later because it was located farthest from the first vehicle. In addition, in the rear bumper, the fire spread more quickly than the front bumper because it was relatively close to the seat.

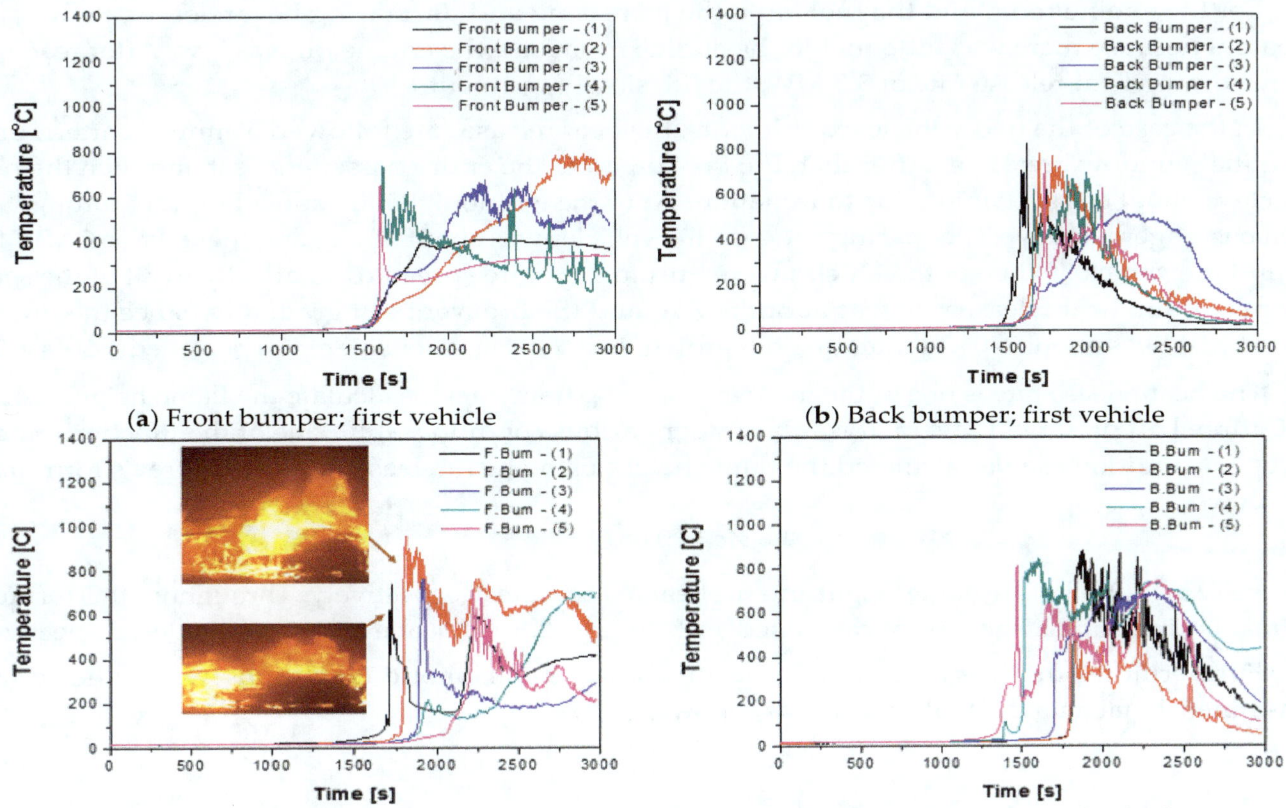

Figure 10. Temperature distribution at bumpers for one and two vehicles: (**a**) Temperature of front bumper in first vehicle, (**b**) temperature of back bumper in first vehicle, (**c**) temperature of front bumper in second vehicle, and (**d**) temperature of back bumper in second vehicle.

3.2. Heat Release Rate

The heat release rates considering the unsteady fire phenomenon are represented in Figure 11. As can be seen in Figure 6, the heat release rate increased rapidly around 180 s (3 min) after ignition because the fire spread into the vehicle interior. Further, when the windshield broke around 300 s (5 min), the heat release rate increased sharply to about 2.3 MW. However, it continued to increase because the fire spread from the passenger seat to the driver seat. The passenger-seat cushion spontaneously ignited and then temporarily decreased. Because the cushion was composed of a composite material, the heat release rate increased and decreased repeatedly until it reached its maximum.

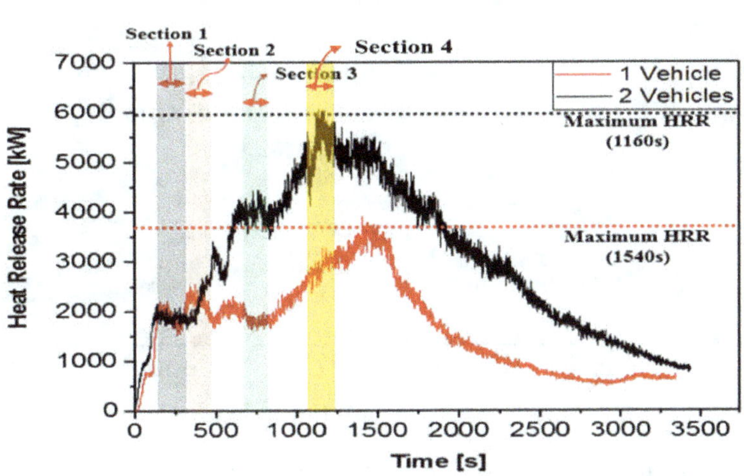

Figure 11. Heat release rate for one and two vehicles as a function of time.

For the engine room and the fuel tank, the increasing and decreasing heat release rate was not apparent because there was little fuel in the engine room and the engine oil was low. After reaching the maximum heat release rate of 3.5 MW, the fire slowly went out.

In the case of the two-vehicle experiments, the heat release rate followed a similar pattern until the initial window breakage. After that, the fire spread to the entire passenger seat and then the next vehicle at about 600 s (10 min), due to radiation. After the ignition, the fire spread rapidly through the open passenger window. The spacing between the vehicles was about 50 cm, and there was a 2 MW fire around the vehicle. When both vehicles were burning, the fire spread dramatically to all of the seats. The maximum heat release rate was about 6 MW, and the fire went out gradually. Since the time to reach 1 MW was about 240 s (4 min), the fire growth in a vehicle is considered to be a medium-fast fire.

The heat release rate is one of the most important parameters to calculate the flame height [24,25]. McCaffrey [24] presented several formulas under various conditions, but, one of the functions under the open condition can be calculated the flame height using heat release rate. McCaffrey's formula is as follows:

$$Zc = 0.08\,Q^{2/5} \tag{1}$$

where Zc (m), Q (kW) are flame height and the heat release rates, respectively. Throughout this formula, the maximum flame heights were calculated as 2.1 m and 2.6 m for one and two vehicles, respectively.

Furthermore, the height of the flame in normal atmospheric conditions was indicated by non-dimensional analysis [25], and it is as follows:

$$L_f = 0.235 \cdot Q^{2/5} - 1.02D \tag{2}$$

where L_f(m), Q (kW), D (m) are flame height, the heat release rates, and diameter of pool, respectively. In case of the function suggested by Heskastad [25], the flame height can be calculated using the heat release rate and pool diameter. Throughout this formula, the maximum flame heights were calculated 4.3 m and 7.6 m, 1.35 m and 2.83 m, and 3.49 m and 4.97 m for one and two vehicles for varying diameters, respectively. Because the width and length of the vehicle have different diameters, the flame heights were calculated with the diameters at 1.8 m width, 4.7 m length, and 2.6 m hydraulic diameter of the vehicle. Flame heights are represented as a function of time and maximum flame heights in Figure 12 and Table 3, respectively. However, measuring the change of flame heights was affected significantly under various conditions.

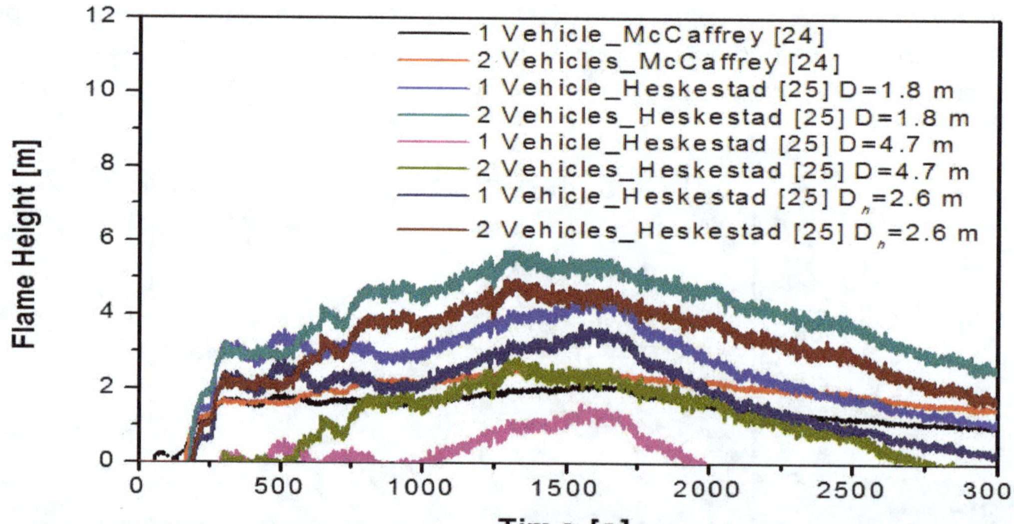

Figure 12. Flame height calculated by Equation 1 (McCaffrey [24]) and Equation 2 (Heskestad [25]) for varying diameter as a function of time.

Table 3. Comparison of flame heights under the open condition.

Number of Vehicle	Maximum Flame Height (m)			
	McCaffrey [24]		Heskestad [25]	
	-	$D = 1.8$	$D = 4.7$	$D_h = 2.6$
One	2.1	4.3	1.35	3.49
Two	2.6	7.6	2.83	4.97

The fire growth can be evaluated using the following generic fire-growth curve:

$$Q_{max} = a \cdot (t - t_i)^2 \quad (3)$$

where a is the fire-growth coefficient (kW/s^2), t means time (s), and t_i means the time of ignition (s). In this study, the time of ignition can be taken as zero. The fire-growth coefficients for vehicle fires up to 1 MW are represented in the Appendix A. Reducing the flame height and fire growth were decisive parameters for suppressing a fire; a fire extinguisher should be considered, especially in enclosed spaces, e.g., tunnels and indoor parking lots.

It is generally well acknowledged that fire experiments are really difficult to do repeated experiments, due to time consumption and expensive time. Thus, various fire articles did not contain a detailed uncertainty analysis. Melcher et al. [26] suggested the impact of random deviations that may occur in a single experiment. Mass loss rates and heat release rates for one and two vehicles were represented with error range in Figure 13. Yellow and lavender indicate the error ranges that occurred in one and two vehicles, respectively.

Applying the actual heat release rates obtained from this study to the numerical analysis study, Park et al. published a study on the effect of a vehicle accident on the evacuation in various tunnel aspect ratio [27].

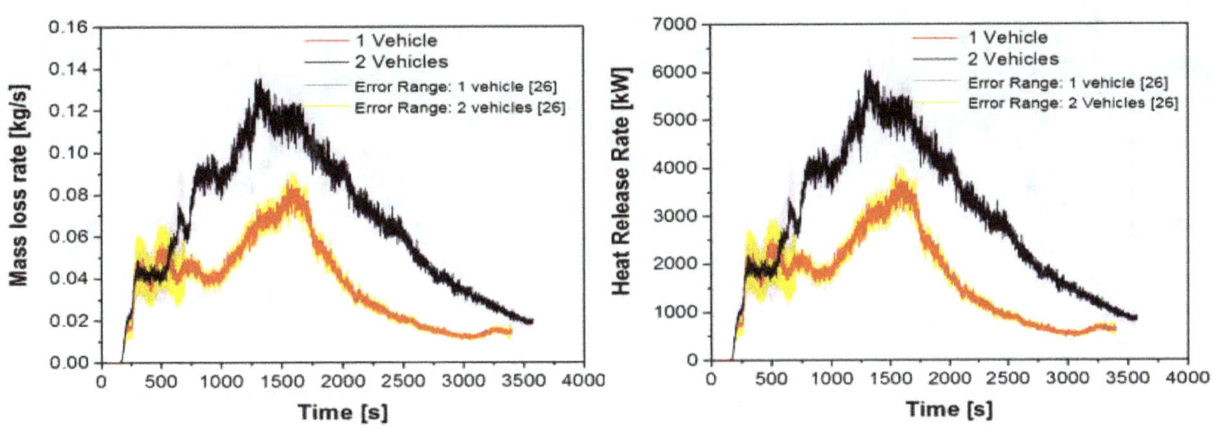

(**a**) Mass loss rate (**b**) Heat Release Rate

Figure 13. Mass loss rate and heat release rate for one and two vehicles as a function of time, error range estimated following Melcher et al. [26]: (**a**) Mass loss rate and (**b**) heat release rate.

The maximum heat release values obtained by other studies are shown in Table 4 and compared with the results obtained through this experiment. As a rule of thumb, the maximum heat release rates were normally represented 2.5 MW to 5 MW. Okamoto et al. [8] found the heat release rates as a function of time in real-scale experiments, and the vehicles used in this experiment were similar to those used in this study. The experimental conditions were quite similar, such as conducting the vehicle fire test in the absence of fuel, but only with different experimental measuring equipment. In addition, the maximum heat release rate was 3.5 MW, the same as this study. However, in the case of research done by Shipp et al. [28], the maximum heat release rate was significantly higher than other studies since the vehicle tests involved fuel spill from the petrol tank. Furthermore, Ingason [29] conducted the vehicle fire experiments in a tunnel so that the maximum heat release rates were different because of ceiling temperatures, ventilation system on maximum heat release rate, and fire growth rates.

Table 4. Comparison of maximum heat release rates throughout large-scale experiments.

Type of Vehicles	Maximum Heat Release Rate (MW)			
	Okamoto et al. [6]	Shipp et al. [28]	Ingason [29]	Park et al.
Small passenger car	3.5	8	2.5	3.5
Large passenger car	4.2	-	<5	-
2 passenger cars	-	-	3.5–10	6

3.3. Toxic Substances

The changes in the amounts of carbon dioxide and carbon monoxide as a function of time are represented in Figure 14. These values represent the toxic gases obtained through the LSC from the vehicle as it burned. The main changes on carbon monoxide and carbon dioxide as a function of time are represented in Tables 5 and 6, respectively. Furthermore, it showed similar trends to the distribution of change of heat release rate. However, judging from the influence on the human body, it is obvious that these values might be not correct, because of the many differing conditions in fire phenomena, e.g., enclosed or open spaces, scale of space, ventilation systems, sprinklers, and so on.

With fires in open spaces, the smoke and toxic gases will be released into the air. In enclosed spaces, e.g., tunnels or indoor parking lots, the smoke will accumulate continuously, and the concentration will greatly increase. However, the effects on the human body can be analyzed based on concentrations of CO and CO_2 from the vehicle itself [30].

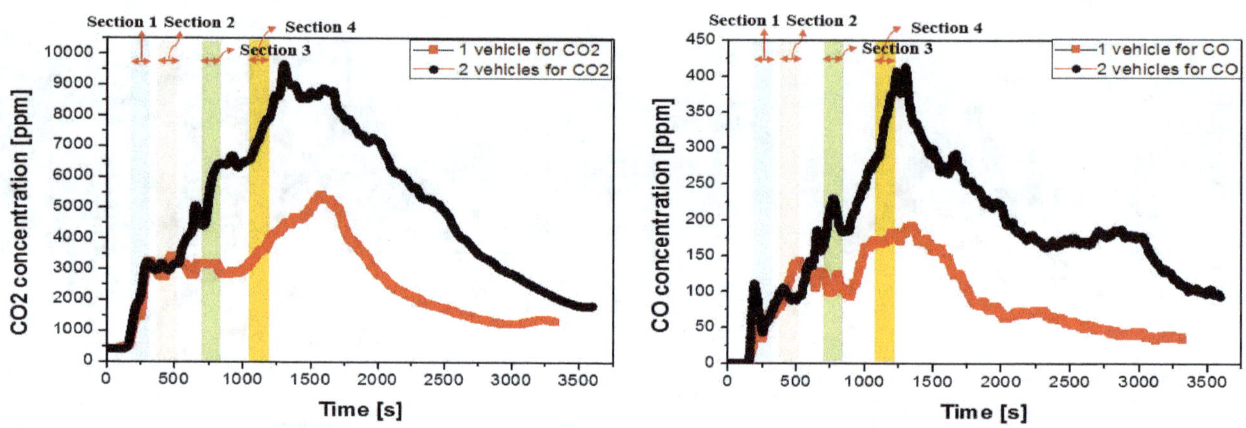

Figure 14. Toxic-substance concentration results: (**a**) Carbon dioxide and (**b**) carbon monoxide.

Table 5. Carbon-dioxide concentration at the major change periods, and influences on the body based on value.

Section	CO_2 Concentration (ppm)		Effect on Health [30]	
	One Car	Two Cars	One Car	Two Cars
Section 1	1000–3000		General condition and mild headaches	
Section 2	3000		Poor air condition and headaches	
Section 3	3000	4500–6200	Poor air condition and headaches	
Section 4	3000–3500	6500–8000	Headaches	Respiratory, circulatory, and cerebral impairment

Table 6. Carbon-monoxide concentration at the major change periods, and influences on the body based on value.

<table>
<tr><th rowspan="2">Section</th><th colspan="2">CO Concentration (ppm)</th><th colspan="2">Effect on Health [30]</th></tr>
<tr><th>One Car</th><th>Two Cars</th><th>One Car</th><th>Two Cars</th></tr>
<tr><td>Section 1</td><td>0–50</td><td>0–100</td><td colspan="2">Slight headaches</td></tr>
<tr><td>Section 2</td><td colspan="2">75–150</td><td colspan="2">Headaches</td></tr>
<tr><td>Section 3</td><td>100–125</td><td>150–225</td><td>Headaches</td><td>Dizziness, nausea, fatigue, headaches</td></tr>
<tr><td>Section 4</td><td>150–175</td><td>300–425</td><td>Dizziness, nausea, fatigue, headaches</td><td>Headache and nausea; life threatening in 3 h</td></tr>
</table>

Based on the values obtained from the LSC, the CO_2 generated from one vehicle will give a person a slight headache. However, in a two-vehicle fire, the slight influence occurs before 1000 s (16 min 40 s); however, after that, it will affect the respiratory system and nervous system, and cause cerebral impairment. Furthermore, only a slight headache can be felt about 500 s (8 min 20 s) after ignition. However, after that, the area should be evacuated. In the worst case, life is threatened, and evacuation may be difficult because of the high concentration of CO over 1200 s (20 min). The higher the heat release rate is, the higher the concentration of CO and CO_2. Therefore, the larger the fire is, the faster the area should be evacuated. Based on the CO and CO_2 concentrations absorbed through the LSC, the impact on human health applies when people were directly exposed. It is apparent that the values obtained through the LSC will vary depending on the location and conditions of the fire inside of buildings, but it can provide indirect guidance on ventilation and evacuation in compartment space while presenting the toxic concentration occurring in the vehicle itself.

4. Conclusions

The aim of this study was to investigate how fire spreads through a vehicle using an actual vehicle-fire test, and to analyze the heat release rate for two vehicles using a large-scale calorimeter. In addition, the influence of the toxic gases generated from the vehicle fires was analyzed. An analysis of the experimental results provided the following conclusions.

(1) In actual vehicle tests, the fire spread from the initial ignition location to the rear seat, engine room, fuel tank, and bumper in regular sequence. In the two-vehicle situation, a similar tendency was observed, and the fire spread to the next vehicle after about 500 s (8 min 20 s).

(2) The fire rose sharply after 200 s (3 min). The maximum heat release rates of one and two vehicles were represented as 3.5 MW and 6 MW, reached at about 1540 s (25 min 40 s) and 1160 s (19 min 20 s), respectively. Since the time to reach 1 MW was about 240 s (4 min) before and after, the fire growth in a vehicle fire is considered to be a medium-fast fire phenomenon.

(3) The evacuation should be totally completed within 20 min after the fire starts, because of the high concentrations of carbon dioxide and carbon monoxide in an enclosed space. This may vary, depending on the size of the location where the fire occurs.

(4) In this study, uncertainty analysis cannot be included because of financial problems like other related studies [6–8], however, we place emphasis on fire-spreading characteristics inside of a vehicle and toxic gases.

Author Contributions: Conceptualization, methodology, investigation, formal analysis, Y.P. and J.R.; writing—original draft preparation, Y.P.; writing—review and editing, J.R.; project administration, supervision, H.S.R.

Appendix A. Fire-Growth Coefficients for Vehicle Fires

Table A1. Fire-Growth Coefficients for one and two vehicle fires.

Time	One Vehicle	Two Vehicles	Time	One Vehicle	Two Vehicles
55	0.000122845	0	111	−0.00003228	0
56	0.000110589	0	112	−0.000100544	0
57	0.000111826	0	113	−0.000102829	0
58	0.000119181	0	114	−0.000125789	0
59	0.000151472	0	115	−0.000112598	0
60	0.000143496	0	116	−0.000120963	0
61	0.000141763	0	117	−0.000118931	0
62	0.000142229	0	118	−0.000123675	0
63	0.00014096	0	119	−0.000136618	0
64	0.00013857	0	120	−0.000132197	0
65	0.000130619	0	121	−0.000123733	0
66	0.000143594	0	122	0.000045496	0
67	0.000153154	0	123	5.5872E-06	0
68	0.000143394	0	124	4.16544E-05	0
69	0.000115387	0	125	3059376E-05	0
70	0.000326219	0	126	3.63424E-05	0
71	0.000223675	0	127	3.67568E-05	0
72	0.000277621	0	128	2.67056E-05	0
73	0.000272314	0	129	−1.11936E-05	0
74	0.000274293	0	130	1.77728E-05	0
75	0.000288267	0	131	9.4848E-06	0
76	0.000282707	0	132	1.01104E-05	0
77	0.000298621	0	133	−3.34768E-05	0
78	0.000276741	0	134	0.00000116	0
79	0.000281992	0	135	0.000144683	0
80	0.000264557	0	136	0.000136411	0
81	0.000300158	0	137	0.000158899	0
82	0.000287592	0	138	0.000142838	0
83	0.000257986	0	139	0.000129723	0
84	0.000279203	0	140	0.000139946	0
85	0.000259619	0	141	0.000131667	0
86	0.000282691	0	142	0.000127328	0
87	0.000290432	0	143	0.000127696	0
88	0.000129774	0	144	0.000236862	0
89	0.000130891	0	145	0.000222347	0
90	0.000116182	0	146	0.000211238	0
91	0.000112029	0	147	0.000220477	0
92	0.000117714	0	148	0.000215518	0
93	0.000104933	0	149	0.00025139	0
94	0.00010421	0	150	0.00021375	0
95	0.000104573	0	151	0.000351069	0
96	0.000099011	0	152	0.000364856	0
97	0.0000978	0	153	0.000371894	0
98	0.00009746	0	154	0.00035011	0
99	0.0000985872	0	155	0.000354869	0
100	0.000088056	0	156	0.000278704	0
101	0.000102904	0	157	0.000365112	0
102	0.0000951808	0	158	0.00045428	0
103	0.000102624	0	159	0.00038917	0
104	−1.50448E-05	0	160	0.000341139	0
105	−9.12816E-05	0	161	0.000341352	0
106	−0.000100579	0	162	0.000358957	0
107	−9.69392E-05	0	163	0.000454278	0
108	−9.98464E-05	0	164	0.000407691	0
109	−2.65664E-05	0	165	0.000914008	0
110	−3.36192E-05	0	166	0.000862808	0

Table A1. *Cont.*

Time	One Vehicle	Two Vehicles	Time	One Vehicle	Two Vehicles
167	0.001054307	0.000220694	203	0.0100165	0.0120997
168	0.001063165	0.000611402	204	0.0110054	0.0114384
169	0.00104391	0.00068039	205	0.0110963	0.0124061
170	0.001228091	0.000880261	206	0.0101954	010114076
171	0.001635272	0.000812598	207	0.0104325	0.0122057
172	0.001834352	0.001029643	208	0.0106026	0.0121818
173	0.001916245	0.001240034	209	0.0111004	0.0124914
174	0.00199748	0.002309933	210	0.0111064	0.0135382
175	0.001945301	0.002533862	211	0.0126027	0.0129667
176	0.002285653	0.002699115	212	0.0107972	0.01428483
177	0.002352554	0.002952645	213	0.011416	0.0135615
178	0.002293226	0.002970237	214	0.0117953	0.0138855
179	0.003488533	0.003338856	215	0.0112167	0.0142959
180	0.003595838	0.003579187	216	0.0117955	0.0135957
181	0.003715666	0.00413451	217	0.0123651	0.0139167
182	0.004403666	0.0041032	218	0.011948	0.0153101
183	0.004578416	0.004038008	219	0.0121137	0.013121
184	0.00505969	0.005178826	220	0.0118276	0.0146671
185	0.004863582	0.006157565	221	0.0121345	0.0160241
186	0.005264023	0.007279477	222	0.0121715	0.0140538
187	0.005925605	0.007346544	223	0.0121353	0.0150714
188	0.005557421	0.006510378	224	0.0125515	0.0146725
189	0.006212605	0.007681355	225	0.0120098	0.0150361
190	0.006590218	0.007854469	226	0.0126213	0.0147933
191	0.005883032	0.007422917	227	0.0113123	0.0146584
192	0.003584963	0.007919211	228	0.0113451	0.0153618
193	0.0066938	0.0087581	229	0.0112896	0.0158492
194	0.0082796	0.0090616	230	0.0117224	0.0150186
195	0.008573	0.0095076	231	0.0121374	0.0151767
196	0.0081337	0.010333	232	0.0122526	0.0159671
197	0.0084333	0.0105719	233	0.0122324	0.0157199
198	0.0092459	0.0096946	234	0.0120766	0.0158552
199	0.0098028	0.0101741	235	0.0121153	0.0149351
200	0.0101057	0.0114833	236	0.0119933	0.0169582
201	0.009347	0.011416	237	0.0112681	0.0161582
202	0.0100612	0.0113177	238	0.0123996	0.016821

References

1. Li, Y.; Spearpoint, M. Analysis of vehicle fire statistics in New Zealand parking buildings. *Fire Technol.* **2007**, *43*, 93–106. [CrossRef]
2. Babrauskas, V. The Cone Calorimeter. In *SFPE Handbook of Fire Protection Engineering*; Springer: New York, NY, USA, 2016; pp. 952–980.
3. Wang, Z.; Hu, X.; Jia, F.; Galea, E.R. A two-step method for predicting time to flashover in room corner test fires using cone calorimeter data. *Fire Mater.* **2012**, *37*, 457–473. [CrossRef]
4. Lai, C.M.; Ho, M.C.; Lin, T.H. Experimental Investigations of Fire Spread and Flashover Time in Office Fires. *J. Fire Sci.* **2010**, *28*, 279–302. [CrossRef]
5. Hakkarainen, T. Rate of Heat Release and Ignitability Indices in Predicting SBI Test Results. *J. Fire Sci.* **2001**, *19*, 284–305. [CrossRef]
6. Katsuhiro, O.; Norimichi, W.; Yasuaki, H.; Tadaomi, C.; Ryoji, M.; Hitoshi, M.; Satoshi, O.; Hideki, S.; Yohsuke, T.; Kimio, H.; et al. Burning behavior of sedan passenger cars. *Fire Saf. J.* **2009**, *44*, 301–310. [CrossRef]
7. Mangs, J.; Keski-Rahkonen, O. Characterization of the Fire Behaviour of a Burning Passenger Car. Part1: Car Fire Experiments. *Fire Saf. J.* **1994**, *23*, 17–35. [CrossRef]

8. Katsuhiro, O.; Takuma, O.; Hiroki, M.; Masakatsu, H.; Norimichi, W. Burning behavior of minivan passenger cars. *Fire Saf. J.* **2013**, *62*, 272–280. [CrossRef]
9. Li, D.; Zhu, G.; Zhu, H.; Yu, Z.; Gao, Y.; Jiang, X. Flame spread and smoke temperature of full-scale fire test of car fire. *Case Stud. Therm. Eng.* **2017**, *10*, 315–324. [CrossRef]
10. Loladze, I. Rising atmospheric CO_2 and human nutrition: Toward globally imbalanced plant stoichiometry? *Trends Ecol. Evol.* **2002**, *17*, 457–461. [CrossRef]
11. Bailey, J.E.; Argyropoulos, S.V.; Kendrick, A.H.; Nutt, D.J. Behavior and Cardiovascular Effects of 7.5% CO_2 in Human Volunteers. *Med. J. Aust.* **2005**, *21*, 18–25. [CrossRef]
12. Mayr, F.B.; Spiel, A.; Leitner, J.; Marsik, C.; Germann, P.; Ullrich, R.; Wagner, O.; Jilma, B. Effects of Carbon Monoxide inhalation during Experimental Endotoxemia in Humans. *Am. J. Respir. Crit. Care Med.* **2005**, *171*, 354–360. [CrossRef] [PubMed]
13. Truchot, B.; Fouillen, F.; Collet, S. An experimental evaluation of toxic gas emissions from vehicle fires. *Fire Saf. J.* **2018**, *97*, 111–118. [CrossRef]
14. Deckers, X.; Haga, S.; Sette, B.; Merci, B. Smoke control in case of fire in a large car park: Full-scale experiments. *Fire Saf. J.* **2013**, *57*, 11–21. [CrossRef]
15. Yan, Z.; Guo, Q.; Zhu, H. Full-scale experiments on fire characteristics of road tunnel at high altitude. *Tunn. Undergr. Space Technol.* **2017**, *66*, 134–146. [CrossRef]
16. Blanchard, E.; Boulet, P.; Desanghere, S.; Cesmat, E.; Meyrand, R.; Garo, J.P.; Vantelon, J.P. Experimental and numerical study of fire in a midscale test tunnel. *Fire Saf. J.* **2012**, *47*, 18–31. [CrossRef]
17. Lonnermark, A.; Ingason, H. Fire Spread and Flame Length in Large-Scale Tunnel Fires. *Fire Technol. J.* **2006**, *42*, 283–302. [CrossRef]
18. Horvath, I.; Beeck, J.V.; Merci, B. Full-scale and reduced-scale tests on smoke movement in case of car park fire. *Fire Saf. J.* **2013**, *57*, 35–43. [CrossRef]
19. Ingason, H. Model scale railcar fire tests. *Fire Saf. J.* **2007**, *42*, 271–282. [CrossRef]
20. Lee, S.R.; Ryou, H.S. An experimental study of the effect of the aspect ratio on the critical velocity in longitudinal ventilation tunnel fires. *Fire Sci. J.* **2005**, *23*, 119–138. [CrossRef]
21. Lee, S.R.; Ryou, H.S. A numerical study on smoke movement in longitudinal ventilation tunnel fires for different aspect ratio. *Build. Environ. J.* **2006**, *41*, 719–725. [CrossRef]
22. Baek, D.; Bae, S.; Ryou, H.S. A numerical study on the effect of the hydraulic diameter of tunnels on the plug-holing phenomena in shallow underground tunnels. *J. Mech. Sci. Technol.* **2017**, *31*, 2331–2338. [CrossRef]
23. Baek, D.; Sung, K.H.; Ryou, H.S. Experimental study on the effect of heat release rate and aspect ratio of tunnel on the plug-holing phenomena in shallow underground tunnels. *Int. J. Heat Mass Transf.* **2017**, *113*, 1135–1141. [CrossRef]
24. McCaffrey, B. Flame Height. In *SFPE Handbook of Fire Protection Engineering*, 2nd ed.; National Fire Protection Assoc.: Quincy, MA, USA, 1995.
25. Heskestad, G. Fire Plumes. In *SFPE Handbook of Fire Protection Engineering*, 2nd ed.; National Fire Protection Assoc.: Quincy, MA, USA, 1995.
26. Melcher, T.; Zonke, R.; Trott, M.; Krause, U. Experimental investigations on the repeatability of real scale fire test. *Fire Saf. J.* **2016**, *82*, 101–114. [CrossRef]
27. Park, Y.; Lee, Y.; Na, J.; Ryou, H.S. Numerical Study on the Effect of Tunnel Aspect Ratio on Evacuation with Unsteady Heat Release Rate Due to Fire in the Case of Two Vehicles. *Energies* **2019**, *12*, 133. [CrossRef]
28. Shipp, M.; Spearpoint, M. Measurements of the severity of fires involving private motor vehicles. *Fire Mater.* **1995**, *19*, 143–151. [CrossRef]
29. Ingason, H. *An Overview of Vehicle Fires in Tunnels*; SP Swedish National Testing and Research Institute: Boras, Sweden, 2001.
30. Agency for Toxic Substances and Disease Registry (ATSDR). Available online: www.atsdr.cdc.gov/2012 (accessed on 27 February 2019).

7

Investigation on Water Hammer Control of Centrifugal Pumps in Water Supply Pipeline Systems

Wuyi Wan *, Boran Zhang and Xiaoyi Chen

Department of Hydraulic Engineering, College of Civil Engineering and Architecture, Zhejiang University, Hangzhou 310058, China; zhangboran@zju.edu.cn (B.Z.); zjuchenxy@zju.edu.cn (X.C.)

* Correspondence: wanwuyi@zju.edu.cn

Abstract: Water hammer control in water supply pipeline systems is significant for protecting pipelines from damage. The goal of this research is to investigate the effects of pumps moment of inertia design on pipeline water hammer control. Based on the method of characteristics (MOC), a numerical model is established and plenty of simulations are conducted. Through numerical analysis, it is found that increasing the pumps moment of inertia has positive effects both on water hammer control as well as preventing pumps rapid runaway speed. Considering the extra cost of space, starting energy, and materials, an evaluation methodology of efficiency on the increasing moment of inertia is proposed. It can be regarded as a reference for engineers to design the moment of inertia of pumps in water supply pipeline systems. Combined with the optimized operations of the valve behind the pumps, the pipeline systems can be better protected from accident events.

Keywords: transient analysis; pumps; moment of inertia; water hammer; pipe flow

---

1. Introduction

Due to the non-uniformity on spatial and temporal distribution of water resources, long-distance water supply systems are required to deliver water from humid regions to arid regions, and from the wilderness to the cities [1]. Water hammer is an undesired hydraulic phenomenon in this kind of engineering, especially in pressured pipelines. It is a pressure wave caused when the running condition of the system is changed, for example, a sudden valve closure at the end of a pipeline system, causing a pressure wave to propagate in the pipe. Extreme water hammer can cause pipe and device damage by high overpressure or low negative pressure [2–5]. Operations such as rapid opening and closing of the valves, starting and shutting down the pumps, and other sudden changes on devices along the pipeline will cause water hammers. To reduce harmful extreme water hammers, designers usually use pressure control devices such as surge tanks [6–8], air valves [9,10], and pressure vessels [11–13], as well as changing the operation time of valves [14–16], to keep the extreme pressure in a safe range. Usually, it is the water hammer occurring place that will load the most extreme pressure. For example, the valve at the end of a pipeline will bear the highest pressure in the whole pipeline in a water hammer after a rapid closing of it, and the outlet tube of a pump will suffer the lowest negative pressure in a water hammer caused by a pump stop event [17–19]. The magnitude of extreme pressure of the worst case is related to the wave speed of water hammer and the time for valve closing according to the Joukowsky formula [20], and it can be significantly reduced by expanding the closing time of the valves. The pressure control devices can protect the part of the pipeline on the other side of it, opposite to the pressure wave source [21]. Thus, the pressure control devices are normally recommended to be set somewhere most likely to produce a water hammer, if the geographical conditions permit this [22]. Nevertheless, pressure control devices set at other special positions may have a great positive influence too. For instance, in-line valves are selected to isolate parts of the system from a high-pressure source to a low-pressure water hammer event [14], in this way greatly reducing or eliminating any return surge.

Sometimes, the pressure control devices are limited by the geographical conditions. For instance, surge tanks provide a free surface to reflect the pressure wave and store energy, reducing the reflection time of the system and limits the influence of the pressure wave, which helps to protect the upstream section from higher pressure. Meanwhile, there is a problem with the limited heights, caused by mass oscillations between the reservoir and the surge tanks, which can be limited by throttles in the surge tanks [23–25]. On the other hand, when the pipeline is buried deep underground, the air valve is difficult to install. Except for adding pressure control devices in the pipeline systems, there are other methods to reduce the extreme pressure. Li et al. [26] proposed an improved bypass pipe which overcomes the limitations of conventional bypass pipes and can significantly help to reduce the maximum extreme pressure in a pumped water supply system. Triki [27,28] investigated other water hammer control strategies using an in-line polymeric short-section, and a branched polymeric penstock, respectively, and found both have a significant effect on mitigating the pressure increase and decrease induced by water hammer waves. Zeng et al. [29] analyzed different guide vane closing schemes on reducing water hammer intensity. Considering controlling the pulsating pressure and runaway speed during the transient processes, a three-phase valve closing scheme showed an obvious advantage, and is validated by correspondingly conducted experiments. Ballun [30] investigated advantages of different kinds of valves and proposed a methodology for designers to predict and ultimately prevent the valve slam. Hur et al. [31] used a frequency response analysis to study the homologous relationships between the rotational speed of a pump and its head or discharge, and derived the transfer functions for the head and discharge between the upstream and downstream pumps. Wan et al. [32], on the other hand, presented an optimal collaborative scheme to prevent inverse rotation and overpressure during the startup process.

In this research, another strategy is proposed to control the transient processes by adjusting the pumps moment of inertia in pumped water supply systems. Compared to other strategies, it can be accessed directly by changing the size of the flywheel of a pump and does not require further artificial operations. Through 1D pressured water supply pipeline model establishment and plenty of numerical simulations, the principles of adding a moment of inertia are analyzed. It is found that this strategy has a positive effect on both extreme pressure and pumps runaway speed control. Combining the operations on valves, more cases are studied, and a kind of evaluation methodology is proposed to judge the improvement. The results show that an optimized strategy of staging the valve closing period, and an appropriate increase of pumps moment of inertia can significantly reduce the water hammer intensity and protect the devices.

2. Materials and Methods

2.1. Method of Characteristics

In this study, the simulation was conducted by programming in FORTRAN based on MOC and the boundary conditions of the involved devices. To simulate a transient process, there are a lot of numerical methods, including the Finite Difference Method (FDM) [33] and the Finite Element Method (FEM) [34,35]. The FDM method of characteristics (MOC) is one of the most widely used methods recently, because of its convenience and accuracy. In water supply pipeline systems, using the method of characteristics [36], the complete dynamic governing equations can be expressed as Equations (1) and (2). There are some assumptions made in the derivation of the equations. The flow in a pipe is one-dimensional and the control volume is fixed relative to the pipe. The velocity and pressure are uniform at the cross sections of the pipe, and the fluid density is constant. The pipe walls and fluid are linearly elastic, that is, stress is proportional to the strain. And the temperature is assumed to be constant [36]:

$$g\frac{\partial h(x,t)}{\partial x}+v(x,t)\frac{\partial v(x,t)}{\partial x}+\frac{\partial v(x,t)}{\partial t}+\tau_w=0 \tag{1}$$

$$v(x,t)\frac{\partial h(x,t)}{\partial x}+\frac{\partial h(x,t)}{\partial t}-v(x,t)\sin\alpha_{\text{Pipe}-\text{Plane}}+\frac{a^2}{g}\frac{\partial v(x,t)}{\partial x}=0 \tag{2}$$

where x and t are independent variables denoting the location and the time, respectively, and correspondingly, the h and v are dependent variables denoting, respectively, the hydraulic pressure and the flow velocity. $\tau_w=\tau_{ws}+\tau_{wu}$ is the instantaneous wall shear stress, which can be divided into two parts [37]. One is $\tau_{ws}=f\frac{v(x,t)|v(x,t)|}{2D}$, representing the quasi-steady component, where f denotes the Darcy friction factor, which is valued as 0.029 in the simulation of this study, D denotes the diameter of the pipeline, and is the dominant friction factor in a steady flow. τ_{wu}, the unsteady component, has a significant influence on the water hammer intensity reducing process. However, it affects little on the value of the extreme pressure, which is usually the first strike when the water hammer occurs. To simulate the unsteady component, various numerical methods were proposed. Instantaneous acceleration-based (IAB) models and weighting function-based (WFB) are the two popular models [37]. $\alpha_{\text{Pipe}-\text{Plane}}$ represents the angle between the pipeline and the horizontal plane. However, in this research, it is the extreme pressure that we care about, the unsteady friction factor was therefore neglected. Controlling the characteristic relation between wave speed and meshes, which is the essential assumption of MOC, is shown as Equation (3),

$$\frac{dx}{dt}=\pm a \tag{3}$$

the variables at the next time step on the middle node can be inferred from the parameters at the last time step on those three adjacent nodes, expressed as Equation (4) derived from Equation (1)–(3):

$$\begin{cases} C^+: h(i,t+\Delta t)=h(i-1,t)-Rq(i-1,t)|q(i-1,t)|+Bq(i-1,t)-Bq(i,t+\Delta t) \\ C^-: h(i,t+\Delta t)=h(i+1,t)+Rq(i+1,t)|q(i+1,t)|-Bq(i+1,t)+Bq(i,t+\Delta t) \end{cases} \tag{4}$$

in which $B=\frac{a}{gA}$, $R=\frac{f\Delta x}{2gDA^2}$, $h(i,t+\Delta t)$ denotes the water head at section i at time $t+\Delta t$; $q(i,t+\Delta t)$ denotes the discharge at section i at time $t+\Delta t$; $h(i-1,t)$ denotes the water head at section $i-1$ at time t; $q(i-1,t)$ denotes the discharge at section $i-1$ at time t; $h(i+1,t)$ denotes the water head at section $i+1$ at time t, and $q(i+1,t)$ denotes the discharge at section $i+1$ at time t. The access of the variables at the next time step can be expressed as two characteristic lines, shown in Figure 1. The positive characteristic line represents the controlling relationship between the variables of the node at section i at time $t+\Delta t$, and the hydraulic head $h(i-1,t)$ and discharge $q(i-1,t)$ of the node at section $i-1$ at time t. While the negative characteristic line represents the controlling relationship between the variables of the node at section i at time $t+\Delta t$, and the variables of the node at the downstream adjacent section at time t. As the variables of the adjacent nodes at time t are already known, the middle node at the next time step can be solved by combining the two characteristic lines. In this way, the variables can be solved in every inner section except the boundary sections.

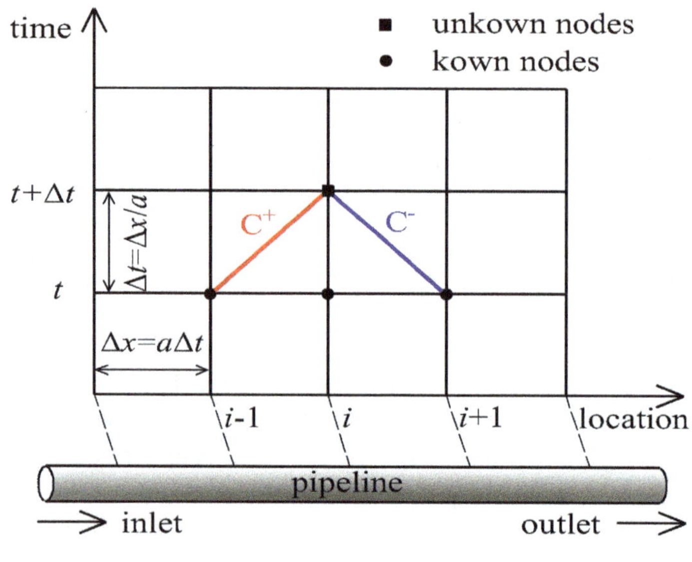

Figure 1. Meshing of the method of characteristics (MOC) in the pipeline.

Furthermore, in a water supply system, there are always many kinds of devices set along the pipelines, such as valves, tanks, pumps, etc. Therefore, to simulate the transient processes in water supply systems, those boundary conditions need to be solved first. In previous research, boundary conditions have been a significant topic in numerical simulations [8,38–41]. Till now, researchers have investigated a series of reliable and effective accesses in numerical simulations of transients in water supply systems.

2.2. Boundary Conditions of Pumps

Pumps are frequently used in hydraulic systems. To simulate a transient process of a hydraulic system with a pump station, the boundary conditions of pumps are necessary to be solved correctly. The running characteristics of a pump depend on several parameters. Four quantities are involved in the characteristics, including the dynamic head H, the flow discharge Q, the shaft torque T, as well as the rotational speed N.

In addition, every single pump has its own complete characteristics curve, which determines the relationship among these four quantities. According to the rated dynamic head H_R, the rated flow discharge Q_R, the rated shaft torque T_R, and the rated rotational speed N_R, the complete characteristics can be expressed in the following form as Equation (5):

$$\begin{cases} WH(x_p) = \frac{H/H_R}{(N/N_R)^2+(Q/Q_R)^2} \\ WB(x_p) = \frac{T/T_R}{(N/N_R)^2+(Q/Q_R)^2} \end{cases} \tag{5}$$

where $WH(x_p)$ and $WB(x_p)$ are two defined dimensionless quantities representing the features of the pumps, the curve of which consists the relationships among the four dominant variables mentioned above. Once two of the variables are determined, the others can be derived using the characteristic curves represented by $WH(x_p)$ and $WB(x_p)$. Figure 2 shows the complete characteristic curve of the pump used in this simulation model. It is an imaginary turbine with the parameters changed according to the example described in reference [2]. The abscissa $x_p = \pi + \tan^{-1}\left(\frac{QN_R}{NQ_R}\right)$ represents the instantaneous position of pump operation. That means only two quantities among them are independent. Along with the complete characteristic curve, a certain operating condition can be concretely expressed numerically.

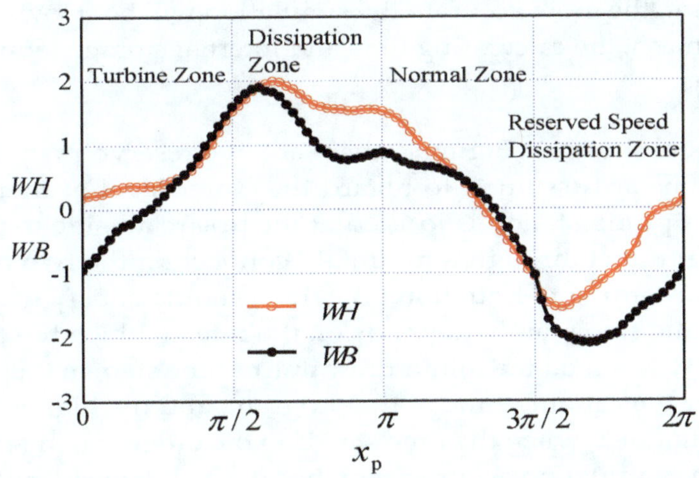

Figure 2. Complete characteristic curve of the pumps. [2].

The characteristic lines of a pump are shown in Figure 3. A working pump provides a water head rise. The water head and discharge on both sides of a pump can be solved using the following set of equations [8], which combine the complete characteristic curve in Figure 3:

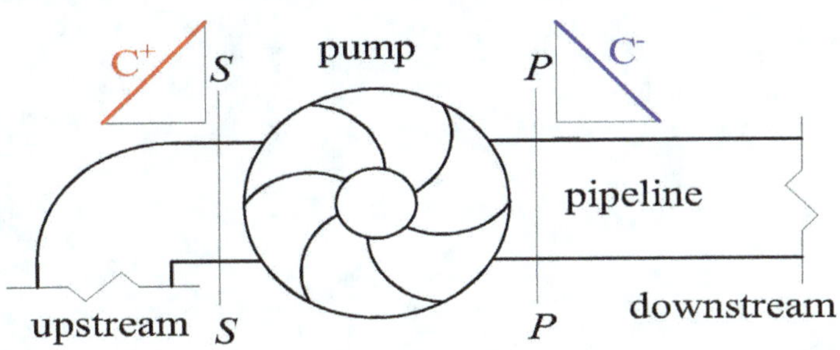

Figure 3. Characteristic lines of a pump.

$$\begin{cases} h(i_{s-s}, t+\Delta t) + \text{tdh} = h(i_{p-p}, t+\Delta t) \\ \text{tdh} = H_R\left(\frac{N}{N_R}^2 + \frac{Q}{Q_R}^2\right)WH(x_\text{p}) \\ \left(\frac{N}{N_R}^2 + \frac{Q}{Q_R}^2\right)WB(x_\text{p}) + \frac{T_0}{T_R} - \frac{WR_g^2}{g}\frac{N_R}{T_R}\frac{\pi}{30\Delta t}\frac{N_0 - N}{N_R} = 0 \\ q(i_{s-s}, t+\Delta t) = q(i_{p-p}, t+\Delta t) \\ h(i_{s-s}, t+\Delta t) = C_P - Bq(i_{s-s}, t+\Delta t) \\ h(i_{p-p}, t+\Delta t) = C_M + Bq(i_{p-p}, t+\Delta t) \end{cases} \quad (6)$$

The solutions have been described in detail in previous research [2,32], which is complex and lengthy. Since it is not the main focus of this study, the details of these solutions are partly omitted here.

3. Model Establishment

3.1. Pipeline Meshing

In a long water supply system, there are various kinds of pipelines, including steel pipes, PCCP pipes and tunnels etc. that divides a water supply system into several zones. Due to the differences in the diameter, roughness, and elasticity, the wave velocity and rubbing effect in the zones are not the same. At the connecting profile of two continuous pipeline sections, usually, it may not be divided just exactly at the abrupt change profile. To deal with the meshing problem, the wave speed or pipe length must be modified in the numerical simulation model, and that inevitably leads to a system error. The best way to limit the simulation system error is to compact the meshing. The closer the adjacent sections are set, the more accurate the simulation will be. However, while increasing the accuracy by using dense mesh, the calculating time and internal storage required in the computer are enhanced at the same time.

In the simplified model shown in Figure 4, there are two reservoirs upstream and downstream, whose boundary conditions are assumed to regard the water level of them kept constant during the transient processes. A pump station is located at the upstream side to provide hydraulic head, and the pipeline is a long tunnel through a mountain connecting the two reservoirs. The meshing is uniformly distributed according to Equation (3), which should satisfy the Courant condition [36]. The maximum value of time step mainly depends on the rate at which the transient process occurs. There is no compulsory restriction on the minimum value of time step, in theory, however, the less the time step is set, the longer the running time of the program, and the higher the cost. From previous investigations [42], it is found that when the time step is in the order of 0.01 s, the detailed reflection of device operations, and the simulation accuracy can be ensured. Considering the running time and accuracy of the calculation, the time step is chosen to be 0.01 s in this numerical model empirically. According to the basic assumption of MOC, Equation (3), which requires the location step to equal the time step multiplying the wave speed. Therefore the mesh density is set every 10 m, as the wave speed is considered as 1000 m/s in the model.

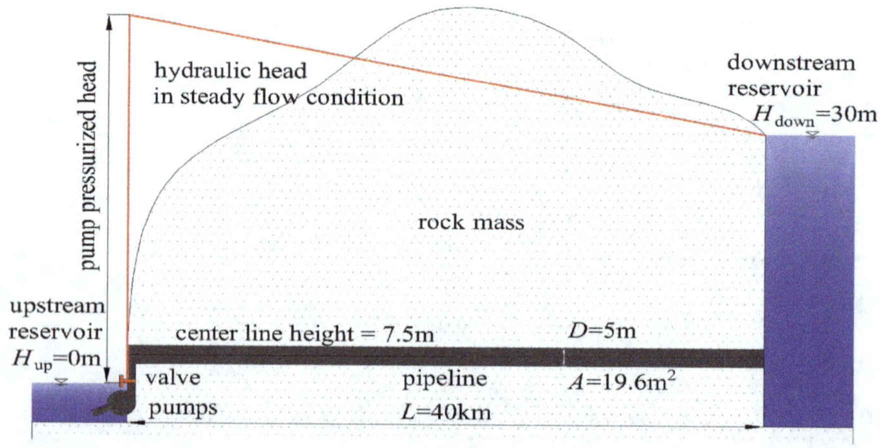

Figure 4. The physical model of a simplified case established for simulation.

3.2. Pumps Moment of Inertia

In a water supply system, pumps are the energy providing devices, pressurizing the delivering liquids meters to tens of meters' water head. Both starting and stopping the pumps can cause water hammer in the pipelines. Starting a pump can cause a positive pressure rising wave delivering downstream, while stopping a pump may lead to a sudden pressure decrease. In a pressured pipeline system, extreme negative pressure is always an undesired phenomenon that may cause the stability failure of pipelines. Determined by the material, the pipelines can tolerate a certain range of negative pressure, which is the available static pressure, similar to its loading capacity of positive pressure. When the intensity of negative pressure is within the tolerable range, it is not severe and can be resisted by the pipeline. However, when the intensity of the negative pressure is beyond the tolerable range of the pipelines, it will be critical.

The moment of inertia is a kind of inherent parameter that determines the rotate speed changing processes, with a dominant influence on the restrictive relationship among the four variables illustrated above, connected with Equation (5). In Equation (7), T is the unbalanced torque, $I = WR_g^2/g$ is the moment of inertia, W is the weight of the rotating part plus entrained liquid, R_g is the radius of gyration of the rotating mass, ω is the angular velocity in radians, and $\frac{d\omega}{dt}$ represents the angular acceleration. It shows that the angular acceleration is inversely proportional to the moment of inertia.

$$T = -I\frac{d\omega}{dt} \tag{7}$$

Figure 5 shows the negative pressure wave producing process in a controlled volume behind the pump after pump stopping. After the pump power off, while the discharge q_{outlet} through section j remains as large as the steady condition, the discharge q_{inlet} through section i decreases with the rotate speed n of the pump. In this way, the water supplied from upstream cannot fulfill the room in the control volume anymore. The faster the angular velocity ω decreases, the worse the negative water hammer will be. Therefore, we prefer the angular acceleration to be less, which demands the moment of inertia of the pump to raise. When the moment of inertia of the pump increases, the acceleration processes of pump start and stop last longer. It should have a positive effect on water hammer control by expanding the transient responding time after a sudden operation.

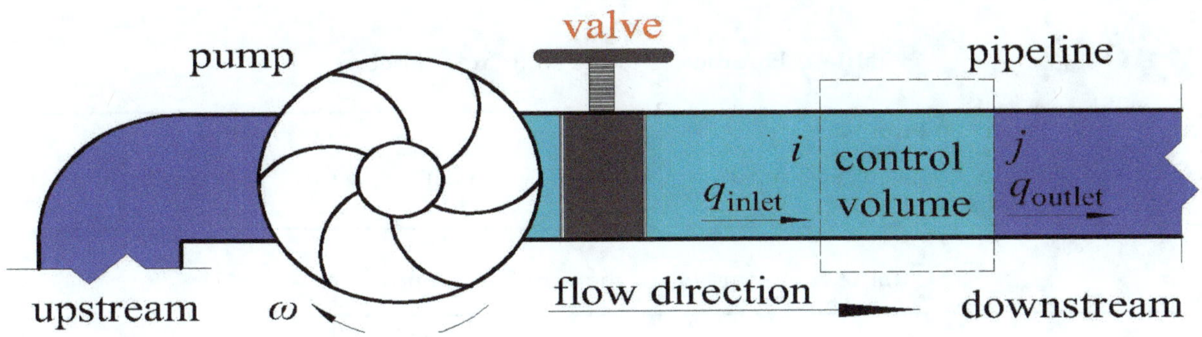

Figure 5. Physical analysis of negative pressure occurring in a pump stopping progress.

4. Simulation and Analysis

4.1. Pump Load Process and Extreme Pressure Along Pipeline

The parameters of the simulated pipeline and pumps are listed in Table 1. The simulated conditions are shown in Table 2, which are changed at the moment of inertia or valve operations to investigate different aspects as follows. t_{V0} is the start time of valve operation, Δt_{V1} is the time period of the first stage of valve operation, Δt_{V2} is the time period of the first stage of valve operation, $\Delta\tau_1$ is the closed ratio of the first stage, and $\Delta\tau_2$ is the closed ratio of the second stage. To investigate the influence of changing pumps moment of inertia on water hammer protection, the first five cases shown in Table 2, which use different moments of inertia without valve operations, are simulated respectively for comparison. In all the simulated cases, the pump failure times are set as 1000 s uniformly. Actually, the model does not need that long to achieve stability, however, it is set at this value to imitate a sudden failure during a normal running condition. As the length of the pipeline is 40 km, and the pressured wave speed is 1000 m/s, the reflection time is therefore $\frac{2L}{a} = 80$ s after the pump's failure. It is the time of the first water hammer response transferring back after the event occurrence. The upstream water level is 0 m, while the downstream water level is 30 m higher, which needs the pumps to provide the head difference and extra energy to overcome the friction along the pipeline. Therefore, after pumps failure, the water will flow backward and push the pumps to inverse if the valve is not closed. In this part of the analysis, the valve is kept open, thus, the backflow would cause the pumps inversion. The operations of the valve will be analyzed in the following section. Increasing the pumps moment of inertia can be easily accessed by adding a flywheel onto the pump's spindles, as shown in Figure 6.

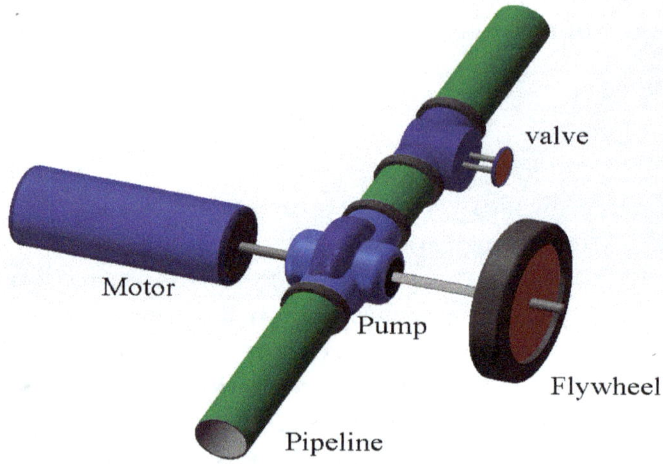

Figure 6. Physical model of a pump with an additional flywheel to increase the moment of inertia.

Table 1. Parameters of the simulated pipeline.

Pumps				Pipeline			
H_R (m)	Q_R (m^3/s)	N_R (r/min)	T_R (N · m)	L (km)	D (m)	a (m/s)	A (m^2)
35.57	8.60	375	84,583	40	5	1000	19.6

Table 2. Parameters of the simulated conditions.

Cases	***I* (kg · m²)**	**Valve Closing**				
		t_{V0}	Δt_{V1}	$\Delta \tau_1$	Δt_{V2}	$\Delta \tau_2$
i_C# (i_C = 1,2, ... ,7)	$i_C \cdot 10^5$	/	/	/	/	/
8#	10^5	1000	600 s	100%	0 s	0%
9#	10^5	1100	600 s	100%	0 s	0%
10#	10^5	1000	60 s	80%	540 s	20%

After a sudden power off, the pumps lose electric force resource to sustain the water supply function, turning gradually slower due to the inertia. In this way, the water discharge drawn through the pumps decreases, while the outlet of the control volume in Figure 5 remains at a high velocity. Therefore, the negative pressure wave occurs as the momentum of the water in the control volume decreases. Figure 7 shows the complete pressure processes behind the pump of the whole simulation under a series of conditions including the pumps' start, normal running condition, pumps' failure, and backflow steady condition. Under different conditions of additional moment of inertia, the start period, normal running, and backflow steady conditions differ little, while the difference in the pump failure period is obvious. For a clear comparison, a detailed view of the pressure changing process in the pump failure period at the location right behind the pumps is shown in Figure 8. After the power off, the hydraulic head behind the pumps decreases dramatically, which produces a negative water hammer. Using pumps with a high moment of inertia, the head dropping process shows a relatively smooth trend compared to other conditions. That leads to a protection effect for the water supply system by reducing the negative water hammer intensity. However, there is an interesting inversion of those pressure curves under different conditions. In this set of curves, the inversion zone appears at about 150 s after the pumps power off. It reveals that the pressure process under a low moment of inertia condition not only decreases rapidly but also has a quicker rise back than the other conditions. Obviously, it can be regarded as a sign of a stronger intensity of the produced negative water hammer.

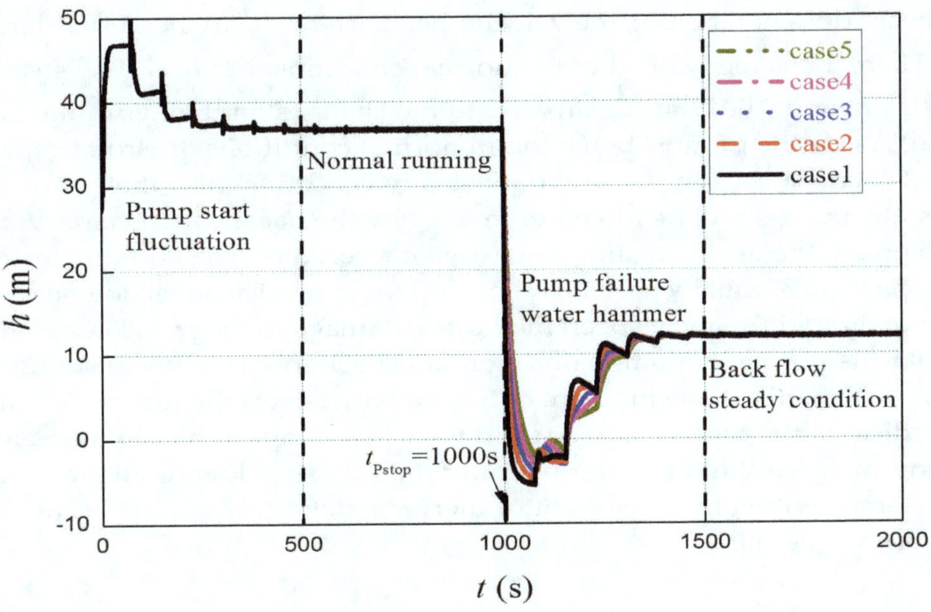

Figure 7. Pressure processes behind the pump of complete time series under different moment of inertia conditions.

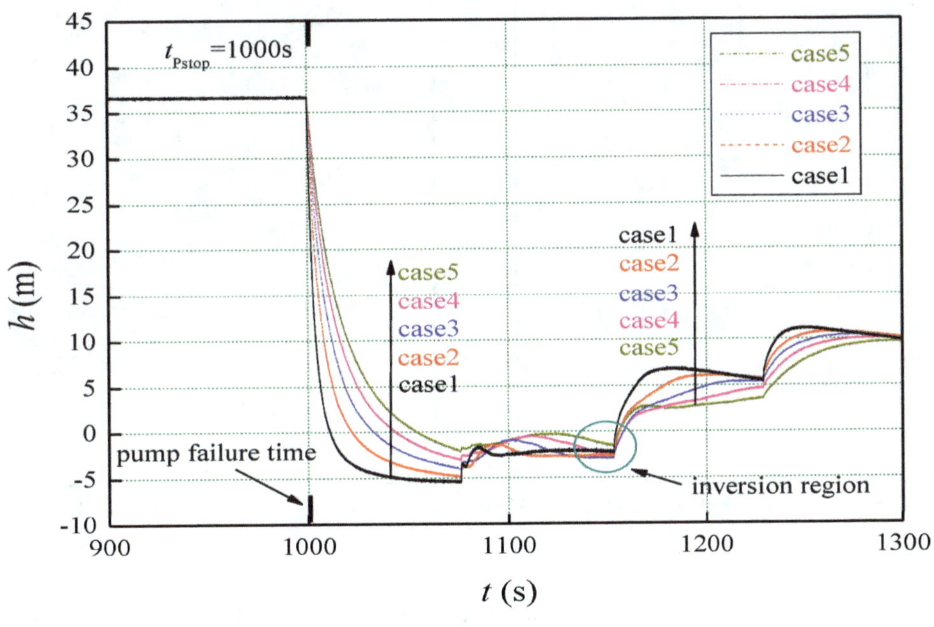

Figure 8. Detailed comparison of pressure processes behind the pump under different moment of inertia conditions.

Figure 9 shows the non-dimensional rotate speed of the pumps and the non-dimensional velocity of the water flowing through the pumps. The non-dimensional rotate speed is using the actual rotate speed divided by the rated rotate speed of the pump, and the non-dimensional velocity is using the actual discharge divided by the rated discharge of the pump. During the period of the normal running condition, the flow discharge is 3.74 m^3/s, the cross area of the pipeline is 19.6 m^2, and the flow velocity is 0.19 m/s. After about 500 s since the pump failure with no operations on the valve, the system reaches another steady condition, where the backflow discharge is -6.1 m^3/s, and the corresponding velocity is -0.31 m/s. The hydraulic head h analyzed in this study, sometimes called the piezometer head, does not include the velocity head. The velocity head is as large as $\frac{v^2}{2g}$, and the velocity $v = \frac{Q}{A}$. Then, the velocity head of the normal condition and backflow steady condition are, respectively, 0.0018 m and 0.0049 m. Compared to the changes on the hydraulic head, the velocity head affects little. As for the pumps' behavior, in normal conditions, the rotate speed is 375 r/min, equal to the rated rotate speed. In the backflow steady condition after the pumps' failure without valve operations, the pumps will be forced to inverse by the backflow force provided by the water level difference between the upstream and downstream reservoirs. The non-dimensional rotate speed of inversion will reach −0.87, finally, which is about −326 r/min. The inversion needs to be controlled, because the extended rapid runaway speed may cause damage to the pumps as well. In comparison, it can be seen that the normal running condition and backflow steady condition do not differ as the moment of inertia changes. The moment of inertia only affects the unsteady transient processes. Apparently, regardless of the rotate speed or fluid velocity, pumps of higher moment of inertia show better stability and produce slighter water hammer than those of low moment of inertia. Thus, it is clear that using pumps with a high moment of inertia is definitely a more secure choice for water hammer control in engineering.

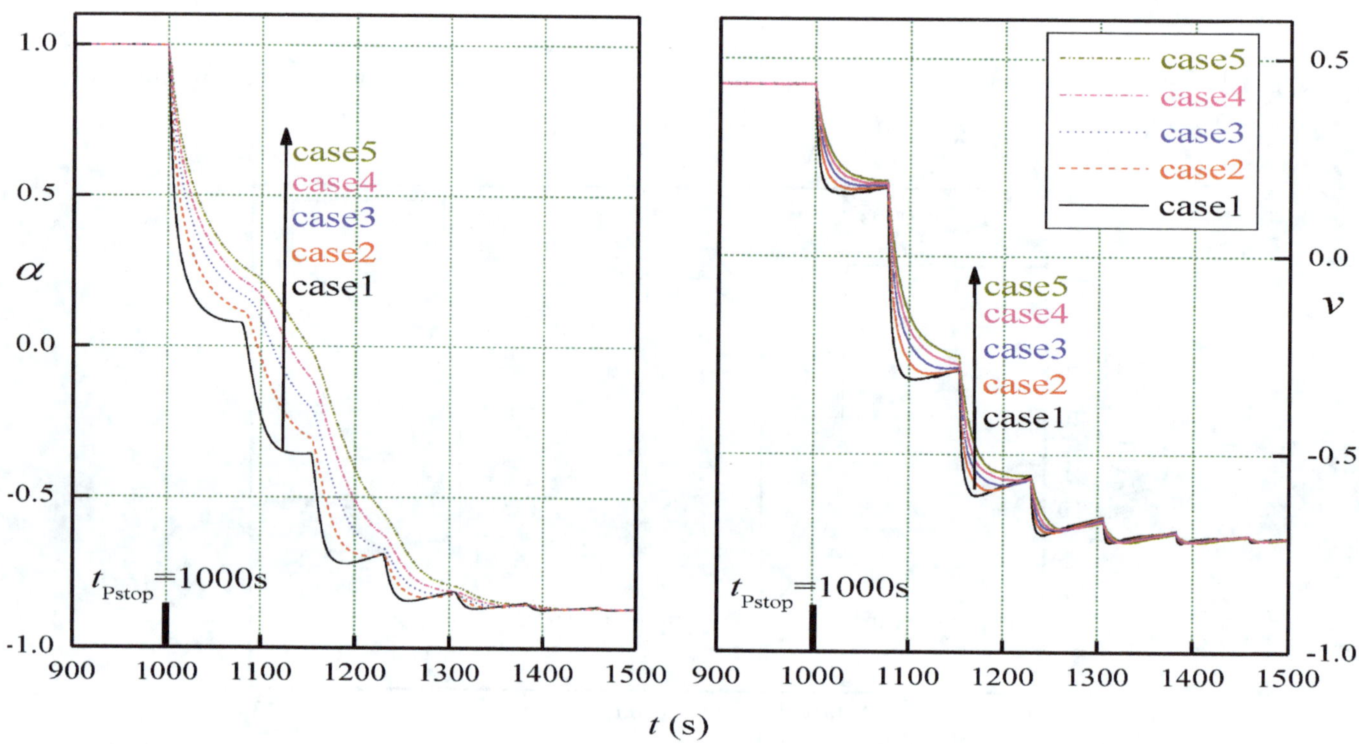

Figure 9. Comparisons of dimensionless rotate speed and dimensionless velocity changing progress under different moment of inertia conditions.

In reality, we are concerned more about the extreme pressure along the pipeline rather than the water hammer process at a certain place. That is because the center pipeline altitude can be changed limited by the topographic conditions. For instance, the pipeline should be deeper underground when crossing a river, and it may rise along with a hillside. Thus, the long-distance water supply pipelines are usually not horizontal. Figure 10 shows the comparison of the extreme pressure curve along the pipeline. The simulation results show that increasing the pumps moment of inertia can not only protect the most severe negative pressure zone but also reduce the water hammer intensity in the whole pipeline system.

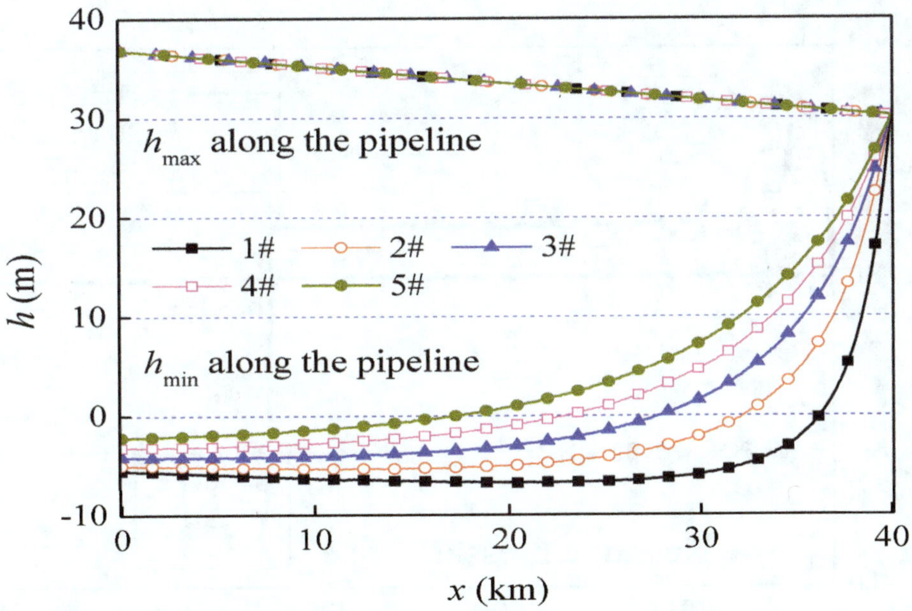

Figure 10. Comparison of the extreme dynamic head along the pipeline under different moment of inertia conditions.

4.2. *Quantity of Additional Moment of Inertia*

The moment of inertia can be accessed by changing the mechanic structure of the pumps. Except for improving the working accuracy by reducing the water hammer intensity, it will accelerate the cost and the occupied space at the same time. Except for adding an additional moment of inertia of pumps, there are still plenty of ways to reduce the water hammer intensity, such as surge tanks, air valves, optimization of operations, etc. Therefore, there is no need to increase the pumps moment of inertia with no limit. That will cost unnecessary economical and space resources. To numerically compare the value of positive effect, two quantities are defined. $\Delta h_{\mathrm{p}\ n1,n2}$ represents the difference of the extreme low pressure at the side of pipe right next to the pumps, where the negative pressure wave first occurs under the conditions of case n1# and case n2#, where n1 and n2 are two integers, representing the order number of two different cases selected from five different cases shown in Table 1. In most conditions, that is the location of the worst zone in a water hammer caused by a pump power off event. Therefore, it is regarded as a typical quantity to measure the protecting effect. However, in the conditions with a low moment of inertia, the rotate speed of the pumps would reduce quickly. The filled water inlet may only relieve the negative pressure of the upstream part but cannot access the downstream part to relieve the negative pressure wave delivered there. Therefore it is possible that the minimum pressure will occur somewhere else in the pipeline. The hydraulic head h, sometimes called piezometer head, equals the pipe center line height adding the relative pressure inside the pipe. When the hydraulic head curve is lower than the altitude of the pipe center line, the pipe will suffer negative pressure. In normal conditions, there is no negative pressure in the pipeline. However, in a water hammer process, the negative pressure wave may bring about a dramatic head decrease, and when the head decrease is greater than the original positive pressure at that point, negative pressure occurs. Therefore, other than additional devices and optimization of operations, reducing the pipeline height can significantly help to prevent negative pressure occurrence in the pipeline, while increasing the positive pressure. Take case 1# for example, the negative pressure wave influence is shown in Figure 11. There are three locations along the pipeline picked out to show the negative pressure delivering process, which are, respectively, $x = 1$ km, $x = 20$ km, and $x = 39$ km. The other one is ΔW, the whole area surrounded by two extreme pressure curves of different conditions, which represents the effect along the whole water supply system. Figure 12 shows the definition of Δh_{p} and ΔW.

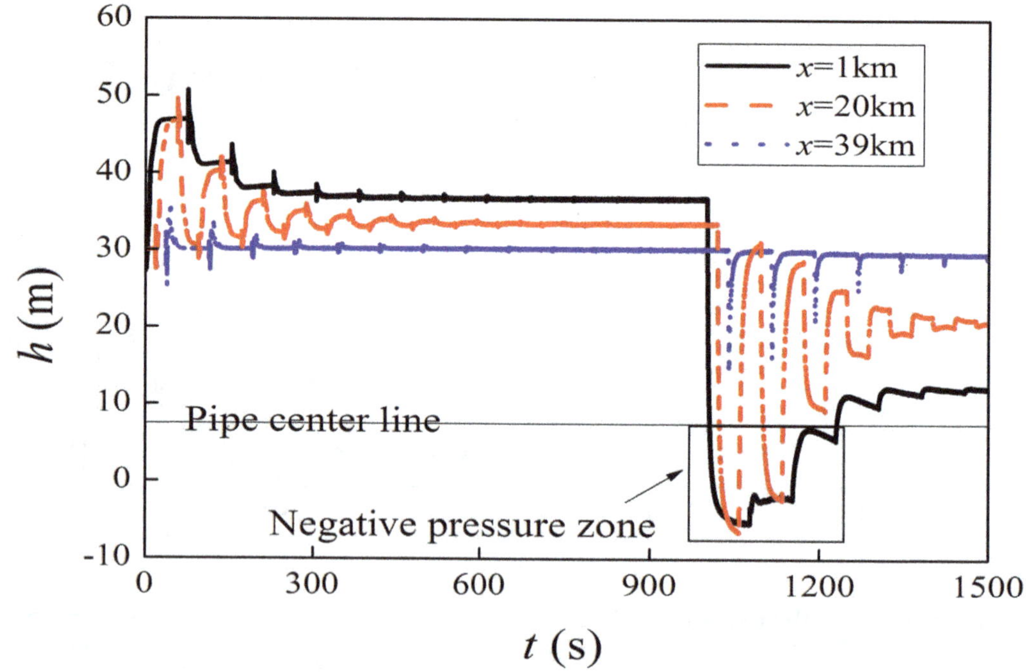

Figure 11. The negative pressure wave delivering process in case 1#.

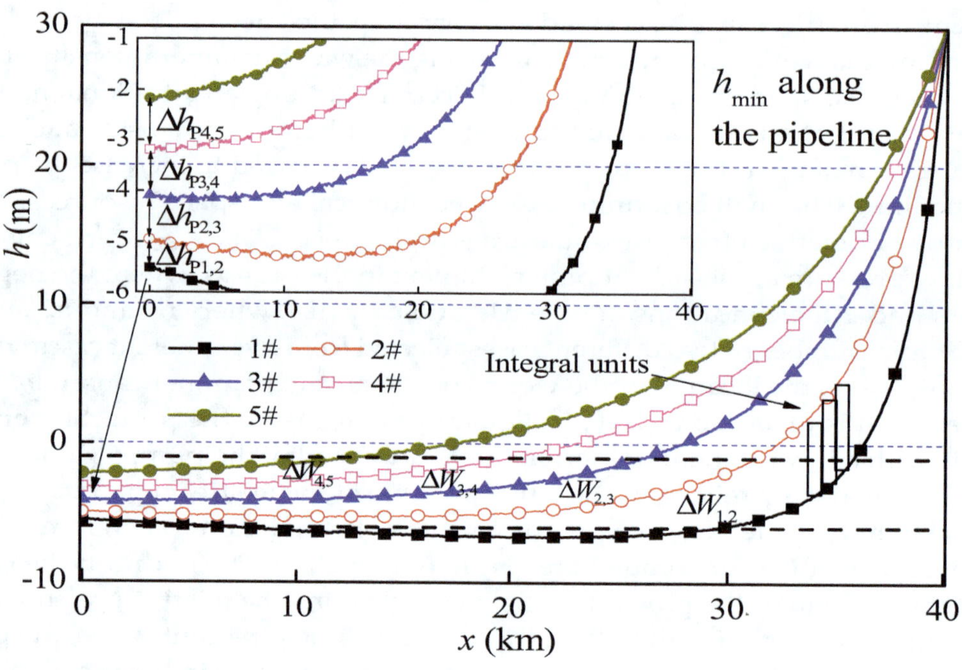

Figure 12. Description of the defined quantities Δh_{P} and ΔW.

To find a rule between the additional moment and the caused positive effect on pressure control, more conditions are simulated, since the original five cases were not adequate to reflect the trend. Seven of them were selected to show the trend of the two defined quantities. Figure 13 shows the relationship between the times of additional moment and the caused positive effect represented by those two quantities of adjacent cases, using the first seven cases listed in Table 2.

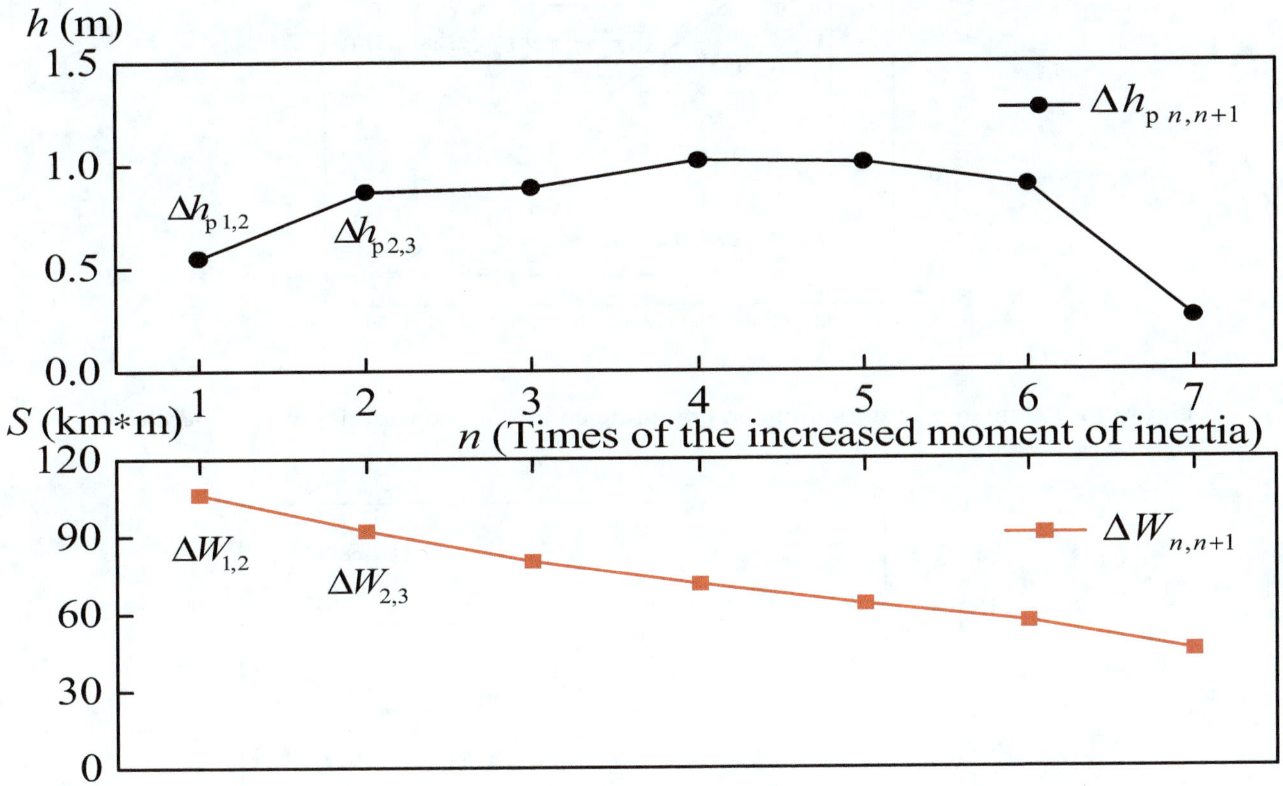

Figure 13. The trend of Δh_p and ΔW changing with increased moment of inertia.

Obviously, as the increase of the additional moment of inertia, the efficiency of the improvement on extreme pressure control effect will not be maintained at a high level. Especially ΔW, which represents the sight of the whole pipeline system, which keeps decreasing while the moment grows. Concerning the most severe zone in the pipeline, which is the point right behind the pump station, the helpful quantity of additional moment is limited below 6 times of the original moment of inertia. However, that can only be a qualitative conclusion, because the concrete numerical value is only suitable for the model established in this study, thus there is a lack of universality.

4.3. Considering the Valve

The simulations conducted above are without considering operations of valves. Normally, to ensure the pumps are not damaged from rapid runaway speed, there is always a non-returning valve set behind the rotors, as shown in Figure 5. Once the pump starts to inverse, the valve should be closed to protect the pump. However, operation of the valve can also bring undesired consequences to the system. If the valve is closed too fast or not at a suitable time, it will cause another water hammer event even more severe than the one caused by pump failure. In order to figure out the effect of putting off the valve closing time, as well as staging the valve operation, three more cases were added to make a comparison. The valve operations' corresponding connection with the system is reflected in Figure 14. Figures 15 and 16 show, respectively, the pressure processes behind the pump and the dimensionless rotate speed, and velocity changing progresses in the conditions considering valve closing listed as cases 8#, 9#, and 10# in Table 2.

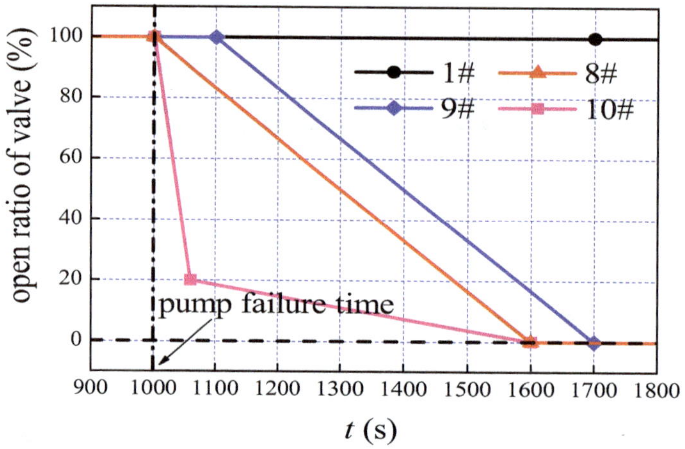

Figure 14. Parameters of the simulated operating conditions considering valve operations.

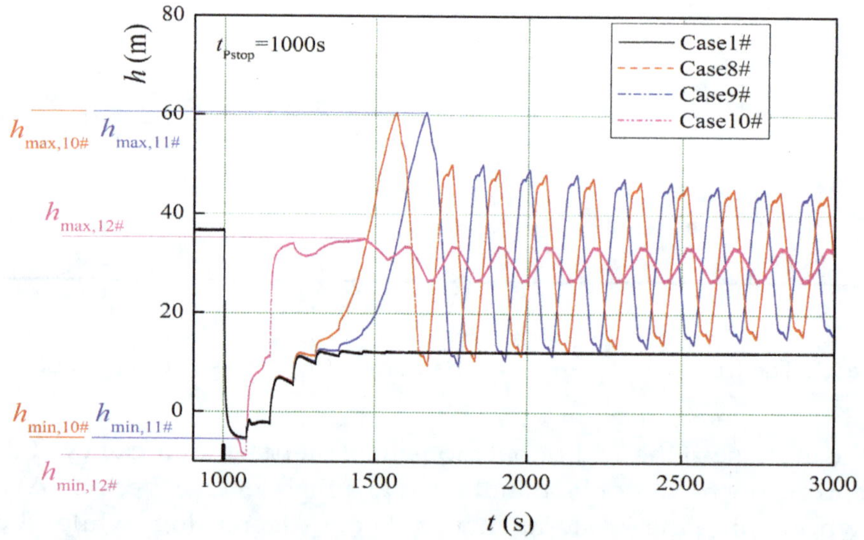

Figure 15. Comparison of pressure processes behind the pump under different valve closing operations.

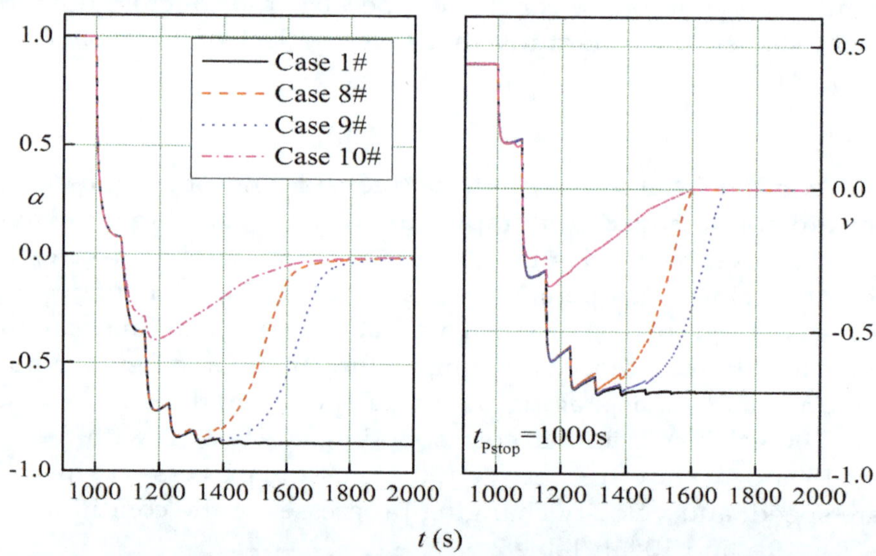

Figure 16. Comparison of dimensionless rotate speed and dimensionless velocity changing progresses under different valve closing operations.

Comparisons among case 8#, case 9#, and case 10# represents two kinds of ways to adjust the operation of the valve, putting off the valve operation time and staging the closing operation. Obviously, putting off the valve closing operation time has a slight influence on pressure control, and a negative effect on controlling the runaway speed. That means the backflow discharge should be controlled by a quick reduction of the valve open ratio before it increases to cause a rapid runaway speed. Though case 10# shows a significant improvement on both runaway speed and pressure control, it magnifies the extreme negative pressure behind the pump at the same time. That is because the rapid close stage I limited the discharge supplement into the negative pressure zone.

Adding pumps moment of inertia, and staging valve closing operation, plenty of more cases are investigated. The remaining other parameters were unchanged, only adjusting the moment of inertia and the time of valve closing in stage I. Ranging moment of inertia from 100,000 to 600,000, and time of valve closing stage I from 60 s to 120 s, the lowest pressure behind the pumps and the maximum runaway speed non-dimensional rotate speed are shown in Figure 17.

As shown, increasing pumps moment of inertia benefits on both aspects. Prolongation of valve closing time can help reduce extreme negative pressure but enlarge runaway speed at the same time. The key region is the period that the flow direction remains forward, which is within about 90 s shown in Figure 9. The best operation is to let the flow go on running to fill the negative pressure zone and close the valve to prevent runaway speed when backflow returns. Staging the time of valve closing is for accessing two aims at the same time. The first is stage I, rapid closing till 20% opening ratio, that is for preventing runaway speed. The other is stage II, slow closing till entirely closed, to avoid another water hammer caused by valve closing.

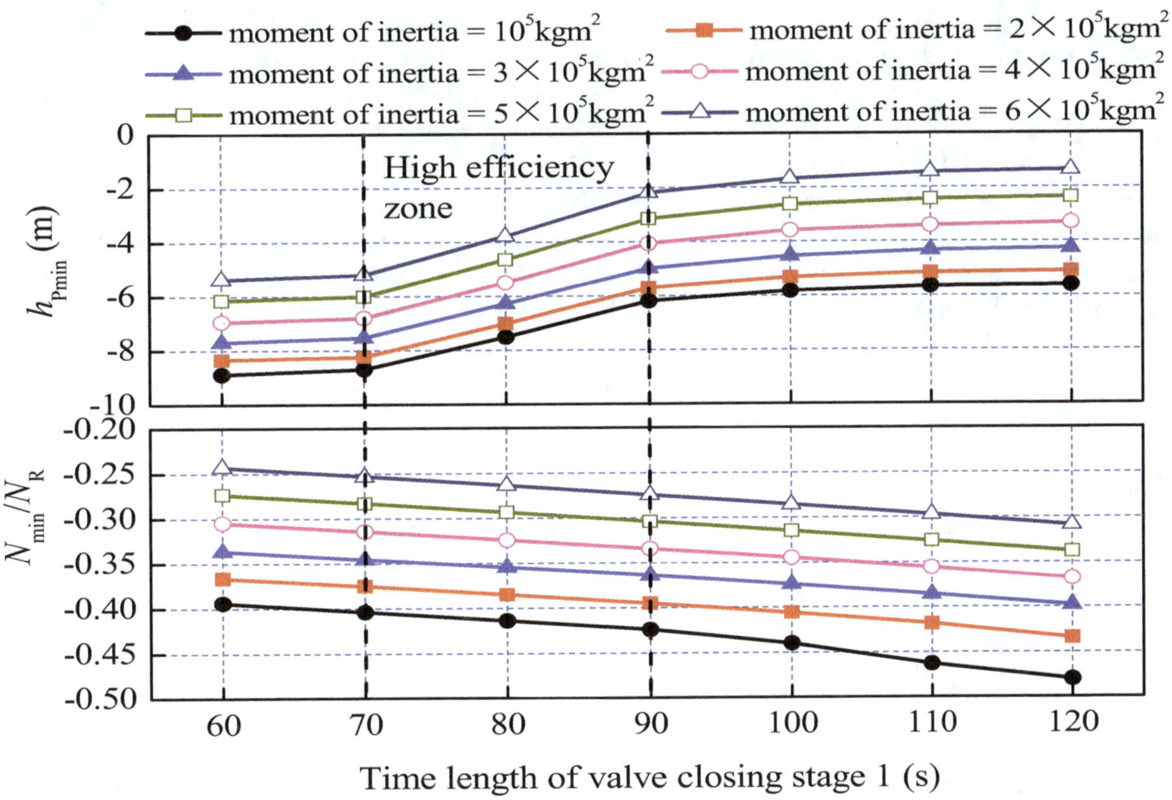

Figure 17. Comparisons of pressure processes behind the pump and dimensionless rotate speed under different additional moment of inertia when combining different valve closing operations.

5. Discussion

Surge tanks, air valves, and pressure vessels are widely used to the control intensity of water hammer. They are usually of good value to improve the system running security. However, the pipeline arrangement of a long-distance water supply system is subject to the geographical conditions. For instance, to cross a river, the pipeline is usually put under the riverbed, or above the water surface like a bridge, and surge tanks are usually limited by the height of the covered thickness above the pipeline. Indeed, the height of the surge tanks can be increased by building a tower upon the land but it leads to much labor and financial cost at the same time. Therefore, reasonable adjustment on devices is a relatively convenient and effective way to control extreme pressure in water hammers. Especially in the projects when the altitude of the pipeline is higher than the pumps. The rising zone of the pump's outlet pipe cannot be protected by surge tanks or other additional devices. The pressure control ability primarily depends on the characteristics of pumps and the operations of valves. Optimizing the operations processes and adding pumps moment of inertia can also help to control the extreme pressure along the pipeline and protect the pumps from fast runaway speed. The comparison among case 1# to 7# indicates that additional moment of inertia has a positive effect on negative pressure control and there is a most effective quantity of it. Furthermore, comparison among case 1#, and 8# to 10#, show the best cooperation of the valve to control the pump's runaway speed is staging time and to avoid the forward flow period. The simulated cases in this research are under specific limitations on pipeline design and pump parameters, the conditions of actual water supply engineering may have complex situations on the devices and pipeline design, including faster flow velocity and larger water hammer intensity. Nevertheless, the advantages of the additional moment of inertia remain the same.

To claim the assumptions used in this research, expect for the physical model simplification, the numerical model used for simulation has some assumptions. The first is the basic assumption of MOC, which requires the relationship between meshing grid and pressure wave speed, which is illustrated in Equation (3). The second is the neglect of unsteady friction factor, which actually has little influence on the prediction of extreme pressure value and does not affect the investigation results either. Besides, the flow in the pipe is one-dimensional and the control volume is fixed relative to the pipe. The velocity and pressure are uniform at the cross sections of the pipe, and the fluid density is constant. The pipe walls and fluid are linearly elastic, that is, stress is proportional to strain. And the temperature is assumed to be constant. Increasing the pumps moment of inertia can be easily accessed by adding a flywheel to the pump's spindles. It makes the response of pumps to accident events slower and slighter, which is normally a helpful change for device protection. It has to be pointed out that the efficiency of improvement will decline as the additional moment increases. For different engineering, there are corresponding valued limits of the additional moment of inertia. In addition, an added moment of inertia requires extra materials, space, and energy consumption. However, it does not need extra power to keep running in steady running conditions. Considering the security improvement on accident events in water supply systems, it is a recommendation to design the pumps with a reasonably high moment of inertia. Though, it has to be clarified that there are still some limitations, the effect of additional pump moment of inertia has just a relieving function for negative pressure when pumps fail. It has little benefit for water hammer control caused by downstream valve rapid closing or other devices operations inside the pipeline. In further research, the investigation on other parameters of pumps and pipeline design will be conducted, such as the complete characteristic curves of the pumps and draft tube design of the pump station.

6. Conclusions

In this study, a 1D numerical model containing a pump station, valves, reservoirs, and pipeline was established. After investigations on the effect of changing the pumps moment of inertia and operations of valves behind the pumps after a power off event, the following conclusions were drawn:

In water supply pipe systems, except for normal pressure control devices, additional pumps moment of inertia can help prevent extreme negative pressure occurring behind the pumps in accident events such as electric power off. Moreover, the extreme pressure curve along the pipeline can also be reduced. Considering the construction cost and the efficiency, there is always an efficient zone of the additional moment of inertia according to different engineering.

The pumps design and optimized valve operations are significant to protect pipe and pump from damage when accident events occur. Through various cases of numerical simulations based on the method of characteristics, it was found that increasing the moment of inertia benefits both negative pressure control after pumps power off and preventing pumps runaway speed. Closing the valve is for protecting the pumps from rapid runaway speed.

Staging the valve closing time can significantly reduce the water hammer caused by valve closing. To limit the negative pressure intensity, the dominant part of valve closing should begin after the period that the pipe flow direction remains forward. As when the flow direction is still forward, the filling water can help reduce the negative pressure caused by pump failure. The best operation is to let the flow go on running to fill the negative pressure zone and close the valve to prevent runaway speed when backflow returns.

Author Contributions: Conceptualization, W.W., and B.Z.; methodology, B.Z.; software, B.Z. and X.C.; validation, B.Z.; formal analysis, B.Z.; investigation, B.Z.; resources, W.W.; data curation, B.Z. and X.C.; writing-original draft preparation, B.Z.; writing-review and editing, W.W.; visualization, W.W.; supervision, W.W.; project administration, W.W.; funding acquisition, W.W.

Acronyms

MOC	method of characteristics

Nomenclature

g	Acceleration of gravity (m/s^2)
h	Pressure head (m)
x	Distance along pipe from the inlet (m)
t	Time, as subscript to denote time (s)
v	Flow velocity (m/s)
τ_w	Instantaneous wall shear stress
$\alpha_{\text{Pipe-Plane}}$	The angle between pipe and the horizontal plane.
τ_{ws}	The quasi-steady component
τ_{wu}	The unsteady component
f	Darcy–Weisbach friction factor
D	Main pipe diameter (m)
a	Wave speed of water hammer (m/s)
Δt	Time step (s)
B	A defined constant parameter of a pipeline
R	A defined constant parameter of a pipeline
i	Serial number of nodes (s)
q	Discharge (m^3/s)
A	Area of section (m^2)
Δx	length of element, space interval step (m)
H	Head of the pump (m)
Q	Discharge through the pump (m^3/s)
T	Torque of the pump (Nm)
N	Rotate speed of the pump
WH	A defined dimensionless variable representing a pump's characteristic
WB	A defined dimensionless variable representing a pump's characteristic
x_{p}	Instantaneous position of pump operation
H_R	Rated head of the pump (m)
N_R	Rated speed of the pump
Q_R	Rated discharge of the pump (m^3/s)
T_R	Rated torque of the pump (Nm)
π	Constant
I	The moment of inertia (kgm^2)

W	The weight of rotating parts plus entrained liquid (kg)
R_g	The radius of gyration of the rotating mass (m)
ω	The angular velocity in radians
n	Non-dimensional rotate speed of the pump
j	Section number
L	The length of pipeline (m)
t_{Pstop}	Time of the pump powering off (m^2)
$h_{\max}$	The maximum hydraulic head (m)
$h_{\min}$	The minimum hydraulic head (m)
Δh	The difference of hydraulic head (m)
i_C	The number of cases
ΔW	The whole area surrounded by two extreme pressure curves
λ	Non-dimensional coefficient
$h_{\text{Psteady},2}$	The hydraulic head at pump in the second steady condition (m)
h_{Pmin}	The minimum hydraulic head at pump in a transient process (m)
$h_{\text{Psteady},1}$	The hydraulic head at pump in the first steady condition (m)

References

1. Kummu, M.; de Moel, H.; Ward, P.J.; Varis, O. How Close Do We Live to Water? A Global Analysis of Population Distance to Freshwater Bodies. *PLoS ONE* **2011**, *6*. [CrossRef] [PubMed]
2. Wylie, E.B.; Streeter, V.L. *Fluid Transients*; McGraw-Hill International Book Co.: New York, NY, USA, 1978.
3. Kwan, E.S.K.; Heilman, C.B.; Shucart, W.A.; Klucznik, R.P. Enlargement of basilar artery aneurysms following balloon occlusion—"water-hammer effect" Report of 2 cases. *J. Neurosurg.* **1991**, *75*, 963–968. [CrossRef] [PubMed]
4. Brunone, B.; Karney, B.W.; Mecarelli, M.; Ferrante, M. Velocity profiles and unsteady pipe friction in transient flow. *J. Water Resour. Plann. Manag.-Asce* **2000**, *126*, 236–244. [CrossRef]
5. Bergant, A.; Simpson, A.R.; Tijsseling, A.S. Water hammer with column separation: A historical review. *J. Fluids Struct.* **2006**, *22*, 135–171. [CrossRef]
6. Xu, B.; Chen, D.; Zhang, H.; Wang, F. Modeling and stability analysis of a fractional-order Francis hydro-turbine governing system. *Chaos Solitons Fractals* **2015**, *75*, 50–61. [CrossRef]
7. Fang, H.; Chen, L.; Dlakavu, N.; Shen, Z. Basic Modeling and simulation tool for analysis of hydraulic transients in hydroelectric power plants. *IEEE Trans. Energy Convers.* **2008**, *23*, 834–841. [CrossRef]
8. Wan, W.; Zhang, B. Investigation of Water Hammer Protection in Water Supply Pipeline Systems Using an Intelligent Self-Controlled Surge Tank. *Energies* **2018**, *11*, 16. [CrossRef]
9. Stephenson, D. Effects of air valves and pipework on water hammer pressures. *J. Transp. Eng.-ASCE* **1997**, *123*, 101–106. [CrossRef]
10. Bergant, A.; Kruisbrink, A.; Arregui, F. Dynamic Behaviour of Air Valves in a Large-Scale Pipeline Apparatus. *Strojniski Vestn.-J. Mech. Eng.* **2012**, *58*, 225–237. [CrossRef]
11. Wan, W.; Huang, W.; Li, C. Sensitivity Analysis for the Resistance on the Performance of a Pressure Vessel for Water Hammer Protection. *J. Press. Vessel Technol. Trans. Asme* **2014**, *136*. [CrossRef]
12. Stephenson, D. Simple guide for design of air vessels for water hammer protection of pumping lines. *J. Hydraul. Eng. Asce* **2002**, *128*, 792–797. [CrossRef]
13. De Martino, G.; Fontana, N. Simplified Approach for the Optimal Sizing of Throttled Air Chambers. *J. Hydraul. Eng. Asce* **2012**, *138*, 1101–1109. [CrossRef]
14. Karney, B.W.; Simpson, A.R. In-line check valves for water hammer control. *J. Hydraul. Res.* **2007**, *45*, 547–554. [CrossRef]
15. Tian, W.; Su, G.; Wang, G.; Qiu, S.; Xia, Z. Numerical simulation and optimization on valve-induced water hammer characteristics for parallel pump feedwater system. *Annals Nucl. Energy* **2008**, *35*, 2280–2287. [CrossRef]
16. Bazargan-Lari, M.R.; Kerachian, R.; Afshar, H.; Bashi-Azghadi, S.N. Developing an optimal valve closing rule curve for real-time pressure control in pipes. *J. Mech. Sci. Technol.* **2013**, *27*, 215–225. [CrossRef]
17. Zhou, J.; Xu, Y.; Zheng, Y.; Zhang, Y. Optimization of Guide Vane Closing Schemes of Pumped Storage Hydro Unit Using an Enhanced Multi-Objective Gravitational Search Algorithm. *Energies* **2017**, *10*, 911. [CrossRef]

18. Zhou, D.; Chen, H.; Zhang, L. Investigation of Pumped Storage Hydropower Power-Off Transient Process Using 3D Numerical Simulation Based on SP-VOF Hybrid Model. *Energies* **2018**, *11*, 1020. [CrossRef]
19. Nagode, K.; Skrjanc, I. Modelling and Internal Fuzzy Model Power Control of a Francis Water Turbine. *Energies* **2014**, *7*, 874–889. [CrossRef]
20. Carratelli, E.P.; Viccione, G.; Bovolin, V. Free surface flow impact on a vertical wall: A numerical assessment. *Theor. Comput. Fluid Dyn.* **2016**, *30*, 403–414. [CrossRef]
21. Moghaddas, S.M.J.; Samani, H.M.V.; Haghighi, A. Transient protection optimization of pipelines using air-chamber and air-inlet valves. *Ksce J. Civ. Eng.* **2017**, *21*, 1991–1997. [CrossRef]
22. Kim, H.; Hur, J.; Kim, S. The Optimization of Design Parameters for Surge Relief Valve for Pipeline Systems. In Proceedings of the Applied Mathematics and Computer Science Conference, Rome, Italy, 27–29 January 2017.
23. Richter, W.; Zenz, G.; Schneider, J.; Knoblauch, H. Surge tanks for high head hydropower plants—Hydraulic layout—New developments/Wasserschlösser für Hochdruck-Wasserkraftanlagen—Hydraulische Auslegung—Neue Entwicklungen. *Geomech. Tunn.* **2015**, *8*, 60–73. [CrossRef]
24. Gabl, R.; Righetti, M. Design criteria for a type of asymmetric orifice in a surge tank using CFD. *Eng. Appl. Comp. Fluid Mech.* **2018**, *12*, 397–410. [CrossRef]
25. Adam, N.J.; De Cesare, G.; Nicolet, C.; Billeter, P.; Angermayr, A.; Valluy, B.; Schleiss, A.J. Design of a Throttled Surge Tank for Refurbishment by Increase of Installed Capacity at a High-Head Power Plant. *J. Hydraul. Eng.* **2018**, *144*, 10. [CrossRef]
26. Li, X.; Zhu, M.; Xie, J. Numerical Simulation of Transient Pressure Control in a Pumped Water Supply System Using an Improved Bypass Pipe. *Strojniski Vestn.-J. Mech. Eng.* **2016**, *62*, 614–622. [CrossRef]
27. Triki, A. Water-Hammer Control in Pressurized-Pipe Flow Using a Branched Polymeric Penstock. *J. Pipeline Syst. Eng. Pract.* **2017**, *8*. [CrossRef]
28. Triki, A. Water-hammer control in pressurized-pipe flow using an in-line polymeric short-section. *Acta Mech.* **2016**, *227*, 777–793. [CrossRef]
29. Zeng, W.; Yang, J.; Hu, J.; Yang, J. Guide-Vane Closing Schemes for Pump-Turbines Based on Transient Characteristics in S-shaped Region. *J. Fluids Eng.-Trans. Asme* **2016**, *138*. [CrossRef]
30. Ballun, J.V. A methodology for predicting check valve slam. *J. Am. Water Work Assoc.* **2007**, *99*. [CrossRef]
31. Hur, J.; Kim, S.; Kim, H. Water hammer analysis that uses the impulse response method for a reservoir-pump pipeline system. *J. Mech. Sci. Technol.* **2017**, *31*, 4833–4840. [CrossRef]
32. Wan, W.; Li, F. Sensitivity Analysis of Operational Time Differences for a Pump-Valve System on a Water Hammer Response. *J. Press. Vessel Technol.-Trans. Asme* **2016**, *138*. [CrossRef]
33. Chaudhry, M.; Hussaini, M. Second-order accurate explicit finite-difference schemes for waterhammer analysis. *J. Fluids Eng.* **1985**, *107*, 523–529. [CrossRef]
34. Kontzialis, K.; Moditis, K.; Paidoussis, M.P. Transient Simulations of the Fluid-Structure Interaction Response of a Partially Confined Pipe Under Axial Flows in Opposite Directions. *J. Press. Vessel Technol.-Trans. Asme* **2017**, *139*. [CrossRef]
35. Kochupillai, J.; Ganesan, N.; Padmanabhan, C. A new finite element formulation based on the velocity of flow for water hammer problems. *Int. J. Press. Vessels Pip.* **2005**, *82*, 1–14. [CrossRef]
36. Wylie, E.B.; Streeter, V.L.; Suo, L. *Fluid Transients in Systems*; Prentice Hall Englewood Cliffs: Englewood, NJ, USA, 1993.
37. Meniconi, S.; Duan, H.F.; Brunone, B.; Ghidaoui, M.S.; Lee, P.J.; Ferrante, M. Further Developments in Rapidly Decelerating Turbulent Pipe Flow Modeling. *J. Hydraul. Eng.* **2014**, *140*. [CrossRef]
38. Karney, B.W.; McInnis, D. Efficient calculation of transient flow in simple pipe networks. *J. Hydraul. Eng.-Asce* **1992**, *118*, 1014–1030. [CrossRef]
39. Wan, W.; Huang, W. Investigation on complete characteristics and hydraulic transient of centrifugal pump. *J. Mech. Sci. Technol.* **2011**, *25*, 2583–2590. [CrossRef]
40. Wan, W.; Huang, W. Investigation of Fluid Transients in Centrifugal Pump Integrated System With Multi-Channel Pressure Vessel. *J. Press. Vessel Technol.-Trans. Asme* **2013**, *135*. [CrossRef]
41. Hou, Q.; Zhang, L.; Tijsseling, A.S.; Kruisbrink, A.C.H. Rapid filling of pipelines with the SPH particle method. In Proceedings of the International Conference on Advances in Computational Modeling and Simulation, Kunming, China, 14–16 December 2011.
42. Wan, W.; Huang, W. Water hammer simulation of a series pipe system using the MacCormack time marching scheme. *Acta Mech.* **2018**, *229*, 3143–3160. [CrossRef]

8

Analysis of Different POD Processing Methods for SPIV-Measurements in Compressor Cascade Tip Leakage Flow

Lei Shi [1], Hongwei Ma [1,*] and Lixiang Wang [2]

1 School of Energy and Power Engineering, Beihang University, Beijing 100191, China; sladeshi@126.com
2 Midea Corporate Research Center, Foshan 528300, China; wanglixiang1106@163.com
* Correspondence: mahw@buaa.edu.cn

Abstract: Though the proper orthogonal decomposition (POD) method has been widely adopted in flow analysis, few publications have systematically studied the influence of different POD processing methods on the POD results. This paper investigates the effects of different decomposition regions and decomposition dimensionalities on POD decomposition and reconstruction concerning the tip flow in the compressor cascade. Stereoscopic particle image velocimetry (SPIV) measurements in the blade channel are addressed to obtain the original flow field. Through vortex core identification, development of the tip leakage vortex along the chord is described. Afterwards, each plane is energetically decomposed by POD. Using the identified vortex core center as the geometric center, the effects of different decomposition regions with respect to the vortex core are analyzed. Furthermore, the effects of different single velocity-components as well as their combination are compared. The effect of different decomposition regions on the mode 1 energy fraction mainly impacts the streamwise velocity component. Though the addition of W velocity component in the decomposition does change the spatial structures of high-order modes, it does not change the dynamic results of reconstruction using a finite number of POD modes. UV global analysis is better for capturing the kinetic physics of the tip leakage vortex (TLV) wandering.

Keywords: POD; tip leakage flow; decomposition region; decomposition dimensionalities; vortex identification; SPIV

---

1. Introduction

The efficiency, the pressure rise and the stable operating range are three prominent parameters for energy and power machinery, such as axial fans and compressors. The tip leakage flow (TLF), driven by the pressure difference between the pressure side and suction side of the blade in the tip region, plays a significant role in all three parameters mentioned above [1–3]. In recent years, the TLF, especially the tip leakage vortex (TLV), has been found to be an inherently unsteady flow phenomenon, with several distinct unsteady behaviors, including vortex wandering [4], vortex splitting [5,6] and vortex breakdown [7,8]. Previous studies showed that these unsteady behaviors not only have profound effects on the TLF mean characteristics, but also a close connection with some important flow phenomena such as rotating stall [9,10], rotating instability [11], oscillations of the TLV [12] and vortex shedding. Therefore, it is necessary to study the unsteady behaviors of the TLV to obtain a thorough knowledge of it.

Stereoscopic particle image velocimetry (SPIV) has been extensively employed to investigate the characteristics of the TLV for its ability to capture an instantaneous snapshot of tip flow structures at various scales [5,8,13,14]. Additionally, multiple averaging methods are also applied, among which the time-averaged method is frequently used. Several stochastic and deterministic characteristics of the TLV have been obtained through this method and statistics analysis [8,13,15]. Besides the simple averaging (ensemble averaging), some researchers employed a triple decomposition to extract the coherent wandering motion from the SPIV data [16]. Through this method, the TLV cores are collocated by linearly shifting the instantaneous core positions to the mean vortex core. This method was concluded to be superior to the simple averaging and recommended to analyze the characteristics of a concentrated vortex [17]. Oweis and Ceccio [14] used a vortex reconstruction method in which ideal Gaussian vortices were fitted to multiple vortices. As a result, they could reduce the effect of primary vortex wandering and reveal changes in flow variability that are masked by the vortex wandering process. However, the TLF is characterized by large-scale vortex and small-scale turbulence motion together. Many important information of unsteady flow behaviors associated with scale may be ignored using these averaging methods as a result.

The proper orthogonal decomposition (POD) method can decompose the examined flow field into orthogonal modes in space and identify the dominant mode based on energy rank [18,19]. It has been widely applied to both experimental and numerical data to identify some coherent structures, such as impeller flow [20], cylinder engine flow [21], wind turbine wake [22] and Jet and vortex actuator-induced flow [23]. Based on optimizing the mean square of the field variable, the POD method can extract flow structures of different scales. It is a promising key technique to analyze the flow structure, the kinematic and dynamic characteristics of the TLV in different length scales and energy levels. Unfortunately, there is a lack of use of the POD method to analyze the unsteady tip leakage flow. Recently, Li [24] introduced the POD method in turbine tip leakage flow analysis. Using the POD method, dominant flow modes governing the unsteady evolution in the tip region are successfully obtained. It is concluded that the POD method with advanced simulation approach can bring an interesting view of what can be done to better understand the leakage flow. Since the rapid development of numerical simulation and experimental techniques, the POD method is providing a new perspective about the tip flow physics.

However, spatial modes and energy distribution can in practice be obtained by different POD processing methods, such as using different decomposition dimensionalities [25,26] or different decomposition regions [27]. In the tip region, the streamwise velocity component is much stronger than the velocity components in the secondary flow direction. The TLV and wall boundary layer would introduce velocity deficits as well. Thus, the velocity and kinetic energy deviation would be quite diverse in different regions and different directions. This inevitably leads to different POD decomposition and reconstruction results using different decomposition regions and decomposition dimensionalities. Though the effects of many factors (POD algorithms [19], number of snapshots [25], analyzed flow variable [28]) have been assessed extensively, few publications have systematically studied the effects of decomposition regions and decomposition dimensionalities on the POD results. A clarification of the impacts of these two factors on POD decomposition and reconstruction results will help with a better and more efficient application of POD method in the TLF analysis.

In this paper, the effects of different decomposition regions and decomposition dimensionalities on POD decomposition and reconstruction of the tip flow are clarified. First, in the blade channel along a full chord length of a single compressor cascade, a SPIV measurement is addressed to obtain the original flow field for the POD decomposition and reconstruction. Combined with vortex core identification, development of the TLV along the chord is described. Then each plane is energetically decomposed and reconstructed using POD. In order to avoid the influence of main flow disturbance and extract more flow field structures associated with the TLV unsteady characteristics, POD decomposition regions are selected with caution. Using the identified vortex core as geometric center, the effects of different decomposition region with respect to the vortex core on POD decomposition are analyzed. Furthermore, the influences of different single velocity-components as well as their combination in POD decomposition are compared, so that to carry out a systematical studying on the effect of these two factors on the POD decomposition. Finally, combined with vortex statistical analysis, the impacts of decomposition dimensionalities in the reconstructed flow field are discussed. It is expected that findings presented in this paper will inspire other researchers who use POD to analyze the TLF.

This paper is organized as follows: firstly, the experimental apparatus and layout are introduced as follows; secondly, an overview of POD method and vortex core identification is given; in the following parts, the detailed comparison results are analyzed; the last section concludes this work.

2. Apparatus and Techniques

2.1. Experimental Facility and Test Conditions

The experiment is carried out in a low speed wind tunnel. The wind tunnel, as schematically shown in Figure 1, has a rectangular exit section of 250 mm by 120 mm (width × height). The test section of wind tunnel consists of a compressor cascade with seven blades, tip wall, hub wall and profile side walls as indicated in Figure 2. The parameters of the cascade and test conditions are listed in Table 1. During the experiment, the inlet velocity of the compressor cascade is kept at 30 m/s. At this velocity, the turbulence intensity in the mainstream is about 2.6%. We also measured the inlet boundary layer since it has a significant impact on the TLF roll-up procedure by interacting with the incoming main flow (Figure 3). The thickness of the boundary layer on the tip wall is about 3 mm.

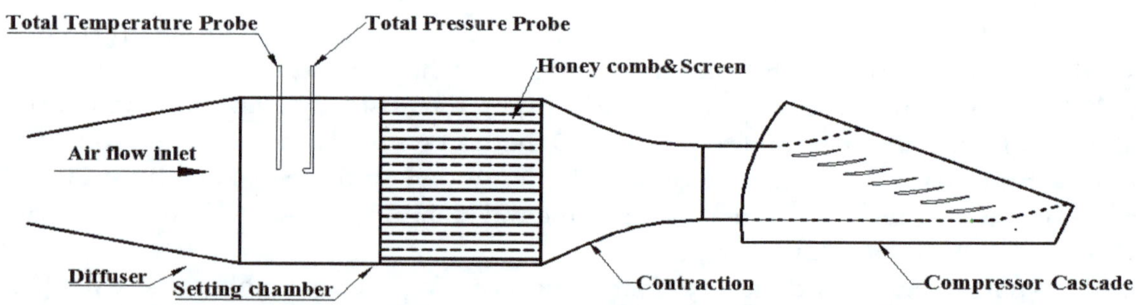

Figure 1. Schematic of the low-speed cascade wind tunnel.

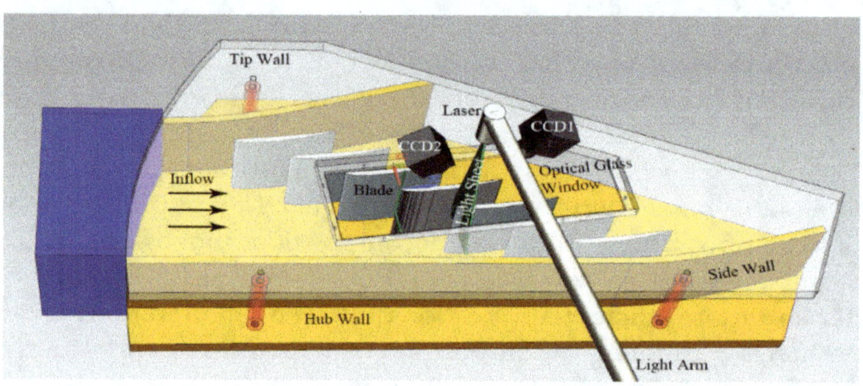

Figure 2. Schematic of compressor cascade and the SPIV measurement configuration.

Table 1. Cascade parameters and test condition.

Items	Details
Number of blades, N	7
Chord length, c	126.8 mm
Span, H	120 mm
Pitch, p	72 mm
Reynolds No., Re	2.81×10^5
Incidence angle	0 degree
Tip clearance/c	5%

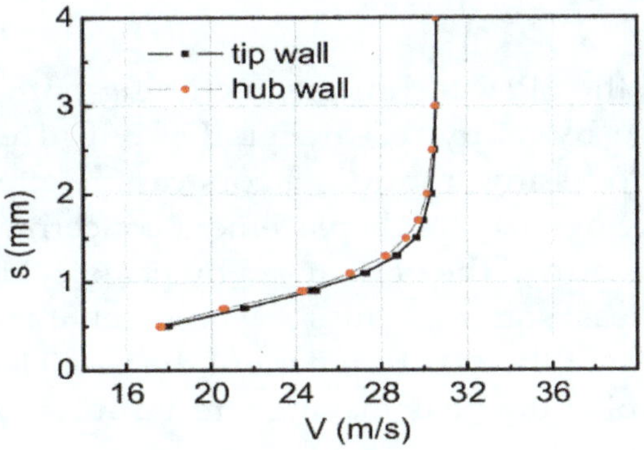

Figure 3. Velocity profiles at the hub and tip walls. Note that the y-axis is the vertical distance of test points to the wall.

2.2. *SPIV Technique*

An advanced commercial SPIV system is employed to measure the TLF field in the blade passage. As shown in Figure 2, the SPIV system is mainly composed of two 2048 × 2048 pixels, 12 bit frame-straddling based CCD cameras, a dual cavity Nd: YAG laser (200 mJ/pulse at a 15 Hz repetition rate) and a laser arm. The system is 2D-3C (two dimensional, three components) with two cameras located at different sides of the light sheet, which enables users to conduct the measurement of all three velocity components of a 3D flow.

The measurement cross sections are set to be perpendicular to the chordwise direction in the passage. 12 measurement locations along the chordwise direction (z-direction) are set, from L/c = 0 to L/c = 1.1 with an interval of 10% chord length, as shown in Figure 4. At each measurement location, at least 800 effective instantaneous images are recorded. The valid size of the cross section is about of 30 × 60 mm^2, which covers nearly half spanwise range of the whole passage. The spatial resolution of the calibrated image is about 0.039 mm/pixel.

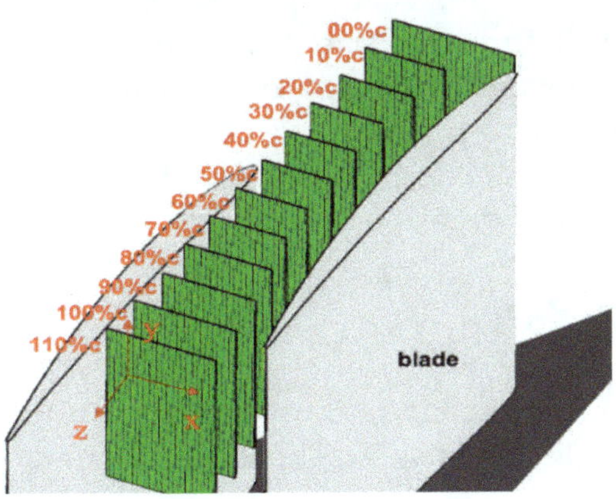

Figure 4. Layout of SPIV measurement cross-sections.

For the data processing, the inter-frame time is set to 10 μs based on the velocity of inlet flow and yields a maximum particle-image displacement of less than 8 pixels. A median-subtraction filter algorithm is applied to all the images to remove non-uniformities in the background light intensity. Subsequently, the particle-image is calculated by the software MicroVec V3 with the two-passes PIV interrogation algorithm. The dimensions of the interrogation region are 64 × 64 pixels in the first pass, and 32 × 32 pixels in the second pass. As a result, the spatial resolution of a single velocity vector is 1.26 mm × 1.26 mm. Once a vector field is calculated, vector validation algorithms must be used to eliminate spurious vectors. In the present work, the relative bias from the average vector is used as a post-processing criterion for eliminating questionable vectors, and the relative bias is set to be 50%.

2.3. Discussion of Uncertainties

The measurement error of the SPIV technique, including both random errors and bias errors, have been thoroughly discussed by many researchers [29–36]. Treated with caution during both the experiment and the data processing, random errors can be effectively eliminated. Bias errors depend crucially on the accuracy of particles displacement concerning the SPIV measurement in the compressor cascade. Among bias errors, the one caused by peak locking [30,32] should be dominated. In order to reduce this error, the least squares Gaussian sub-pixel fit for peak detection [5] is employed. Figure 5 illustrates the histogram of the displacement of the particles at the L/c = 1.0 measurement plane, which shows a well control of the peak locking effect during the SPIV experiments.

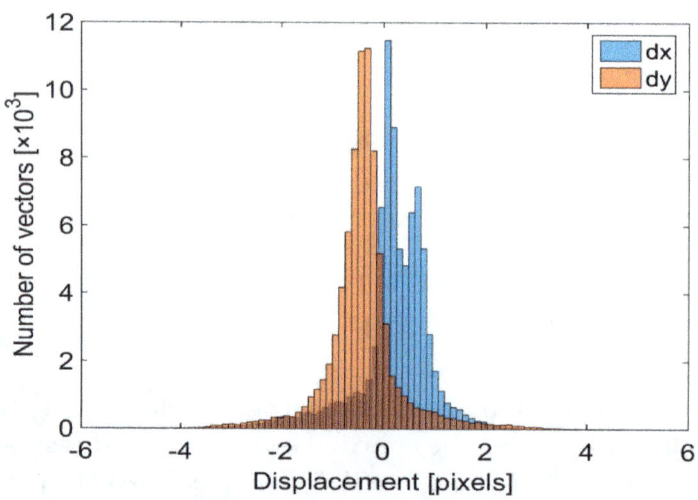

Figure 5. Histograms of the measured particle-image displacement at L/c = 1.0. (79,680 vectors in total, dx and dy are the particle-image displacements in the x and y directions).

According to the analyses of Westerweel [34,36] and Raffle [30], the displacement errors of the in-plane component in the SPIV measurement is about 0.05 pixel. Taking other uncertainties (background noise, acceleration, velocity gradient) into consideration, the displacement errors of the in-plane component are estimated as 0.1 pixels.

Based on the theoretical analysis by Zang and Prasad [37], as shown in Equation (1) the error ratio between the out-of-plane and in-plane components depends on the camera included half angle:

$$\frac{U_z}{U_x} = \frac{U_z}{U_y} = \tan\alpha \quad (1)$$

where α is the camera included half angle; U_x, U_y and U_z are the uncertainties of the x, y and z velocity components. In this paper, this angle is set to 45°. Therefore, the displacement error of the out-of-plane component is 0.1 pixels as well.

The relative uncertainty of instantaneous velocity in the u-component can be estimated from the equation:

$$U_{B,u_{ij}^k} = \frac{\Delta/K}{u_{ij}^k} \quad (2)$$

where K is converge factor related to measured velocity distribution characteristics. It is set to 2.576 for a 99% confidence interval. u_{ij}^k is the corresponding instantaneous velocity at the i, j grid node in the kth snapshot. Δ is error bound calculated by Equation (3). Likewise, the relative uncertainties of instantaneous velocity field in the v- and w-components can also be deduced:

$$\Delta = \frac{d}{m \cdot dt} \quad (3)$$

where d is estimated displacement error, m is image magnification and dt is the inter-frame time.

The measurement uncertainties of the time-averaged velocity, contain both the statistical factor (type A uncertainty) and the factors unrelated with the statistical analysis (type B uncertainty).

Type A relative uncertainty of the time-averaged velocity can be estimated as [38]:

$$U_{A,\overline{u_{ij}}} = \frac{\sqrt{\frac{1}{N(N-1)}\sum_{k=1}^{N}\left(u_{ij}^{k} - \overline{u_{ij}}\right)^2}}{\overline{u_{ij}}} \tag{4}$$

where u_{ij} is the time-averaged velocity in the u-component at the i, j grid node, N is the number of snapshots acquired, u_{ij}^{k} is the corresponding instantaneous velocity in the kth snapshot. Similarly, type A relative uncertainty of the time-averaged mean velocity in the other two components can also be deduced.

In the compressor cascade, the main factor contributing to the type B uncertainty of the time-averaged velocity($U_{B,\overline{u_{ij}}}$) is the time-averaged of the instantaneous velocity deviation shown in Equation (2).

Thus, the combined standard relative uncertainty of the time-averaged velocity in the u-component can be given as:

$$U_{\overline{u_{ij}}} = \sqrt{\left(U_{A,\overline{u_{ij}}}\right)^2 + \left(U_{B,\overline{u_{ij}}}\right)^2} \tag{5}$$

The uncertainties of the time-averaged velocity can be calculated by the propagation of error formula:

$$U_{velocity} = \sqrt{\frac{\left(\overline{U}\cdot\sigma_U\right)^2 + \left(\overline{V}\cdot\sigma_V\right)^2 + \left(\overline{W}\cdot\sigma_W\right)^2}{\left(\overline{U}^2 + \overline{V}^2 + \overline{W}^2\right)^2}} \tag{6}$$

where $\sigma_U, \sigma_V, \sigma_W$ are the uncertainties of the time-averaged velocity in u-, v- and w-component.

The uncertainties of instantaneous velocity and time-averaged velocity are shown in Figure 6. Since the introduction of type A uncertainty, the uncertainty in the time-averaged velocity field is elevated. It can be seen that the uncertainty of time-averaged velocity is about 0.5–2% in the mainstream region, and it can achieve 6% near the tip wall. Moreover, it should be pointed out that in the tip region the accuracy of the in-plane components is about 0.5–3% and for the out-of-plane component is about 5–10%, and these results are not shown here for simplicity. Though the uncertainty of the out-of-plane component is higher, the statistical results are worthful for qualitative analyses of the unsteady characteristics in the compressor cascade, and constructive conclusions can be drawn from these results.

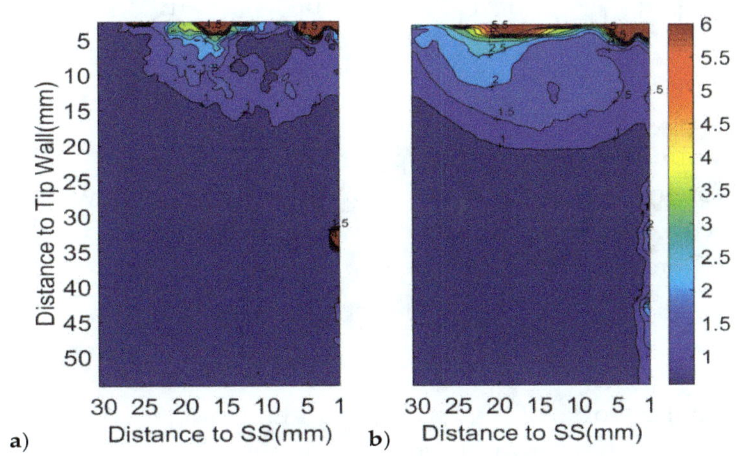

Figure 6. Distribution of measurement relative uncertainties in percentage terms at L/c = 1.0, (**a**) the instantaneous velocity, (**b**) the time-averaged velocity. The velocity is the resultant velocity of components in the x-, y- and z- direction.

2.4. PODMethod

In this paper, a set of instantaneous SPIV measured velocity field $V^{(k)}$ (snapshots) with total number of K is decomposed into a linear combination of K spatial orthonormal basis function (POD modes, φ_m) and the corresponding coefficients $C_m^{(k)}$:

$$V^{(k)} = \sum_{m=1}^{K} C_m^{(k)} \varphi_m \tag{7}$$

with the constraints that the following function is minimized [39]:

$$\sum_{k=1}^{K} \left\| V^{(k)} - \sum_{m=1}^{M} C_m^{(k)} \varphi_m \right\|^2 \rightarrow \min \tag{8}$$

Considering the instantaneous snapshots $V^{(k)}$ made available from the SPIV measurement:

$$V^{(k)} = \left(u_{i,j}, v_{i,j}, w_{i,j}\right)^{(k)} \tag{9}$$

where k is the snapshot index, u, v, w are the velocity components in x, y and z direction respectively, i, j are the index of spatial position coordinates in the velocity distributions for a collocated setup, as illustrated in Figure 7.

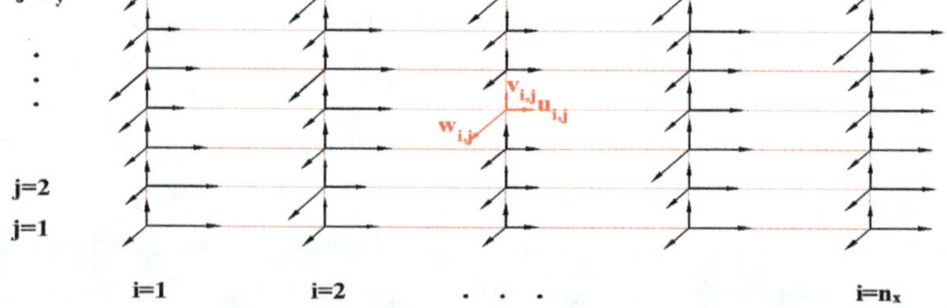

Figure 7. Example of discrete SPIV data on a uniform collocated grid.

Then the velocity component of every single snapshot $V^{(k)}$ is collated as following:

$$U = \begin{bmatrix} U^{(1)} \\ U^{(2)} \\ \vdots \\ U^{(K)} \end{bmatrix} = \begin{bmatrix} u^{(1)}_{i=1,j=1} & u^{(1)}_{i=1,j=2} & \cdots & u^{(1)}_{i=1,j=n_y} & u^{(1)}_{i=2,j=1} & \cdots & u^{(1)}_{i=n_x,j=n_y} \\ u^{(2)}_{i=1,j=1} & u^{(2)}_{i=1,j=2} & \cdots & u^{(2)}_{i=1,j=n_y} & u^{(2)}_{i=2,j=1} & \cdots & u^{(2)}_{i=n_x,j=n_y} \\ \cdots & \cdots & \cdots & \cdots & \cdots & \cdots & \cdots \\ u^{(K)}_{i=1,j=1} & u^{(K)}_{i=1,j=2} & \cdots & u^{(K)}_{i=1,j=n_y} & u^{(K)}_{i=2,j=1} & \cdots & u^{(K)}_{i=n_x,j=n_y} \end{bmatrix} \tag{10}$$

where $n_x \times n_y$ is the total number of the spatial positions.

The other two component-wise velocities are processed the same way and matrix V and matrix W are obtained, respectively. Then, simultaneously using both u and v velocity components and neglecting w, most researchers tend to define the spatial correlation matrix as follows [21,40]:

$$C = \frac{1}{K}\left(UU^T + VV^T\right) \tag{11}$$

However, according to the works of Arányi [25] and Lengani [26], the results would be noticeably different if a multi-dimensional field is analyzed component-wise instead of globally. In this paper, we make a deeper quantitatively comparison of the decomposition results using different analysis dimensionalities to clarify whether decomposition dimensionalities impact the whole modes or just specific modes. Moreover, as the aim of most POD analyses is to reconstruct approximately the vector field, we also compare the reconstruction results based on two reconstruction criteria. Therefore, the spatial correlation matrices (C) are defined according to different analysis dimensionalities (AD) as shown in Table 2.

Table 2. Five spatial correlation matrices for different decomposition dimensionalities.

Processing Methods	Global Analysis		Component-Wise Analysis		
AD	UVW-simultaneously	UV-simultaneously	U	V	W
C	$\frac{1}{K}(UU^T + VV^T + WW^T)$	$\frac{1}{K}(UU^T + VV^T)$	$\frac{1}{K}UU^T$	$\frac{1}{K}VV^T$	$\frac{1}{K}WW^T$

The minimization problem described in Equation (8) is realized by solving the eigenvalue problem of correlation matrix C:

$$C\beta_m = \lambda_m \beta_m \tag{12}$$

where Eigenvalues λ_m are arranged in descending order according to the magnitude of eigenvalues, the basis function φ_m are obtained by projecting the snapshots onto the eigenvector β_m.

The POD coefficients $c_m^{(k)}$ are obtained by projecting the snapshots onto the basis function φ_m and $\frac{1}{2}\left(c_m^{(k)}\right)^2$ represents the energy contribution by the mth mode to the kth snapshot.

The K $\times$ K coefficient matrix $c_m^{(k)}$:

$$c_m^{(k)} = \begin{bmatrix} c_1^{(1)} & c_2^{(1)} & \cdots & c_K^{(1)} \\ c_1^{(2)} & c_2^{(2)} & \cdots & c_K^{(2)} \\ \vdots & \vdots & \vdots & \vdots \\ c_K^{(K)} & c_K^{(K)} & \cdots & c_K^{(K)} \end{bmatrix} \tag{13}$$

The whole kinetic energy from all the snapshots captured by the mth mode is:

$$KE = \frac{1}{2}\sum_{m=1}^{K}\left(\sum_{k=1}^{K}\left(C_m^{(k)}\right)^2\right) \tag{14}$$

The fraction of the energy captured by the mth mode is:

$$ke_m = \frac{1}{2}\sum_{k=1}^{K}\left(C_m^{(k)}\right)^2 / KE \tag{15}$$

By setting the amplitude of higher-order modes to zero, a low order flow field contains the most energetic modes is reconstructed:

$$V^{(k)} = \sum_{m=1}^{M} c_m^{(k)} \varphi_m \tag{16}$$

where M is the reconstruction order less than K.

2.5. TLV Vortex Core Identification

The TLV is a concentrated vortex before splitting into several small vortices or breakdown. In order to identify the TLV core correctly, some criteria, such as Q criterion, λ_2 criterion and Δ criterion, have been developed. The ability of the three criteria is similar to each other and none of them can be applied without error to all situations. Since the reduced $-\lambda_2$ criterion has been employed in compressor TLV many times and can separates the TLV from the background high shear layer, here the reduced $-\lambda_2$

criterion is employed [41]. This reduced criterion is effective to identify the TLV core and is given by Equation (17):

$$\lambda_2 = \left(\frac{\partial V_x}{\partial x}\right)^2 + \left(\frac{\partial V_y}{\partial y}\right)^2 + 2 \times \frac{\partial V_y}{\partial x}\frac{\partial V_x}{\partial y} \tag{17}$$

where x and y are coordinate axes in the measured cross section as demonstrated in Figure 4, and V_x and V_y are the velocity components in the corresponding directions. In this paper, zero threshold of λ_2 is selected to identify the border of the TLV core.

After the identification, the circulation and radius of the TLV core are quantitatively analyzed. The circulation, which represents the strength of the vortex, is obtained by calculating the line integral of velocity along the border. As shown in Figure 10, a red ellipse is used to fit the TLV core and the center and the radius of the TLV core are calculated from this fitting ellipse. The five-pointed star represents the center of the TLV core identified. The x-axis in the figure is the distance of the TLV core center to the blade suction surface. The y-axis is the distance of the center to the cascade tip wall. Detailed information about the identification program may be found in our previous study [42].

3. Results

3.1. Tip Flow Characteristics

Figure 8 lays out the time-averaged SPIV measurement results of the normalized streamwise velocity and λ_2. The trajectory of the TLV cores can be seen clearly from the λ_2 distribution. The concentrated TLV can be found after L/c = 0.5, which is an ambiguous result for the deficiency of SPIV spatial resolution in the z-direction.

As the TLV propagates downstream, the TLV core radius and the absolute value of vortex circulation expands rapidly (Quantitatively shown in Figure 9a). Figure 9a also quantitatively illustrates that the TLV moves slowly away from the blade tip suction surface in the x-direction (Schematically shown in Figure 9b from a bird's eye view). In the y-direction, the TLV moves slowly away from the tip wall. The variance tendencies of the positions of the TLV in the time-averaged flow fields conform well with the SPIV results obtained in a laboratory-scale compressor at the design condition [5,6,42].

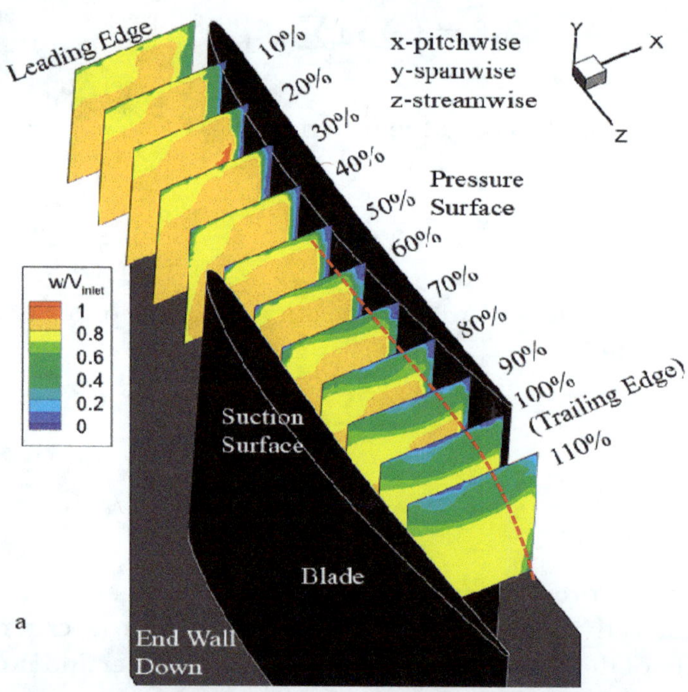

Figure 8. *Cont.*

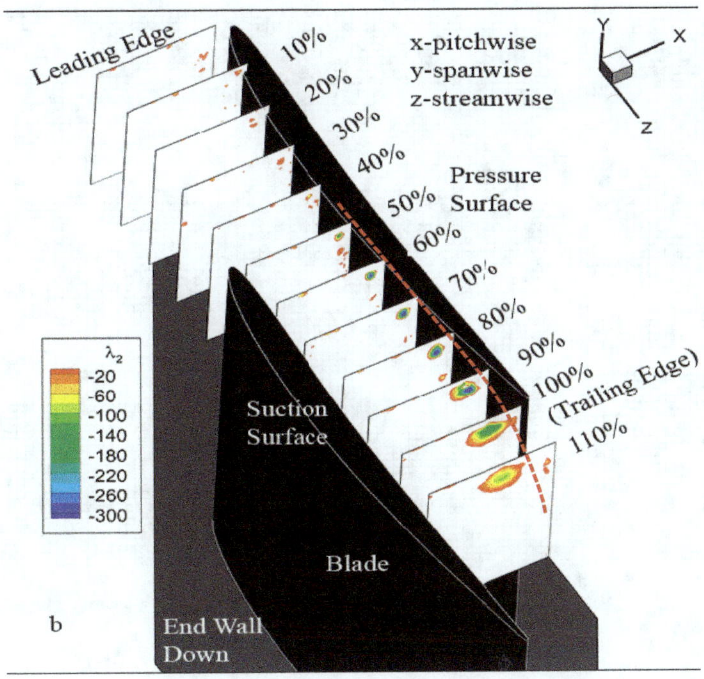

Figure 8. Combined maps of time-averaged SPIV measurement results in the blade passage. (**a**) Normalized Streamwise Velocity; (**b**) λ_2. The red dash lines on both maps sketch the trajectory of the TLV cores.

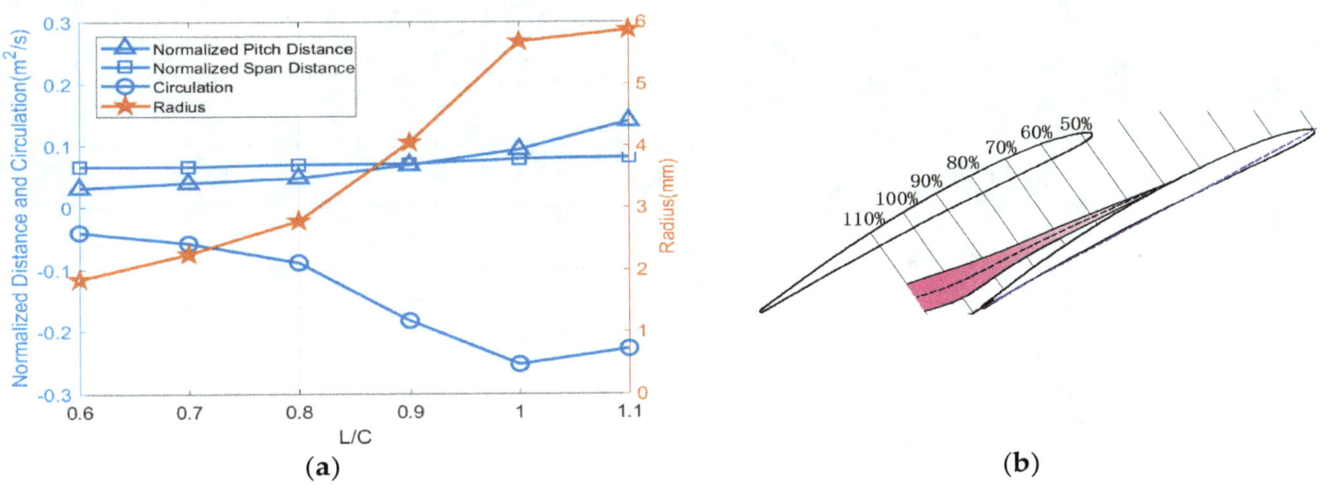

Figure 9. (**a**) Statistics of identified TLV time-averaged parameters; (**b**) Schematic of TLV core trajectory. Normalized pitch distance is the distance of the TLV core center to the blade suction surface in the x-direction normalized by the span, normalized span distance is the distance of the TLV core center to the tip wall in the y-direction normalized by the span.

In Figure 10 the instantaneous vortex cores from L/c = 0.7 to L/c = 1.1 in all the SPIV snapshots are illustrated by single points. The vortex core center, border and border fitting ellipse of the time-averaged vortex are also shown in Figure 10 for comparison. At L/c = 0.7 most instantaneous TLV are near the core center of the time-averaged TLV. The position and circulation distribution of the TLV is in a relatively concentrated form. With the development of the TLV core to the trailing edge, the pitchwise and spanwise distribution range of the vortex expands. From L/c = 0.9 to L/c = 1.1, when the TLV is shedding form the blade suction side, wandering of the TLV cores becomes significantly stronger and the TLV motions cover a much wider area.

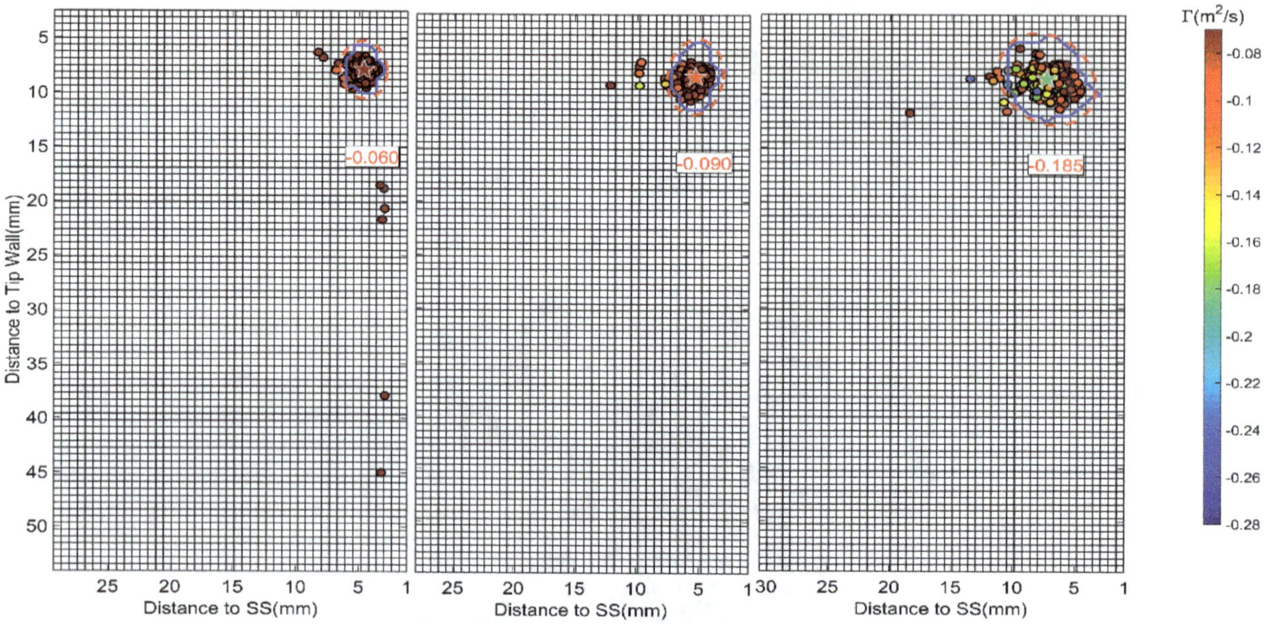

(a) L/c = 0.7 (b) L/c = 0.8 (c) L/c = 0.9

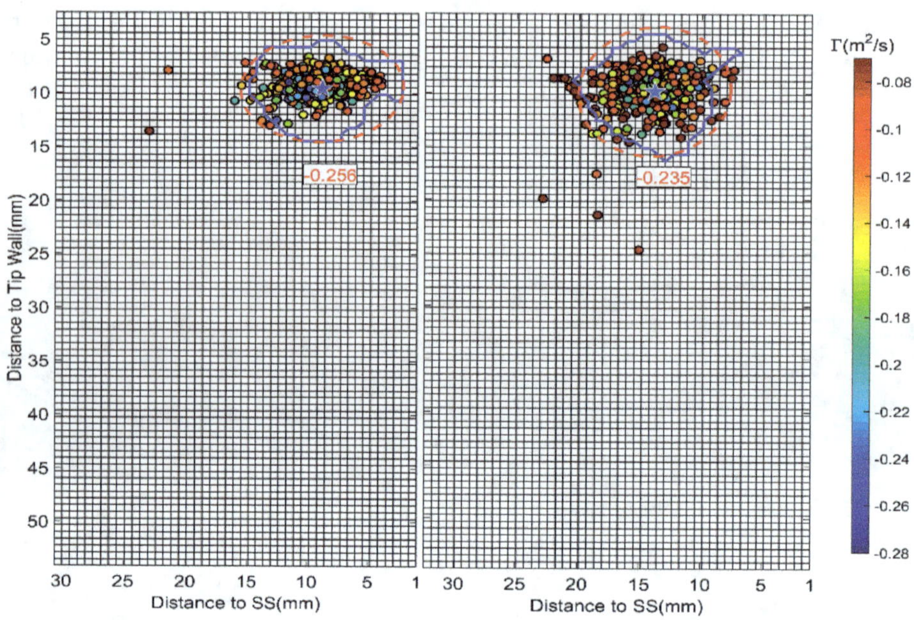

(d) L/c = 1.0 (e) L/c = 1.1

Figure 10. Statistical distribution of the locations, area and circulation of identified TLV cores. The color of the points shows the vortex circulation.

Figure 11 shows the standard deviation of normalized SPIV measured velocity. The flow velocity in every single snapshot is normalized by the inlet average velocity. The standard deviation is calculated as Equation (18) to show the unsteadiness of the TLV:

$$\sigma = \sqrt{\frac{1}{K-1}\sum_{i=1}^{K}\left(\frac{Vs_i - \overline{V}s}{V_{inlet}}\right)^2} \tag{18}$$

where Vs_i is the instantaneous velocity in every single snapshot, $\overline{V}s$ is the time-averaged velocity.

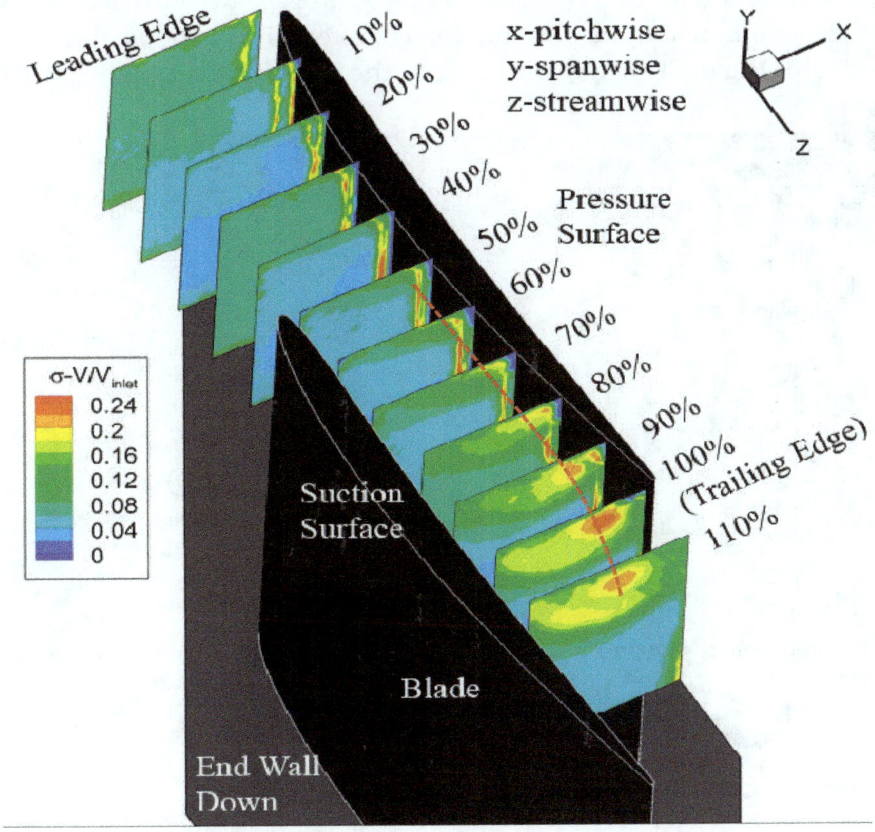

Figure 11. Standard deviation of normalized SPIV measured velocity in the blade passage. The velocity is the resultant velocity of components in the x-, y- and z- direction.

As the vortex propagates downstream, the standard deviation of the normalized SPIV measured velocity near the TLV core increases. In this paper, the unsteadiness of the TLV is mainly originated from the TLV wandering as the vortex maintains a concentrate vortex and does not show any evidence of vortex splitting or breakdown.

The aforementioned experimental results demonstrated the main characteristics of the TLV. In the tip region, streamwise velocity component is much higher than velocity components in secondary flow directions. The TLV and wall boundary layers would introduce velocity deficits as well. Thus, in the compressor cascade, we should keep in mind that the decomposition regions and decomposition dimensionalities would have significant impacts while use the POD method to further analyze the TLV wandering characteristics. Therefore, the two factors should be investigated with caution to get reasonable POD results.

3.2. Effect of Decomposition Region on the Energy Fraction of Mode 1

Since the TLV is a concentrated vortex, to focus on the unsteady characteristics caused by the TLV wandering, circular decomposition regions are deliberately chosen. The decomposition regions are concentric circles with the time-averaged TLV core center as their common center. The radius R_{dr} of the decomposition regions are multiples of the time-averaged TLV core radius R_v. The borders of decomposition regions are shown as dotted line in Figure 12.

Figure 12 b shows the instantaneous TLV cores center, the time-averaged TLV core center and the time-averaged TLV core border. Figure 12c,d shows two instantaneous velocity snapshots chosen randomly. The minimum decomposition region ($R_{dr}/R_v = 2$) covers most instantaneous vortex cores centers. This selection of decomposition regions would keep most information of the TLV wandering and turbulence fluctuation around the TLV while removes the interference of the main flow. While the decomposition region converge to the TLV core center with R_{dr}/R_v reducing from 7 to 2, the effect of main flow on the POD results would be eliminated. Therefore, it is beneficial to extract more flow field structure associated with the unsteady wandering of the TLV and turbulence fluctuation around it.

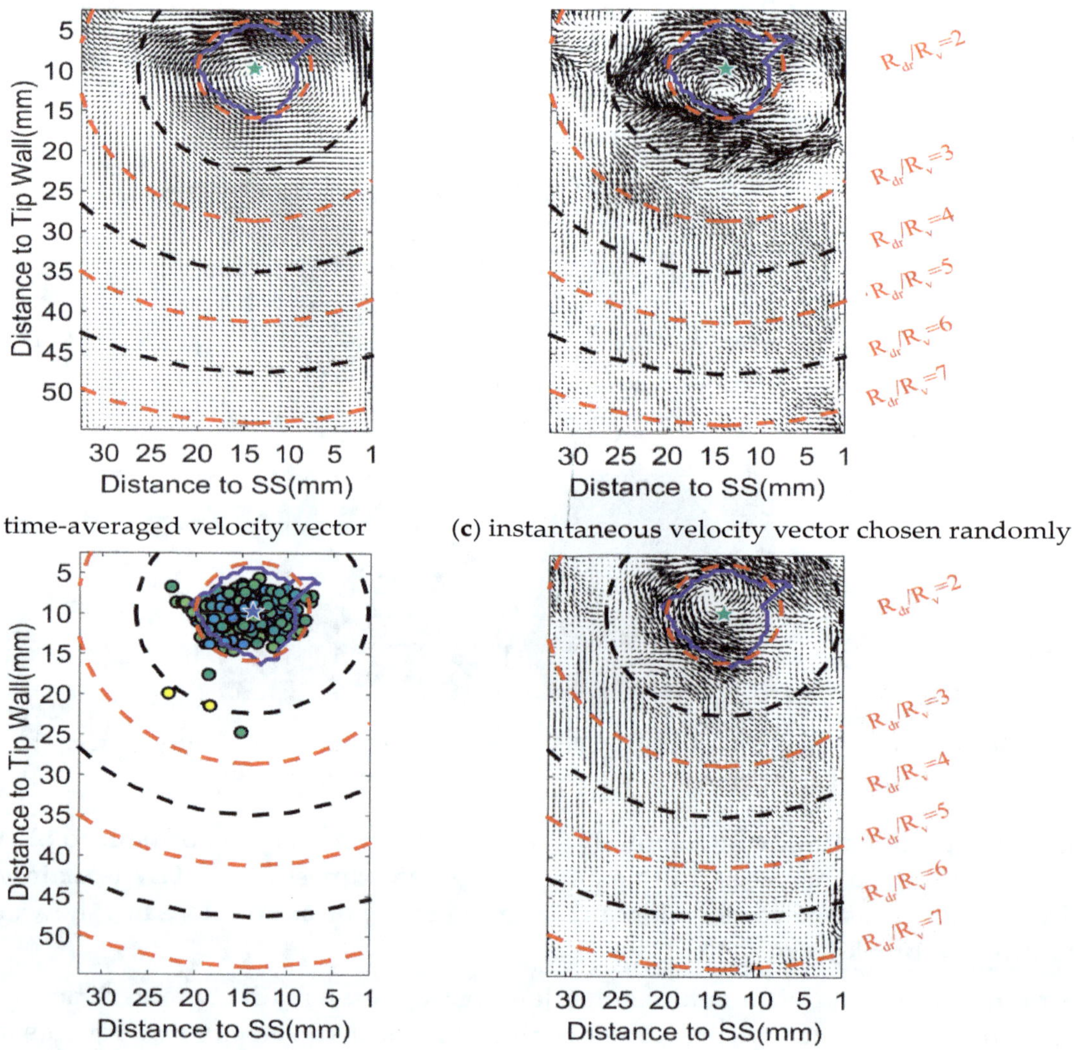

(**a**) time-averaged velocity vector (**b**) instantaneous vortex cores scatter (**c**) instantaneous velocity vector chosen randomly (**d**) instantaneous velocity vector chosen randomly

Figure 12. Decomposition regions at L/c = 1.1.

Before the POD decomposition is performed, many researchers have addressed the difference in POD results between subtracting the time-averaged flow field and not subtracting [25,43,44]. As will be confirmed in the following, while the average is not removed, mode 1 is representative of the time-averaged velocity field. Since the POD modes are all orthogonal, the higher-order modes would be well estimates of the fluctuation. In this paper, the POD analysis is conducted without subtracting the average in the snapshots.

The energy rank is the first information provided by the POD eigenvalues λ_m [45]. Among the energy rank, the energy fraction of mode 1 is the most important. The total energy fraction of all other higher-order modes can be obtained by Equation (19):

$$\frac{\sum_{m=2}^{K} \lambda_m}{\sum_{m=1}^{K} \lambda_m} = 1 - \frac{\lambda_1}{\sum_{m=1}^{K} \lambda_m} \tag{19}$$

Therefore, the energy fraction of mode 1 reflects the relative contribution of fluctuation to the whole energy of the flow field. The energy fraction of mode 1 can represent the level to which extent the instantaneous fields approach the time-averaged result.

From L/c = 0.6 to L/c = 1.1, as the fluid propagates downstream, the energy fraction of mode 1 decreases as shown in Figure 13. It should be noted that the results are from POD decompostion using three velocity vector components U, V and W simultaneously. The POD highlights that the flow posses a higher energy in fluctuation as the vortex propagates downstream, consistent with the results of the standard deviation previously observed in Figure 11. Figure 13 also shows that the deviation of mode 1 energy fraction between L/c = 0.6 and L/c = 1.1 becomes evident as the decomposition region converge to the TLV core. Indeed, the TLV experiences the process of vortex shedding from the blade suction side and wanders much stronger at the downstream near the trailing edge.

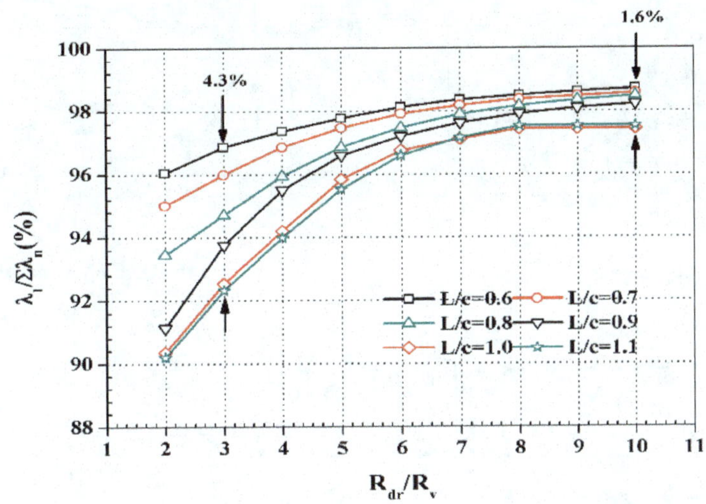

Figure 13. Mode 1 energy fraction at different decomposition region sizes and different chord position.

At the same chordwise position, as the decomposition region size decreases, the energy fraction of mode 1 decreases as well. The energy fraction of fine scale fluid structure, normally contained in the higher-order modes, is weakened by the introduction of main flow. It proves that the decomposition region chosen is beneficial to focus on the unsteady characteristics analysis of the TLV and turbulence fluctuation around it.

To further investigate the effect of decomposition regions on energy distribution, four more different analysis dimensionalities, such as UV global analysis, U component-wise analysis, V component-wise analysis and W component-wise analysis, are used.

3.3. *Effect of Decompostion Dimensionalities on the Energy Fraction of Mode 1*

Figure 14 demonstrates the energy fraction of mode 1 at different decomposition region sizes and different chord positions using four different analysis dimensionalities.

Note that the decomposition region sizes have marginal impact on the results associated with U component-wise analysis, V component-wise analysis and UV global analysis. However, the impact become considerable on the results associated with W component-wise and UVW global analysis while both of the results have the same variability tendency. This corresponding to the fact that the dominant flow direction is z-direction (streamwise) in this cascade while x-direction and y-direction are the secondary flow directions where velocity amplitude is relatively low.

As the flow travels downstream, the energy fraction of mode 1 decreases dramatically. From L/c = 0.6 to L/c = 1.1, the deviation of the energy fraction of mode 1 is nearly 10% in the U component-wise analysis, and higher than 20% in V component-wise analysis and UV global analysis as shown in Figure 14. While the W component is included, the deviation reduced to less than 5%. Therefore, the POD without the W component highlights the development of unsteady characteristics along the streamwise direction. The figure also shows that, using POD without the W component, mode 1 energy fraction at L/c = 0.9 deviate from that at L/c = 1.1 by twice as much as the deviation from L/c = 0.6 to L/c = 0.9. Near the trailing edge, the TLV is shedding from the blade suction surface, the instantaneous fields diverge considerably from the time-averaged result. UV global analysis captures this physics as the energy transferred from mode 1 expands significantly from L/c = 0.9 to L/c = 1.1.

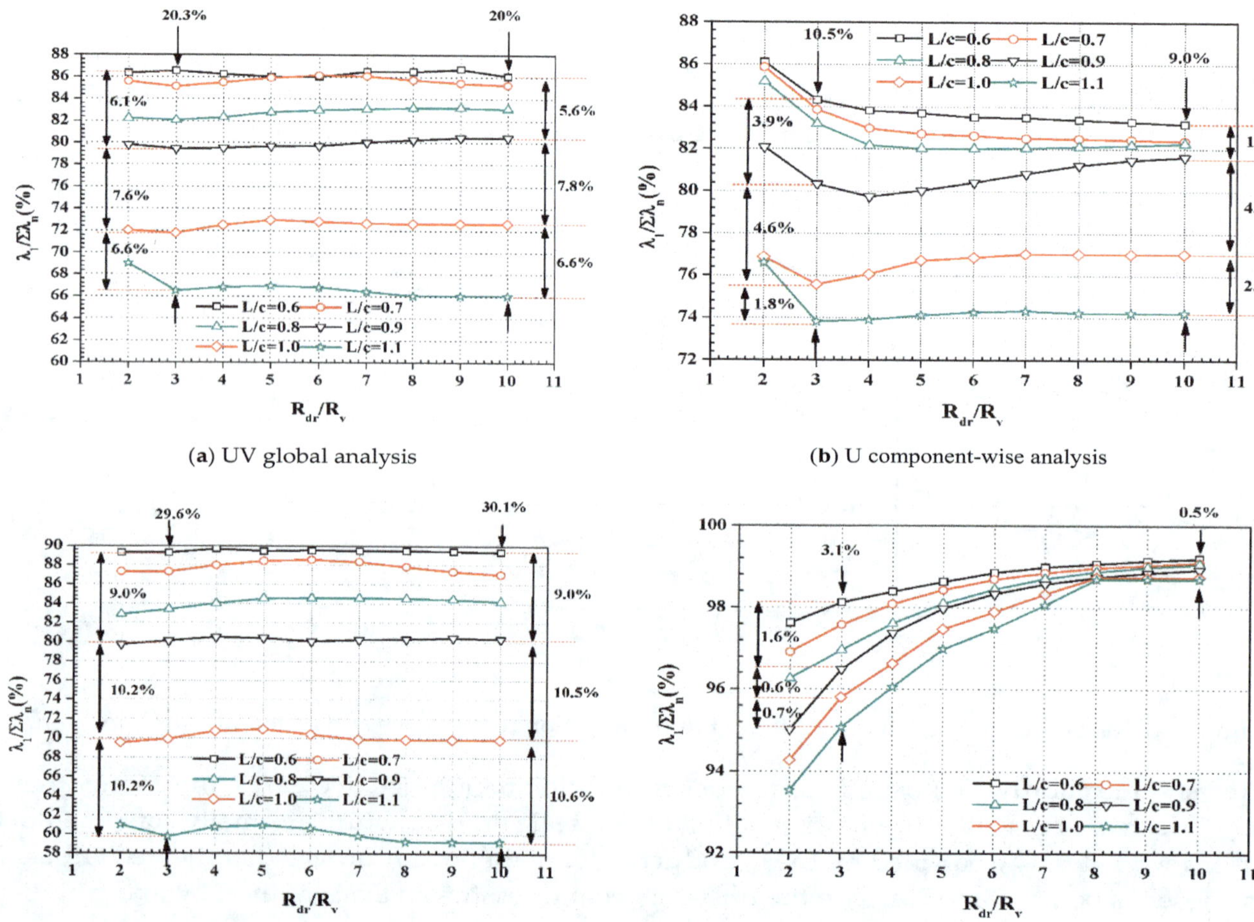

Figure 14. Mode 1 energy fraction at different decomposition region sizes and different chord position.

Therefore, the effect of decomposition region on the mode 1 energy fraction mainly impacts the streamwise velocity component. When the decomposition region converges to the TLV core, the fraction of streamwise velocity deficit region increases, the impact of main flow with uniform streamwise velocity reduces, the unsteadiness caused by vortex wandering and turbulence fluctuation becomes evident.

Thus, the decomposition region size in the following work is all $R_{dr}/R_V = 3$, in order to extract more flow field structure associated with vortex wandering and turbulence fluctuation and to avoid the effect of main flow disturbance on POD. The results of five decomposition dimensionalities in the selected decomposition region are shown in Figure 15. It is interesting to note that the introduction of W velocity component also changes the absolute amplitude of mode 1 energy fraction. The amplitudes of mode 1 energy fraction are nearly equal in the results of UVW global analysis and W component-wise analysis and they are much higher than the result in the other three analyses.

Based on the aforementioned discussions, we can draw a conclusion that there is a marked difference among mode 1 energy fractions using different decomposition dimensionalities. If the aim of the analysis is to approximately reconstruct the flow field, the effects of decomposition dimensionalities on the spatial structures of POD modes and energy distribution should be urgent to clarify.

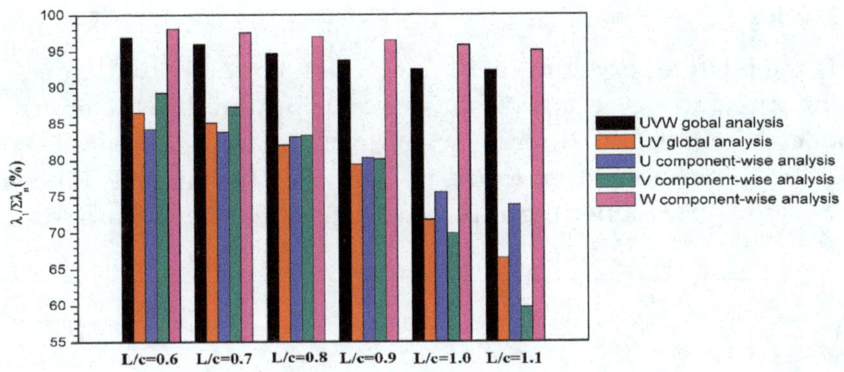

Figure 15. Mode 1 energy fraction at different retained dimensionalities and different chord position when the decomposition region size is Rdr/Rv = 3.

3.4. *Effect of Decomposition Dimensionalities on POD Modes*

3.4.1. Mode 1 and Its Relationship with the Time-Averaged Flow Field

To quantify the degree of similarity about two velocity vector fields, the relevance index R_p is used. R_p is obtained by projecting one velocity vector field onto another velocity vector field [43]:

$$R_p = \frac{(M1, M2)}{||M1|| \cdot ||M2||} \quad (20)$$

where the denominator denotes the product of the L2 norm and the numerator is the inner product of two velocity fields. Rp = 0 means the two velocity field are orthogonal, Rp = 1 if the direction of two velocity vector field are identical, and Rp = −1 if the direction of two velocity vector field are exactly opposite.

The flow patterns of mode 1 from all three decomposition dimensionalities are excellent estimates of the time-averaged velocity field (shown in Figure 16). Moreover, the relevance indices R_p of all three mode 1 and the time-averaged velocity field are nearly 1, which also conforms this conclusion quantitatively. The results from other chordwise positions show the same correlation (not shown).

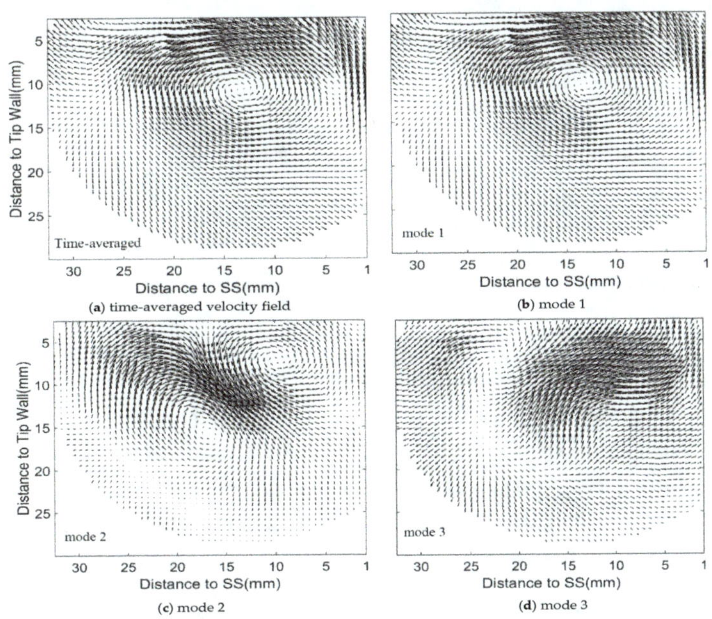

Figure 16. Time-averaged velocity vector field and spatial structures of mode 1, 2 and 3 in UV global analysis at L/c = 1.1.For simplicity, the spatial structures of UVW global analysis and UV component analysis have not shown.

3.4.2. Higher-order Modes

Figure 17 reflects the relative contribution of the modes to the fluctuation energy of the flow field. The modes from W component-wise analysis show a distribution that the energy of the 3rd mode is similar to the 2nd mode. A completely different scenario characterized the POD energy distribution of other four decomposition dimensionalities, especially UVW global analysis. The energy content of the 2nd POD mode appears sensibly higher than that characterizing the 3rd mode.

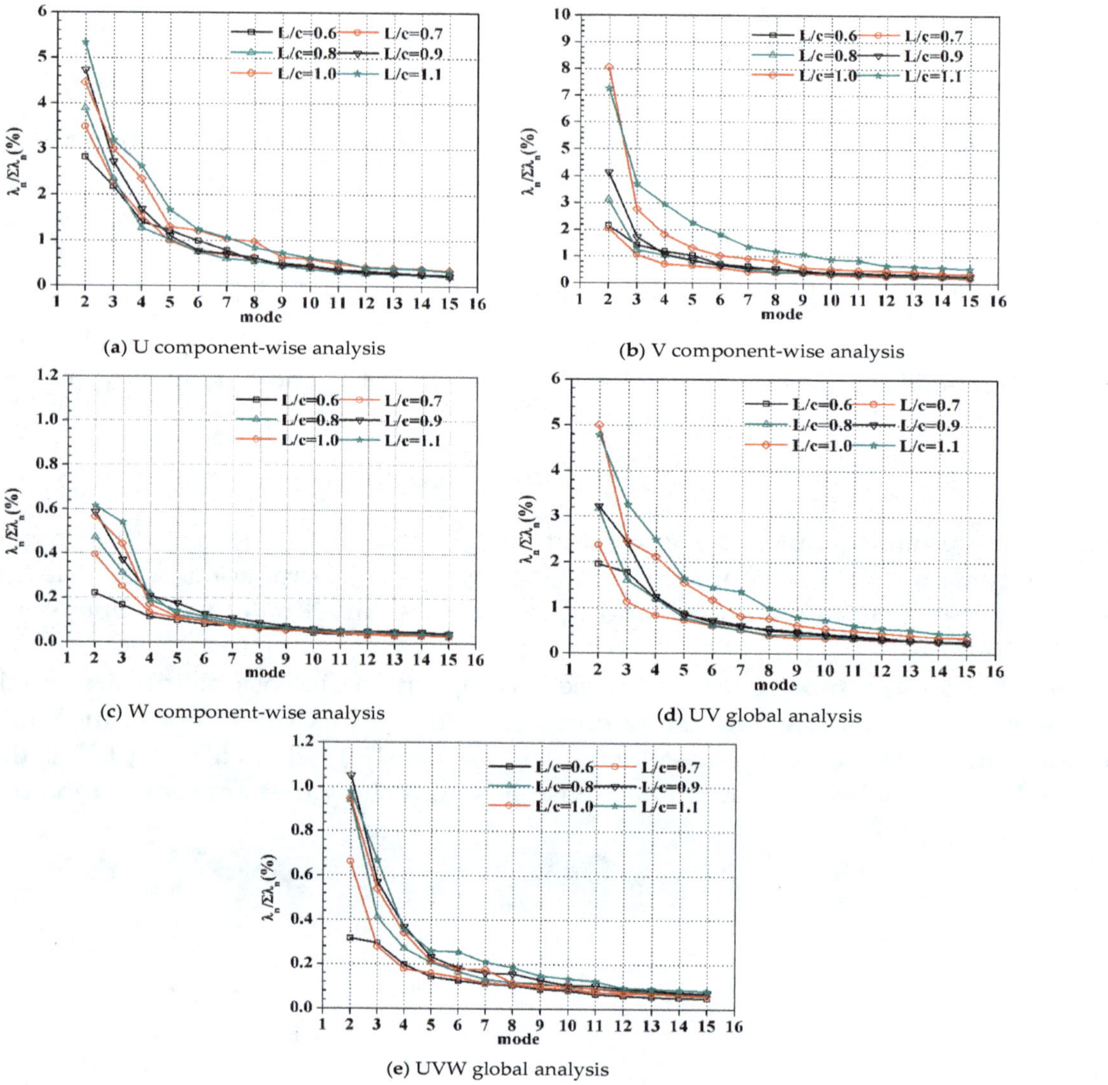

(a) U component-wise analysis
(b) V component-wise analysis
(c) W component-wise analysis
(d) UV global analysis
(e) UVW global analysis

Figure 17. Relative energy of modes higher than 2 from all five different decomposition dimensionalities.

Though the impact of W velocity component in mode 1 energy fraction is dominant as mentioned above, the addition of W velocity component in POD decomposition does not change the relative energy distribution of high-order modes. The 2nd and 3rd POD spatial modes from different decomposition dimensionalities have been checked. No matter which decomposition dimensionality used, two large well-defined vortex-like structures can be observed in the 2nd POD spatial modes (shown in Figure 16). However, the 3rd POD spatial modes are totally different from each other.

To further study the impact of decomposition dimensionalities in the high-order modes, the velocity correlation coefficients account for correlation of each mode from different decomposition dimensionalities are shown in Figure 18. The correlation coefficients of the 2nd POD mode are nearly 1, which verified that the 2nd POD mode is unaffected. The correlation coefficients of modes order higher than 3 are weak. In all, the three different decomposition dimensionalities lead to almost the same 2nd mode while significant difference can be observed in modes which order are higher than 2.

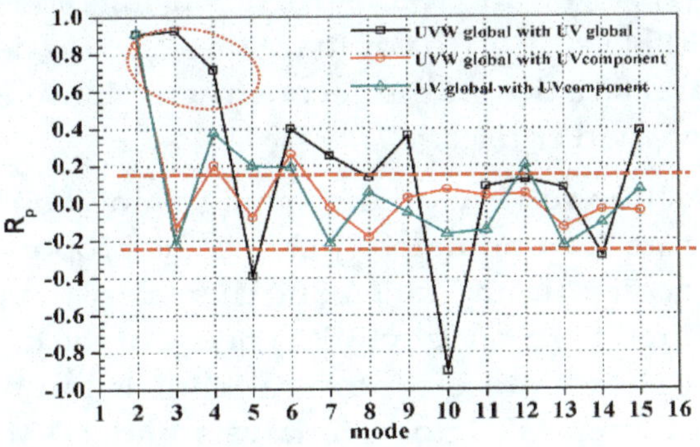

Figure 18. The velocity correlation of high-order modes from different decomposition dimensionalities at L/c = 1.1.

3.5. Effect of Decomposition Dimensionalities on POD Reconstruction

In the tip region, the flow is inherently unsteady. A single POD mode usually cannot bear a significant amount of energy and is not able to completely describe the original flow structures. An ensemble of POD modes is demanded to describe a particular dynamics. To study the dynamics of the TLV wandering, it is of great interest to analyze the vortex distribution characteristics with random noise being removed. In this section, to study the effect of different decomposition dimensionalities on the reconstructed flow field, the modes required to represent 95% energy of the original flow field and vortex distribution characteristics in the reconstructed flow field are compared.

3.5.1. Modes Required to Represent 95% Energy of the Original Flow Field

At the same chordwise position, modes required to represent 95% energy of the original flow field from UVW global analysis and W component-wise analysis are much less than that from other three processing methods. It also proves that the addition of W velocity component in the decomposition changes the energy fraction of every single mode and results in more energy gathering in the low-order modes.

From L/c = 0.6 to L/c = 1.1, an interesting feature in the POD reconstruction is that the modes required are dramatically increased using the U and V velocity components while maintain almost the same using the W velocity component (Figure 19). As the vortex propagates downstream, more and more modes are required to adequately represent the flow field in the x-direction and y-direction. Combining with the discussion in Section 3.3, UV global analysis is better for capturing the kinetic physics of the TLV because it is expected that more high-order modes will be present as the flow travels downstream due to vortex wandering and vortex shedding.

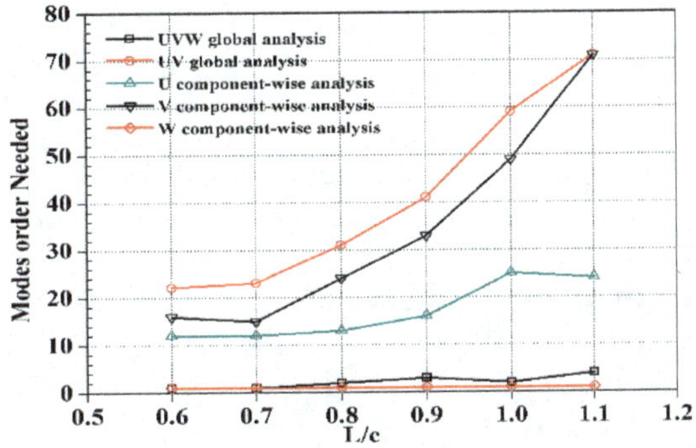

Figure 19. The modes required to represent 95% energy of the original flow field using five different POD processing methods.

3.5.2. Effect of Decomposition Dimensionalities on the Vortex Distribution Characteristics

Two criteria of reconstructing flow field using POD modes is studied in this paper. Reconstruction using a finite number of modes [28,46,47] and reconstruction using modes that represent specific energy portion of the original flow field [45,48,49]. To compare the effect of decomposition dimensionalities on the vortex distribution characteristics in the reconstructed flow field, every single snapshot is reconstructed according to above two criteria.

Results from the reconstruction relying on the same retained energy in the reconstructed flow field are shown in Figure 20. Figure 21 quantitatively shows the proportion of the TLV core parameters. They together demonstrate the geometric and kinematic characteristics of the TLV in the reconstructed flow field. Since the modes required to reconstruct 95% energy of the original flow field is much less in UVW global analysis, the TLV shows more concentrated distribution characteristics in Figure 20. In the reconstructed flow field from UV global analysis and UV component analysis, the TLV cores are decentralized and have the similar distribution characteristics with the original flow field. The histogram of vortex distribution from UVW global analysis is symmetric that the mean value provides a good estimate for the center of the data. The histograms from UV global analysis and UV component analysis have the same shape, and are for a distribution that are skewed left. The reconstruction relying on the retained energy in the reconstructed flow field is sensitive to different decomposition dimensionalities.

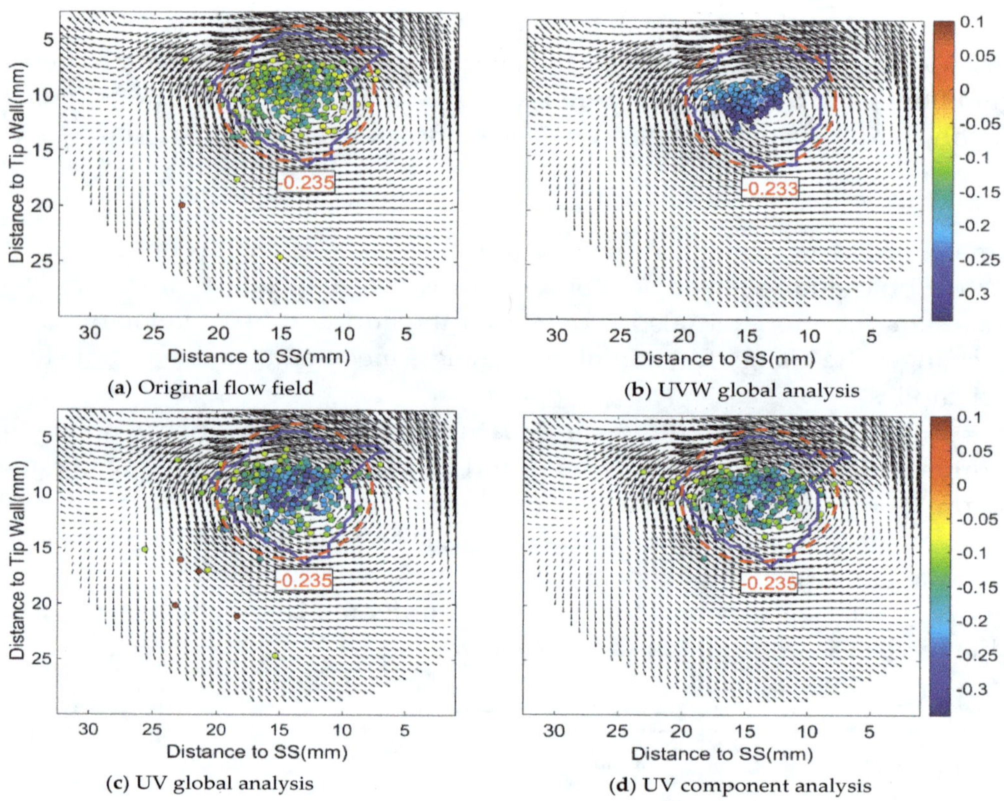

Figure 20. Vortex distribution in the reconstructed flow field that represents 95% energy of the original flow field at L/c = 1.1. The color of the points shows the vortex circulation.

Figures 22 and 23 show the corresponding reconstruction results using the first ten modes. The vortex distributions in the reconstructed flow field using different decomposition dimensionalities show the same concentrated characteristics. The histograms of these three analyses have the same shape as well. Though the addition of W velocity component does change the spatial structures of high-order modes, it does not change the dynamic results of reconstruction using the same number of POD modes.

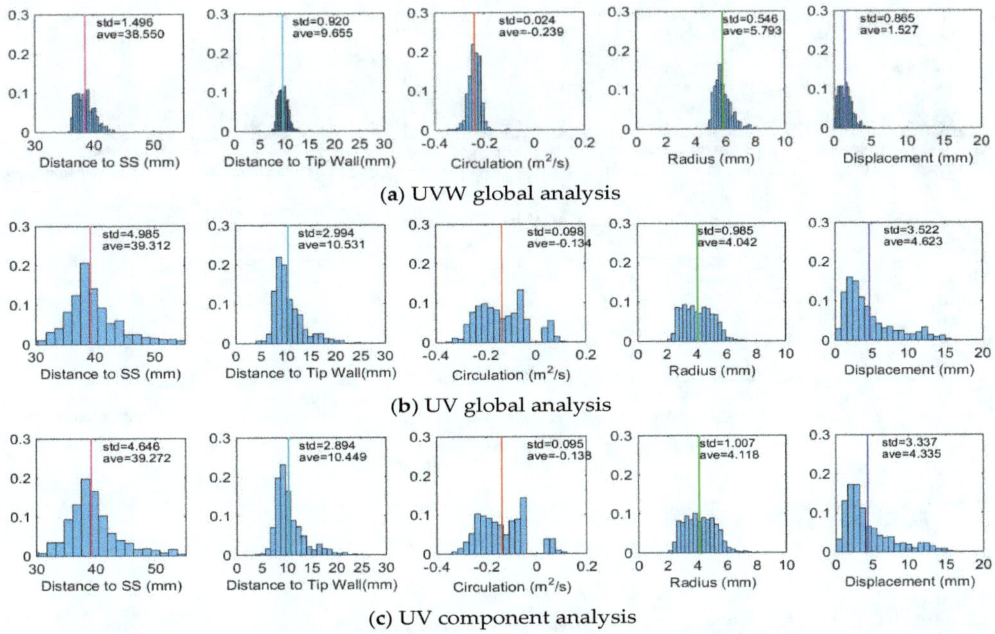

Figure 21. Statistics of instantaneous reconstructed flow field that represents 95% energy fraction of the original flow field at L/c = 1.1. Note that the parameters from left side to the right are the distance of the TLV core center to the suction side, the distance of the TLV core center to the cascade tip wall, the TLV vortex circulation, the TLV vortex core radius and the distance of instantaneous TLV core center to the time-averaged TLV core center respectively. 'std' and 'ave' in the figure show the standard deviation and time-averaged value respectively.

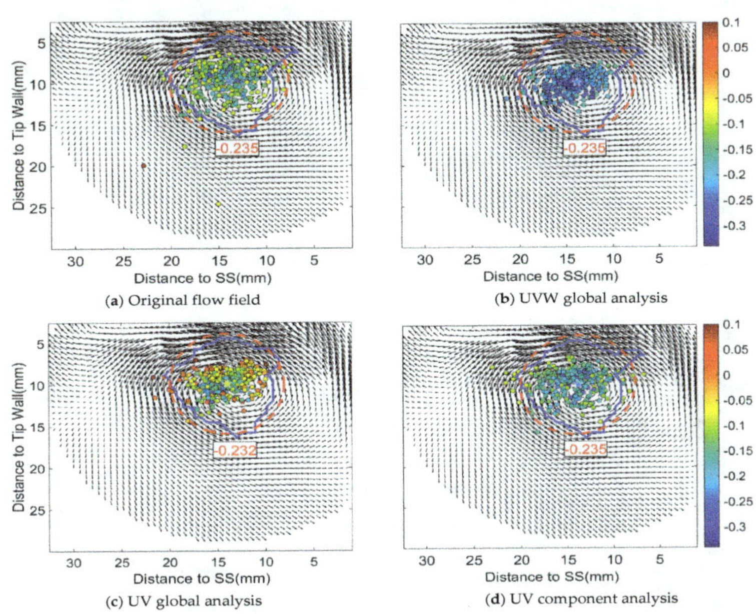

Figure 22. Vortex distribution from reconstructed flow field using first 10 modes at L/c = 1.1.

To further validate the conclusions drawn above, we checked the reconstruction results using different number of modes and different retained energy portions of the original flow field. The standard deviation of the TLV radius and displacement are chosen to quantitatively analysis the difference between the two reconstruction criteria. The results in Figure 24 confirm that reconstructions using the same modes number can obtain nearly the same vortex distribution characteristics while reconstructions relying on the same energy fraction would lead to totally different vortex distribution characteristics.

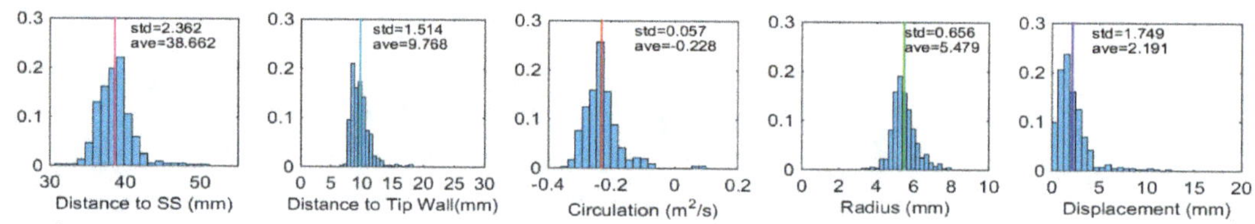

(a) UVW global analysis

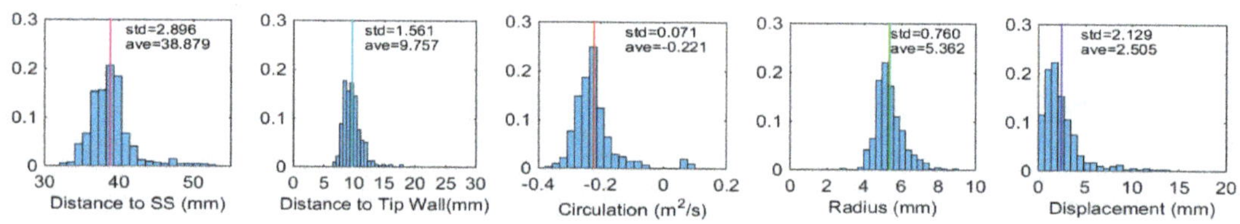

(b) UV global analysis

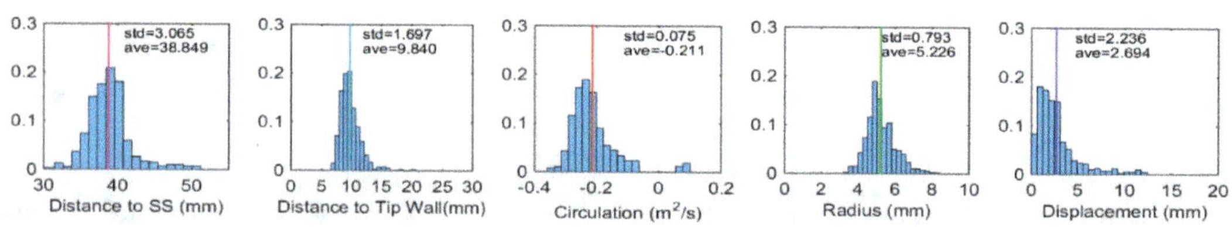

(c) UV component analysis

Figure 23. Statistics of instantaneous reconstructed flow field using first 10 modes at L/c = 1.1.

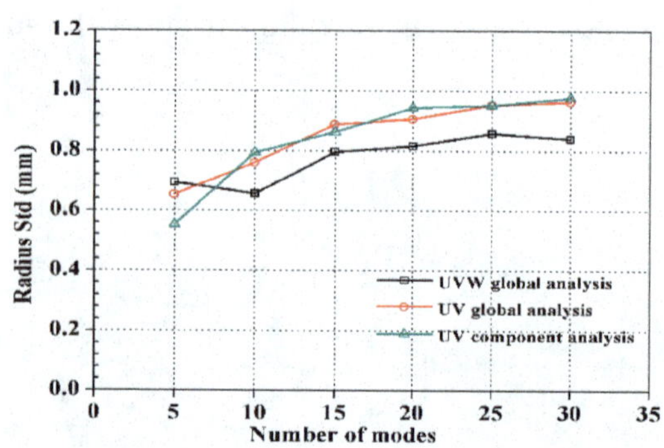

(a) TLV radius Std using the same modes

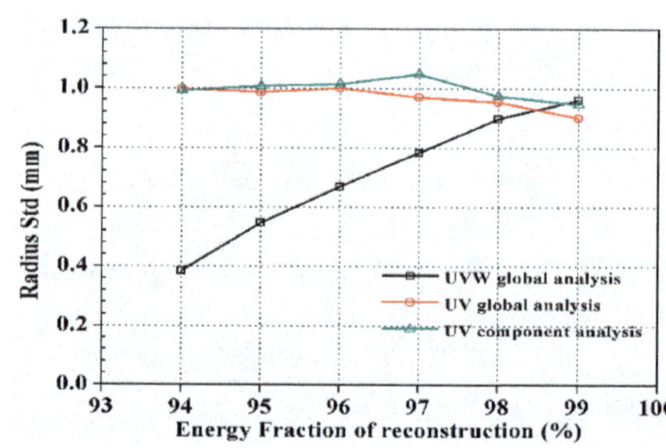

(b) TLV radius Std using the same energy fraction

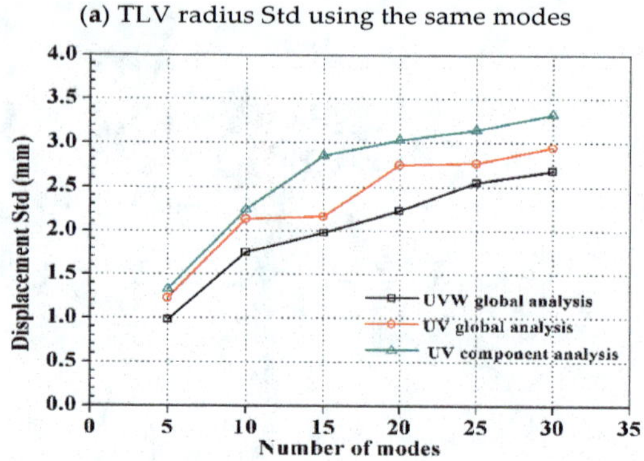

(c) TLV displacement Std using the same modes

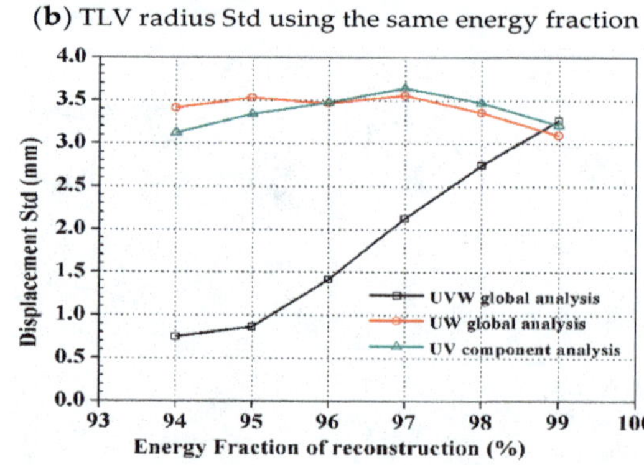

(d) TLV displacement Std using the same energy fraction

Figure 24. Comparison of reconstruction results based on the two reconstruction criteria at L/c = 1.1. The displacement is the distance of instantaneous TLV core center to the time-averaged TLV core center.

4. Conclusions

In this paper, based on the original tip flow field in the compressor cascade obtained by a SPIV measurement, the effects of different decomposition regions and decomposition dimensionalities on POD decomposition and reconstruction have been clarified. Several conclusions can be made as listed below:

(1) From L/c = 0.9 to L/c = 1.1, when the TLV is shedding from the blade suction side, wandering of the TLV core becomes significantly stronger and the TLV motions cover a much wider area. As the TLV propagates downstream, the energy fraction of mode 1 decreases.

(2) The decomposition region sizes have marginal impact on POD decomposition of secondary flow velocity. The impact become considerable on the results associated with W component-wise velocity.

(3) For POD decomposition, using different dimensionalities, energy distributions and modes higher than 2 would be totally different. The dominant POD modes, such as the 1st and 2nd POD modes, stay unaffected. Using POD without the W component, mode 1 energy fraction at L/c = 0.9 deviate from that at L/c = 1.1 by twice as much as the deviation from L/c = 0.6 to L/c = 0.9.

(4) The reconstruction relying on the retained energy in the reconstructed flow field is sensitive to different decomposition dimensionalities. The reconstruction using a finite number of POD modes is unaffected.

(5) UV global analysis is better for capturing the kinetic physics of the TLV. As the flow travels downstream, more high-order modes with higher energy fraction will be present in UV global analysis, which is corresponding to the kinetic physics of the TLV that the unsteadiness of vortex wandering and vortex shedding becomes stronger.

Author Contributions: Conceptualization, L.S. methodology, L.S.; software, L.S.; validation, L.S.; formal analysis, L.S.; investigation, L.S.; resources, L.S.; data curation, L.S.; writing—original draft preparation, L.S.; writing—review and editing, H.M., L.W.; supervision, H.M.; funding acquisition, H.M.

Nomenclature

i,j	The index of the grid points in the velocity distributions
k	Snapshot index
L	SPIV measurement cross section chordwise position
c	Chord length
H	Span
p	Pitch length
Re	Reynolds number of the inlet flow
s	The vertical distance of test points to the wall.
Γ	Circulation
ave	Time-averaged value
Std	Standard deviation
KE	Whole kinetic energy
ke	Energy of specific mode
R_p	Relevance index
M1	Velocity vector fields
R_{dr}	Radius of the decomposition region
R_v	Radius of the TLV core
λ_m	POD eigenvalues
U_x, U_y, U_z	The uncertainties of the x, y and z velocity components
Δ	Error bound
K	Converge factor
σ_U, σ_V, σ_W	The uncertainties of the time-averaged velocity in u-, v- and w-component
φ_m	POD mode
AD	Analysis dimensionalities
C	Correlation matrices

References

1. Denton, J.D. *Loss Mechanisms in Turbomachines*; ASME 1993 International Gas Turbine and Aeroengine Congress and Exposition, 1993; American Society of Mechanical Engineers: New York, NY, USA, 1993.
2. Storer, J.; Cumpsty, N. An approximate analysis and prediction method for tip clearance loss in axial compressors. *J. Turbomach.* **1994**, *116*, 648–656. [CrossRef]
3. Adamczyk, J.; Celestina, M.; Greitzer, E. *The Role of Tip Clearance in High-Speed Fan Stall*; ASME 1991 International Gas Turbine and Aeroengine Congress and Exposition, 1991; American Society of Mechanical Engineers: New York, NY, USA, 1991.
4. Straka, W.A.; Farrell, K.J. The effect of spatial wandering on experimental laser velocimeter measurements of the end-wall vortices in an axial-flow pump. *Exp. Fluids* **1992**, *13*, 163–170. [CrossRef]
5. Liu, B.; Yu, X.; Liu, H.; Jiang, H.; Yuan, H.; Xu, Y. Application of SPIV in turbomachinery. *Exp. Fluids* **2006**, *40*, 621–642. [CrossRef]
6. Xianjun, Y.; Baojie, L.; Haokang, J. Characteristics of the tip leakage vortex in a low-speed axial compressor. *AIAA J.* **2007**, *45*, 870–878. [CrossRef]
7. Furukawa, M.; Inoue, M.; Saiki, K.; Yamada, K. The Role of Tip Leakage Vortex Breakdown in Compressor Rotor Aerodynamics. *J. Turbomach.* **1999**, *121*, 469–480. [CrossRef]
8. Brandstetter, C.; Jüngst, M.; Schiffer, H.-P. Measurements of Radial Vortices, Spill Forward, and Vortex Breakdown in a Transonic Compressor. *J. Turbomach.* **2018**, *140*, 061004. [CrossRef]
9. Day, I. *The Fundamentals of Stall and Surge*; Lecture series-van Kareman Institute for fluid dynamics; Von Kareman Institute for fluid dynamics: Rhode-Saint-Genèse, Belgium, 1996; pp. B1–B27.
10. Inoue, M.; Kuroumaru, M.; Tanino, T.; Yoshida, S.; Furukawa, M. Comparative studies on short and long length-scale stall cell propagating in an axial compressor rotor. *J. Turbomach.* **2001**, *123*, 24–30. [CrossRef]
11. Mailach, R.; Lehmann, I.; Vogeler, K. Rotating instabilities in an axial compressor originating from the fluctuating blade tip vortex. *J. Turbomach.* **2001**, *123*, 453–460. [CrossRef]
12. Zierke, W.; Farrell, K.; Straka, W. Measurements of the tip clearance flow for a high-reynolds-number axial-flow rotor. *J. Turbomach.* **1995**, *117*, 522–532. [CrossRef]
13. Liu, B.; An, G.; Yu, X.; Zhang, Z. Quantitative Evaluation of the Unsteady Behaviors of the Tip Leakage Vortex in a Subsonic Axial Compressor Rotor. *Exp. Therm. Fluid Sci.* **2016**, *79*, 154–167. [CrossRef]
14. Oweis, G.F.; Ceccio, S.L. Instantaneous and time-averaged flow fields of multiple vortices in the tip region of a ducted propulsor. *Exp. Fluids* **2005**, *38*, 615–636. [CrossRef]
15. Wu, H.; Tan, D.; Miorini, R.L.; Katz, J. Three-dimensional flow structures and associated turbulence in the tip region of a waterjet pump rotor blade. *Exp. Fluids* **2011**, *51*, 1721–1737. [CrossRef]
16. Reynolds, W.; Hussain, A. The mechanics of an organized wave in turbulent shear flow. Part 3. Theoretical models and comparisons with experiments. *J. Fluid Mech.* **1972**, *54*, 263–288. [CrossRef]
17. Van der Wall, B.G.; Richard, H. Analysis methodology for 3C-PIV data of rotary wing vortices. *Exp. Fluids* **2006**, *40*, 798–812. [CrossRef]
18. Cordier, L. *Proper Orthogonal Decomposition: An Overview*; Lecture series 2002-04, 2003-03 and 2008-01 on post-processing of experimental and numerical data; Von Karman Institute for Fluid Dynamics: Rhode-Saint-Genèse, Belgium, 2008.
19. Taira, K.; Brunton, S.L.; Dawson, S.T.M.; Rowley, C.W.; Colonius, T.; McKeon, B.J.; Schmidt, O.T.; Gordeyev, S.; Theofilis, V.; Ukeiley, L.S. Modal Analysis of Fluid Flows: An Overview. *AIAA J.* **2017**, *55*, 4013–4041. [CrossRef]
20. Guo, R.; Li, R.; Zhang, R. Reconstruction and Prediction of Flow Field Fluctuation Intensity and Flow-Induced Noise in Impeller Domain of Jet Centrifugal Pump Using Gappy POD Method. *Energies* **2018**, *12*, 111. [CrossRef]
21. El-Adawy, M.; Heikal, M.; A Aziz, A.; Adam, I.; Ismael, M.; Babiker, M.; Baharom, M.; Firmansyah; Abidin, E. On the Application of Proper Orthogonal Decomposition (POD) for In-Cylinder Flow Analysis. *Energies* **2018**, *11*, 2261. [CrossRef]
22. Bastine, D.; Witha, B.; Wächter, M.; Peinke, J. Towards a Simplified DynamicWake Model Using POD Analysis. *Energies* **2015**, *8*, 895. [CrossRef]

23. Cadirci, S.; Gunes, H. *Proper Orthogonal Decomposition Analysis of JaVA Flows for Cross-Flow Conditions*; ASME 2011 International Mechanical Engineering Congress and Exposition, 2011; American Society of Mechanical Engineers: New York, NY, USA, 2011; pp. 871–878.
24. Li, H.; Su, X.; Yuan, X. Entropy Analysis of the Flat Tip Leakage Flow with Delayed Detached Eddy Simulation. *Entropy* **2018**, *21*, 21. [CrossRef]
25. Arányi, P.; Janiga, G.; Zähringer, K.; Thévenin, D. Analysis of different POD methods for PIV-measurements in complex unsteady flows. *Int. J. Heat Fluid Flow* **2013**, *43*, 204–211. [CrossRef]
26. Lengani, D.; Simoni, D.; Ubaldi, M.; Zunino, P. POD analysis of the unsteady behavior of a laminar separation bubble. *Exp. Therm. Fluid Sci.* **2014**, *58*, 70–79. [CrossRef]
27. Maurel1, S.; Borée, J.; Lumley, J.L. Extended Proper Orthogonal Decomposition: Application to Jet/Vortex Interaction. *Flow Turbul. Combust.* **2001**, *67*, 125–136. [CrossRef]
28. Kostas, J.; Soria, J.; Chong, M.S. A comparison between snapshot POD analysis of PIV velocity and vorticity data. *Exp. Fluids* **2005**, *38*, 146–160. [CrossRef]
29. Perret, L.; Braud, P.; Fourment, C.; David, L.; Delville, J. 3-Component acceleration field measurement by dual-time stereoscopic particle image velocimetry. *Exp. Fluids* **2006**, *40*, 813–824. [CrossRef]
30. Raffel, M.; Willert, C.E.; Scarano, F.; Kähler, C.J.; Wereley, S.T.; Kompenhans, J. *Particle Image Velocimetry: A Practical Guide*; Springer: Berlin, Germnay, 2018.
31. Keane, R.D.; Adrian, R.J. Optimization of particle image velocimeters. I. Double pulsed systems. *Meas. Sci. Technol.* **1990**, *1*, 1202. [CrossRef]
32. Huang, H.; Dabiri, D.; Gharib, M. On errors of digital particle image velocimetry. *Meas. Sci. Technol.* **1997**, *8*, 1427. [CrossRef]
33. Hain, R.; Kähler, C. Fundamentals of multiframe particle image velocimetry (PIV). *Exp. Fluids* **2007**, *42*, 575–587. [CrossRef]
34. Westerweel, J. Theoretical analysis of the measurement precision in particle image velocimetry. *Exp. Fluids* **2000**, *29*, S003–S012. [CrossRef]
35. Boillot, A.; Prasad, A. Optimization procedure for pulse separation in cross-correlation PIV. *Exp. Fluids* **1996**, *21*, 87–93. [CrossRef]
36. Westerweel, J. Fundamentals of digital particle image velocimetry. *Meas. Sci. Technol.* **1997**, *8*, 1379. [CrossRef]
37. Zang, W.; Prasad, A.K. Performance evaluation of a Scheimpflug stereocamera for particle image velocimetry. *Appl. Opt.* **1997**, *36*, 8738–8744. [CrossRef] [PubMed]
38. Cantrell, C.D. *Modern Mathematical Methods for Physicists and Engineers*; Cambridge University Press: Cambridge, UK, 2000.
39. Lumley, J. *The Structure of Inhomogeneous Turbulence*; Atmospheric Turbulence and Wave Propagation, Nauka: Moscow, Russia, 1967; pp. 166–178.
40. El-Adawy, M.; Heikal, M.R.; A Aziz, A.R.; Siddiqui, M.I.; Munir, S. Characterization of the Inlet Port Flow under Steady-State Conditions Using PIV and POD. *Energies* **2017**, *10*, 1950. [CrossRef]
41. Jeong, J.; Hussain, F. On the identification of a vortex. *J. Fluid Mech.* **2006**, *285*, 69–94. [CrossRef]
42. Ma, H.; Wei, W.; Ottavy, X. Experimental investigation of flow field in a laboratory-scale compressor. *Chin. J. Aeronaut.* **2017**, *30*, 31–46. [CrossRef]
43. Hao, C.; David, L.R.; Volker, S. On the use and interpretation of proper orthogonal decomposition of in-cylinder engine flows. *Meas. Sci. Technol.* **2012**, *23*, 085302.
44. Liu, K.; Haworth, D.C. *Development and Assessment of POD for Analysis of Turbulent Flow in Piston Engines*; SAE International: Warrendale, PA, USA, 2011.
45. Lengani, D.; Simoni, D.; Pichler, R.; Sandberg, R.; Michelassi, V.; Bertini, F. *On the Identification and Decomposition of the Unsteady Losses in a Turbine Cascade*; ASME Turbo Expo 2018: Turbomachinery Technical Conference and Exposition, 2018; American Society of Mechanical Engineers: New York, NY, USA, 2018; p. V02AT45A013.
46. Hossain, M.S.; Bergstrom, D.J.; Chen, X.B. Visualisation and analysis of large-scale vortex structures in three-dimensional turbulent lid-driven cavity flow. *J. Turbul.* **2015**, *16*, 901–924. [CrossRef]
47. Hellström, L.H.O.; Ganapathisubramani, B.; Smits, A.J. The evolution of large-scale motions in turbulent pipe flow. *J. Fluid Mech.* **2015**, *779*, 701–715. [CrossRef]

48. Raiola, M.; Discetti, S.; Ianiro, A. On PIV random error minimization with optimal POD-based low-order reconstruction. *Exp. Fluids* **2015**, *56*, 75. [CrossRef]
49. Chen, H.; Reuss, D.L.; Hung, D.L.S.; Sick, V. A practical guide for using proper orthogonal decomposition in engine research. *Int. J. Engine Res.* **2012**, *14*, 307–319. [CrossRef]

9

Bubble Size and Bubble Concentration of a Microbubble Pump with Respect to Operating Conditions

Seok-Yun Jeon [1,2], Joon-Yong Yoon [1] and Choon-Man Jang [2,*]

1 Department of Mechanical Engineering, Hanyang University, 55 Hanyangdeahak-ro, Sangnok-gu, Ansan, Gyeonggi-do 15588, Korea; lo21c@hanyang.ac.kr (S.-Y.J.); joyoon@hanyang.ac.kr (J.-Y.Y.)

2 Department of Land, Water and Environment Research, Korea Institute of Civil Engineering and Building Technology, 283, Goyangdae-ro, Ilsanseo-gu, Goyang-si, Gyeonggi-do 10223, Korea

* Correspondence: jangcm@kict.re.kr

Abstract: The present paper describes some aspects of the bubble size and concentration of a microbubble pump with respect to flow and pressure conditions. The microbubble pump used in the present study has an open channel impeller of a regenerative pump, which generates micro-sized bubbles with the rotation of the impeller. The bubble characteristics are analyzed by measuring the bubble size and concentration using the experimental apparatus consisting of open-loop facilities; a regenerative pump, a particle counter, electronic flow meters, pressure sensors, flow control valves, a torque meter, and reservoir tanks. To control the intake, and the air flowrate upstream of the pump, a high precision flow control valve is introduced. The bubble characteristics have been analyzed by controlling the intake air flowrate and the pressure difference of the pump while the rotational frequency of the pump impeller was kept constant. All measurement data was stored on the computer through the NI (National Instrument) interface system. The bubble size and concentration are mainly affected by three operating parameters: the intake air flowrate, the pressure difference, and the water flowrate supplied to the pump. It is noted that the operating conditions that can most effectively generate microbubbles in the range of 20 to 30 micrometers are at the pressure of 5 bar and at the air flowrate ratio of 4.0 percent for the present pump. Throughout the experimental measurements, it was found that the pump efficiency changed by less than 1.2 percent, depending on the intake air supply. The performance characteristics of microbubble generation obtained by experimental measurements are analyzed and discussed in detail.

Keywords: microbubble pump; bubble generation; pump efficiency; bubble size; concentration; particle counter

---

1. Introduction

A microbubble defined as a bubble having the size of 50 micrometers or less and having several unique characteristics resulting from their fine size [1]. It has the characteristics of a large surface area due to its very small diameter, a slow rising velocity inside the liquid due to the low buoyancy force, self-pressurized dissolution, a superior mass transfer efficiency, and a high dissolution rate. Due to these characteristics, microbubbles are broadly applied in various fields; sewage treatment, water purification technology, ozone water disinfection, mineral processing, natural ecology restoration, bio-pharmaceutical production, fine chemical reactions, and so on [2,3]. With further applications of microbubble technology in human life and industry, it has been getting more attention in recent years. In addition to the field of application using the inherent characteristics of the microbubbles above, they have the potential to be applied in applications such as non-detergent cleaning, skin care, sterilization cleaning and anion effect generation.

The formation and growth mechanism of microbubbles is closely related to cavitation which is one of the hydrodynamic phenomena. Although it is able to be used as a method for generating microbubbles in research area, it is not used in large-scale bubble generation applications due to high cost and technical reasons. Therefore, the following methods are generally used for bubble generation.

Typical bubble generation methods include pressurized dissolution type, gas–water circulation type, and Venturi-type. For the pressurized dissolution type bubble generator, microbubbles are generated as the pressure decreases due to injecting pressurized air into a water tank. The method uses a supersaturated condition to dissolve gas under high pressure of 3–4 bars and the gas becomes very unstable and escapes from the water [4]. As a result, microbubbles are generated in a very short time. Kim et al. [5] investigated the effect of swirling chamber and breaker disk on bubble generation by pressurized dissolution method. For the gas–water circulation type bubble generator, Takahashi [6] proved that gas mixed in a vortex flow changes continuously into microbubbles as it is broken down by the vortex. Ikeura et al. [7] found that ozone microbubbles generation by pressurized dissolution type is more efficient than gas–water circulation type to remove the residual pesticide in vegetables. The Venturi-type bubble generator is similar to the pressurized dissolution type in the bubble generation method and has been widely used due to its simple structure. Yoshida et al. [8] found that microbubbles were generated through the collision of a gas with the pressure wall formed by a shock wave. Yin et al. [9] conducted an experimental study on the bubble size distribution in the turbulence flow for the Venturi-type bubble generator and proved that the averaged bubble diameter has a dependence on the Reynolds number. Uesawa et al. [10] researched the mechanism of the bubble breakup phenomenon in a Venturi tube and obtained experimental results that the bubbles expanded once into a divergence region of the Venturi tube and then contracted rapidly and broke up into a great number of fine bubbles. Gordiychuk et al. [11] investigated the effects of the flowrates of air and water on the bubble size for a Venturi tube connected bubble generator. They measured bubble size by picturing the flow and using a post-processing algorithm. Zhao et al. [12] visualized the details of the transportation of individual bubbles in a Venturi tube by using a digital image analysis (DIA) method. An extremely rapid deceleration and intense breakup process was observed at the entrance of the diverging section of the bubble generator.

The bubble generator types described above are based on the assumption that the growth and release of a sphere bubble is formed from nozzles or orifices, and this process consists of bubble expansion stage and collapse stage [13]. This complex process of the bubble formation necessarily involves a variety of additional techniques for precisely controlling the bubble size and amount of production. On the other hand, a microbubble pump generates microbubbles by forming a vortex flow using centrifugal force in the radial direction and reaction force in the circumferential direction by rotation of impeller.

The microbubble pump with a regenerative pump type impeller (hereinafter, the term 'microbubble pump' is used.) has the advantages of continuous micro-sized bubble generation and a being compact single unit. In addition, the microbubble pump is able to be applied to any small-scale application requiring microbubbles, and is easy to maintain, due to its simple structure. The microbubble pump adopted in the present study has a disk-type impeller with many vanes on its periphery rotating in an annular flow path as that is usual in conventional regenerative pumps. Therefore, it follows inherent characteristics of a regenerative pump, such as having high pressure at a low flow capacity [14]. Despite the relatively low mechanical efficiency compared to the other types of pumps having an equivalent impeller diameter, the regenerative pump has advantages, such as versatile capacity and scalability, low production cost, and robust reliability. Until now, researchers on regenerative turbomachinery have been interested in improving efficiency and performance. Karlsen-Davies and Aggidis [15] reviewed overall aspects of regenerative pumps and their basic principle of operation, areas of application, performance challenges, and the improvement of geometrical parameters. Quail et al. [16] carried out parametric design through one- and three-dimensional numerical techniques and Jeon et al. [17] tried to enhance the pump performance using an optimized design. Shimizu et al. [18] experimentally analyzed the flow patterns and the bubble size of methane–water two-phase flows.

The microbubble pump has both functions of a regenerative pump and a microbubble generation in a pressurized condition. The regenerative pump having multi impeller blades generates high suction pressure with the rotation of the impeller, and thus sucks gas through tube installed upstream of the pump. The inhaled gas mixes with water inside the pump and dissolves. It is noted that the microbubble pump is able to remove water-soluble gases effectively in chemical plant environments without complicated processes [19]. Also, it has the potential to replace the bubble generator applied to conventional dissolved air flotation (DAF) systems and has an advantage in terms of operating cost and energy saving [20].

Despite the various advantages described above, research on the microbubble pump has been rare until now. This is especially true when considering the bubble size and the concentration generated from the microbubble pump with respect to operating conditions which are important in evaluating the pump performance.

In the present study, the bubble size and the concentration of a microbubble pump are investigated by experimental measurements. To measure the bubble characteristics with respect to flowrate and pressure conditions, an open-looped measuring facility has been designed. A particle counter having optical sensors and an infrared laser and a tube-linked peristaltic pump are used for analyzing the microbubbles generated from the pump. Bubble characteristics are also measured by controlling the intake air flowrates and the pressure rise of the pump, while the rotational frequency of the pump impeller was kept constant. All measured data is stored on the computer through the NI interface system. The characteristics of generated microbubbles are obtained by experimental measurements and are analyzed and discussed in detail.

2. Experimental Setup

2.1. Microbubble Pump

The microbubble pump introduced in the present study has an open channel impeller, as shown in Figure 1. It has a duct inlet for water and a port for air supply, and a duct outlet for discharging water with microbubbles. The pump is composed of a regenerative impeller with multiple blades, a casing, and a dissolution tank for discharging the pressurized and mixed bubble water. Jang et al. [21,22] conducted a numerical analysis of a regenerative blower and observed that the fluid passes through each impeller blade with recirculation flow. In the same manner as the microbubble pump, the regenerative impeller induces higher a pressure at the inlet than the outlet and the local recirculation flow at each blade passage produces microbubbles by mixing and crushing the water and air supplied.

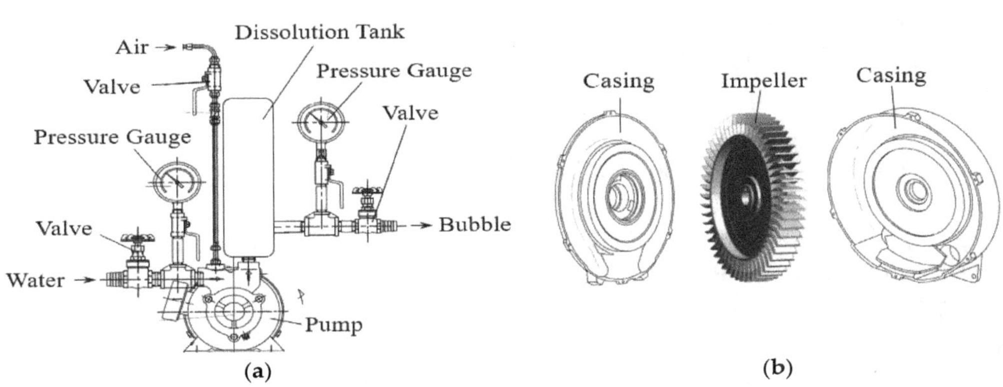

Figure 1. Schematic view of test microbubble pump: (**a**) overall view; (**b**) pump.

The pump is made of stainless steel with capacity of 1.1 kW at rated operating conditions. A mechanical seal is installed between the pump casing and the drive shaft to prevent water leakage at high-pressurized operating conditions. The detailed specifications of the test pump are shown in Table 1. The flow and pressure coefficients of the pump from performance test at rated operating condition are 0.39 and 2.04, respectively. The outer diameter of the impeller is 72.6 mm and the number of blades is 48.

Table 1. Specifications of test microbubble pump.

Parameters	Value
Flow coefficient	0.39
Pressure coefficient	2.04
Rotational speed, rpm	3550
Diameter of impeller, mm	72.6
Number of impeller blades, ea	48

The flow coefficient (ϕ) and pressure coefficient (ψ) of the pump are defined as follows.

$$\phi = \frac{Q}{AU_t} \tag{1}$$

$$\psi = \frac{2\Delta P}{\rho_w {U_t}^2} \tag{2}$$

where Q is the volumetric flowrate of water, ΔP is the pressure rise between the inlet and outlet of the microbubble pump, U_t is the rotational velocity of the pump impeller at the blade tip, ρ_w is the density of water, and A is the cross-sectional area of the side channel of the microbubble pump.

2.2. Experimetal Apparatus and Method

Figure 2 shows the schematic diagram of the experimental apparatus of the microbubble pump. Using the experimental apparatus of the pump, the generation feature of the microbubbles and the bubble sizes are measured and analyzed with respect to the pump's operating conditions. A water tank having the capacity of 1 ton is installed upstream of the pump to stabilize the water supply to the microbubble pump. The experimental apparatus has an open-loop water supply system, so the large water tank is necessary to keep the water supply continuous. Upper and lower reservoirs made of acrylic are installed upstream and downstream of the test pump for the checks of the water level and the visual monitoring of the microbubbles. The water level of the upper reservoir is kept constant by the optimal operation of the pump installed between the water tank and the upper reservoir. Mixed water containing microbubbles flows into the lower reservoir through the pipeline at the bottom of it, and the vertical height from the discharge port to the free surface is 0.42 m.

Two electronic flow meters are installed upstream of the microbubble pump to measure the flowrate of the supplied water and the air sucked to the microbubble pump, respectively. As for measuring the water flowrate supplied to the pump, the electronic flow meter having a range between 7 and 150 LPM (liter per minute) has been used (VVX25, Sika, Kaufungen, Germany). Considering the small air flowrate to the pump, a thermal flowrate sensor with high sensitivity and high response characteristics is introduced. The air flowrate meter has a measuring range between 0 and 2 LPM (CMS0002, Azbil, Tokyo, Japan). Two electronic pressure sensors are installed, one upstream and one downstream of the test pump to measure the pressure difference of the pump. A pressure sensor (model: PSAN-(L)V01C(P)A, Autonics, Busan, Korea) having a range between −101.3 and 0 kPa is installed upstream of the pump. On the other hand, a pressure sensor up to 1000 kPa has been used for measuring high positive pressure (model: PSAN-(L)1C(P)A, Autonics, Busan, Korea). Two valves are installed to control the suction air flowrate and the pressure of the pump.

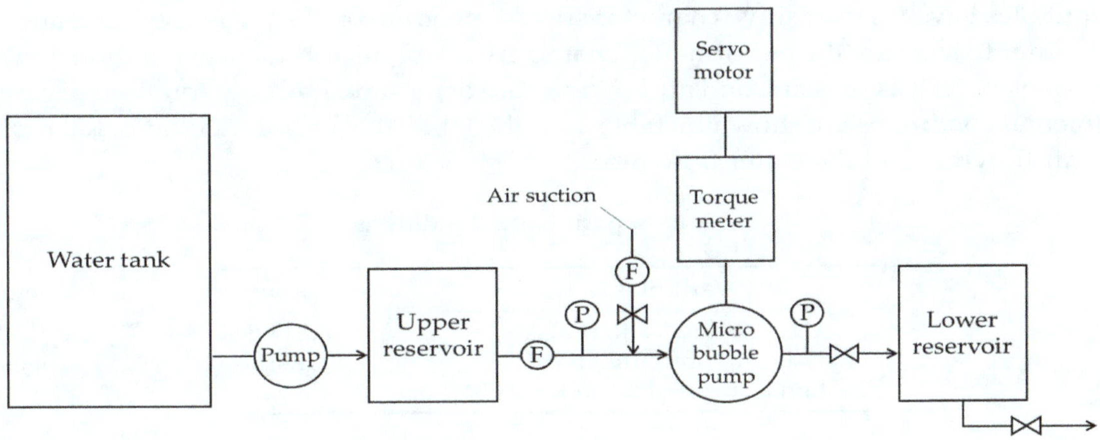

Figure 2. Schematic diagram of a microbubble pump experimental apparatus. (F: electronic flow meter, P: electronic pressure sensor).

The pump driving system is introduced to control the rotational speed of the pump and to measure the pump driving torque. A servo motor (FMA-CN10-AB00, Higen, Changwon, Korea) of 1.0 kW is installed to drive the pump at a constant rotational speed. A torque meter with a tachometer (SS-100, Ono Sokki, Yokohama, Japan) is selected for measuring the rotational speed and the torque of the pump.

A microbubble counting system is used to measure the size and the concentration of the microbubbles that are generated by the microbubble pump in respect to time. As shown in Figure 3, a microbubble counting system consists of a peristaltic pump, a particle counter, and a data processing PC. A particle counter (PC3400, Chemtrac, Norcoss, GA, USA) having optical sensors and processing electronics, including an infrared laser, is introduced. The measuring range of the bubble size is between 2 and 125 μm while the pump flowrate of the microbubble into the particle counter is kept at 75 mL/min. A constant flow pump (JWS-600, JenieWell, Seoul, Korea) is used to send a constant amount of the microbubbles to the counting system through a tube-linked peristaltic pump (JWS-600, JenieWell, Seoul, Korea).

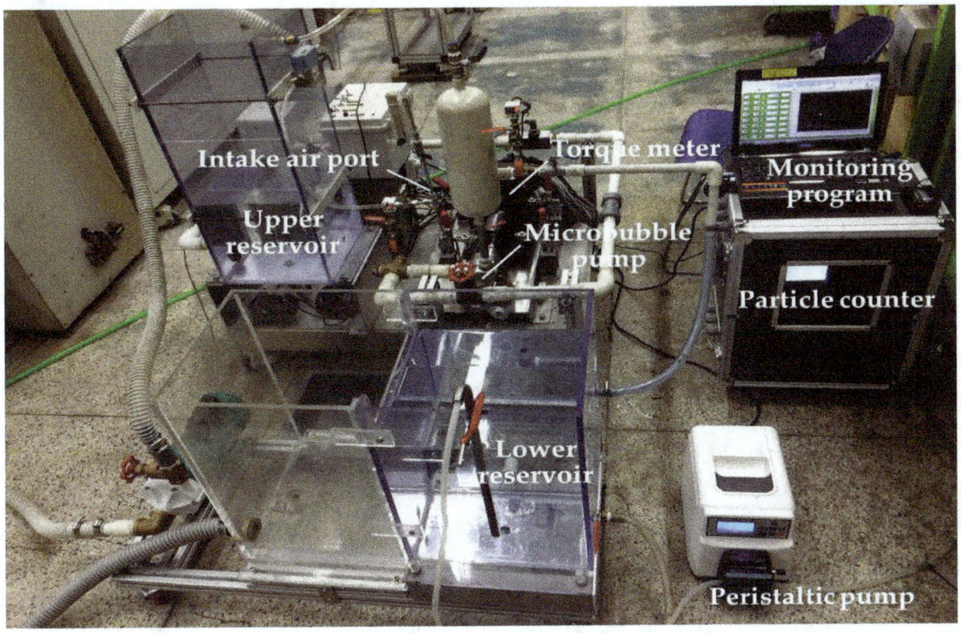

Figure 3. Picture of a microbubble pump experimental apparatus.

Microbubbles have very sensitive characteristics, depending on the pump pressure and suction air flowrate. In order to measure the performance characteristics of microbubble generation, the measuring ranges of variables, such as pressure and air flowrate, are determined through preliminary experiments. The experimental conditions are shown in Table 2. In the table, the air flowrate ratio, Ra, represents the ratio of the air flowrate and the water flowrate.

Table 2. Experimental conditions.

Variables	Value
Air Flowrate Ratio (*Ra*), %	0.8–4.8
Pump pressure, bar	4–6
Rotational speed of impeller, RPM	3550

3. Result and Discussion

3.1. Performance Curve of Test Microbubble Pump

Prior to measuring the microbubbles generated, the pump performance is obtained by an open loop pump test facility, as shown in Figure 4. It is noted that the pump performance is determined by the water supply only, in the absence of air suction. Mechanical efficiency (η) of the pump is derived from following Equation (3).

$$\eta = \frac{Q\Delta P}{T\omega} \tag{3}$$

where T and ω are the torque acting on the rotational axis and the angular velocity of the microbubble pump impeller, respectively.

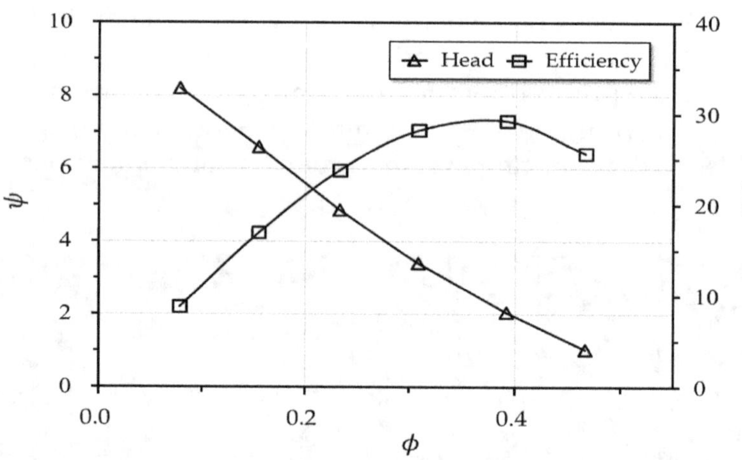

Figure 4. Performance curve of test microbubble pump.

3.2. Bubble Size and Distribution

Since the diameter of microbubbles are usually considered to be 50 micrometers or less, the present study also analyzes the bubble size range. Figures 5–7 show bubble concentration (c) with respect to the bubble diameter (d) at the three different pump pressure conditions, while the air flowrate ratio (Ra) also being changed. The particle counter used for the measurement of the bubble size and concentration automatically converts the total counts of particles to the volume of them and displays the concentration of particles in ppb. That is, the bubble concentration in the unit of ppb means the relative volume ratio of air and water with a scale of one billionth. It is noted that the bubble concentration is determined at the fixed rotational speed of impeller of 3550 rpm. As shown in Figure 6, the bubble concentration is higher when the pump pressure is at 5 bar, than the other two pressures in Figures 5 and 7. A higher bubble concentration is distributed between the 20 and 30 micrometer bubble diameters at a pump pressure of 5 bar, as shown in Figure 6. When considering the microbubble generation by mixing the water and suction air at the blade passage with the rotation of the impeller, it is understood that optimal pump pressure is important to generate a higher concentration of microbubbles. On the other hand, the amount of suction air is also an important factor for bubble generation.

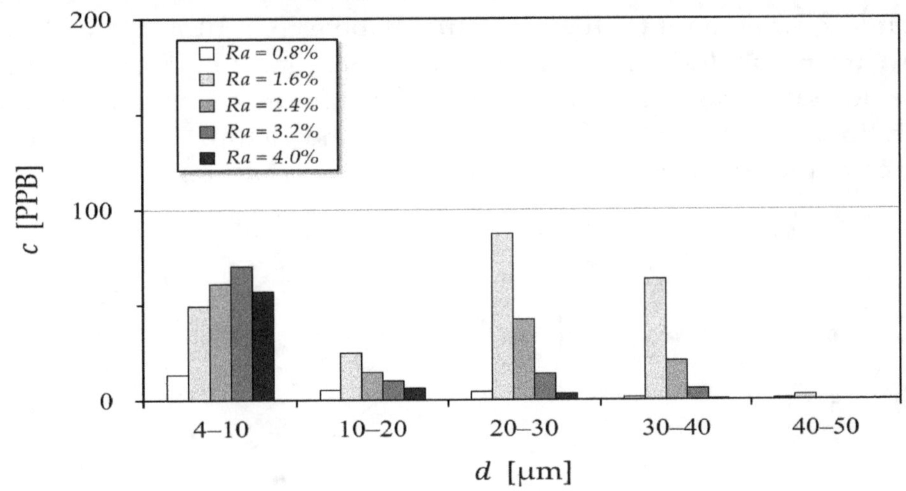

Figure 5. Bubble concentrations at the pump pressure of 4 bar.

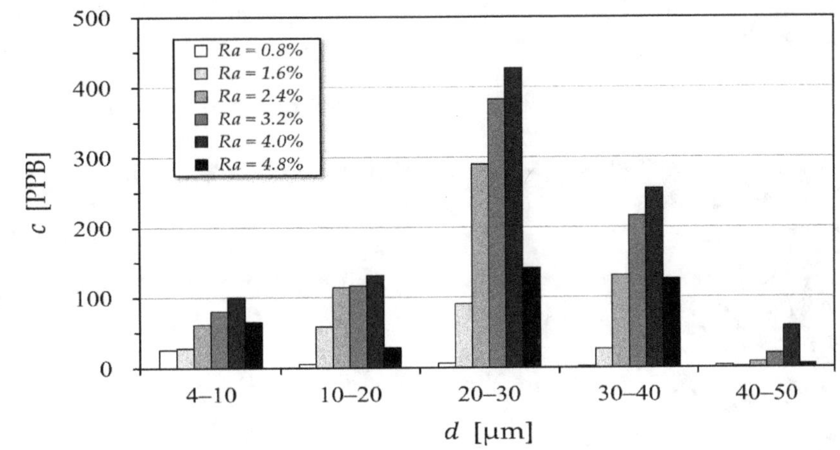

Figure 6. Bubble concentrations at the pump pressure of 5 bar.

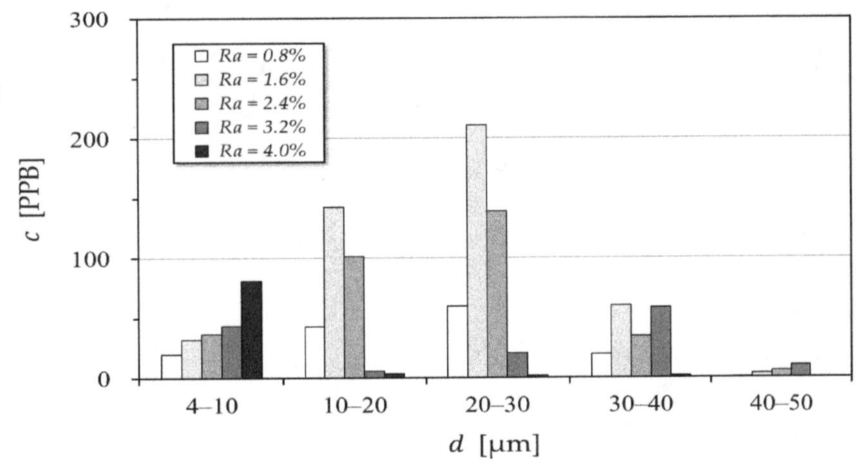

Figure 7. Bubble concentrations at the pump pressure of 6 bar.

Figure 8 shows the bubble concentration with respect to the air flowrate ratios at the pump pressure of 5 bar, which is the optimal operation condition for microbubble generation. As shown in the figure, the microbubble concentration at the range of 20–30 micrometers is the most prevalent when compared to other ranges. It is noted that the most efficient operational condition for microbubble generation is located near the 4.0 percent air flowrate ratio. The bubble concentration increases with the increase of the intake air flowrate ratios from 0.8 to 4.0, and then rapidly decreases at 4.8 percent. Thus, the optimal intake air amount for the maximum bubble concentration can be determined.

Figure 9 shows microbubbles from the maximum generation condition with the pump pressure of 5 bar and the air flowrate ratio of 4.0 percent, and the minimum condition with the pump pressure of 4 bar and the air flowrate ratio of 3.2 percent. Unlike to the picture of Figure 9a, the microbubbles have a gray milky color under the best operational condition in Figure 9b.

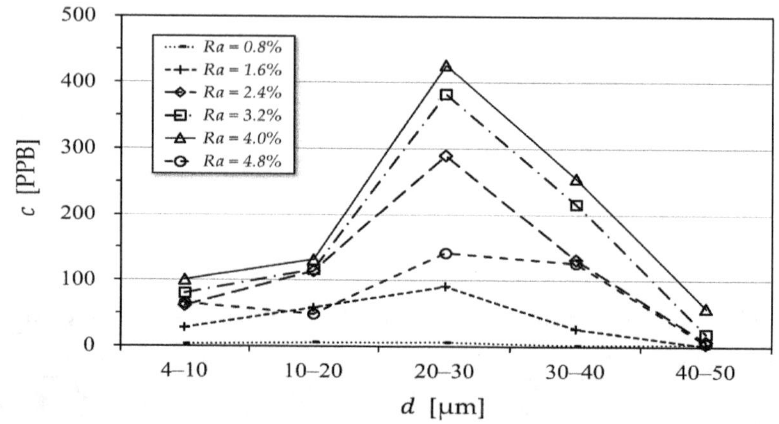

Figure 8. Bubble concentration with respect to the airflow ratios at the pump pressure of 5 bar.

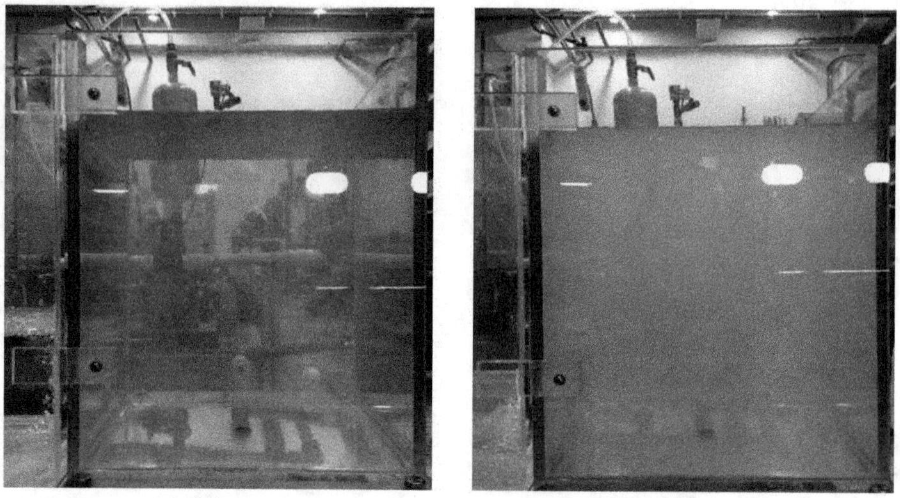

(a) (b)

Figure 9. Picture of microbubbles: (**a**) pump pressure of 4 bar and air flowrate ratio of 3.2 percent; (**b**) pump pressure of 5 bar and air flowrate ratio of 4.0 percent.

3.3. *Effect of Intake Air Flowrate on Pump Performance*

In order to understand the effect of the suction air flowrate on the pump performance, the pump performance is compared at the operational conditions of the pump, with the pressure of 5 bar and the air flowrate ratio of 4.0 percent, as shown in Figure 10. This operation condition corresponds to the maximum generation of microbubbles. As shown in the figure, pressure and efficiency have a similar tendency for both air suction conditions, while the maximum difference of the pump efficiency is less than 1.2 percent. Therefore, the performance difference of the microbubble pump can be neglected depending on the air flowrate ratio.

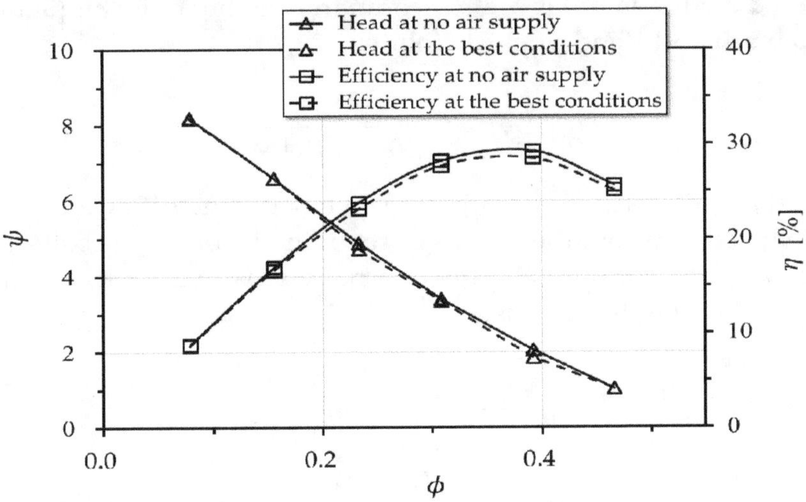

Figure 10. Effect of the suction air flowrate on the pump performance.

3.4. Relationship between Microbubble Generation and Dissolved Oxygen

For measuring the dissolved oxygen concentration during microbubble generation, a dissolved oxygen meter with a range between 0 and 90 mg/L has been used (Orion Star A223, Thermo Scientific, Waltham, MA, USA). Its resolution is 0.01 or 0.1 mg/L and has a relative accuracy is ±0.2 mg/L.

Figure 11 shows the dissolved oxygen and microbubble concentration with respect to time at the pump pressure of 5 bar and the air flowrate of 4.0 percent, which is the maximum microbubble generation condition. The dissolved oxygen concentration is measured during microbubble generation. As shown in the Figure, dissolved oxygen rapidly increases up to 180 s after pump operation, then increases more gradually, subsequently decreasing rapidly after stopping pump operations at 420 s. It can be seen that the tendency of the oxygen dissolved amount changing in resects to time is the same as the generated amount of microbubbles. From these results, it can be said that the time variation of the oxygen concentration and the concentration of microbubbles measured in the present study have a close relationship from generation to extinction.

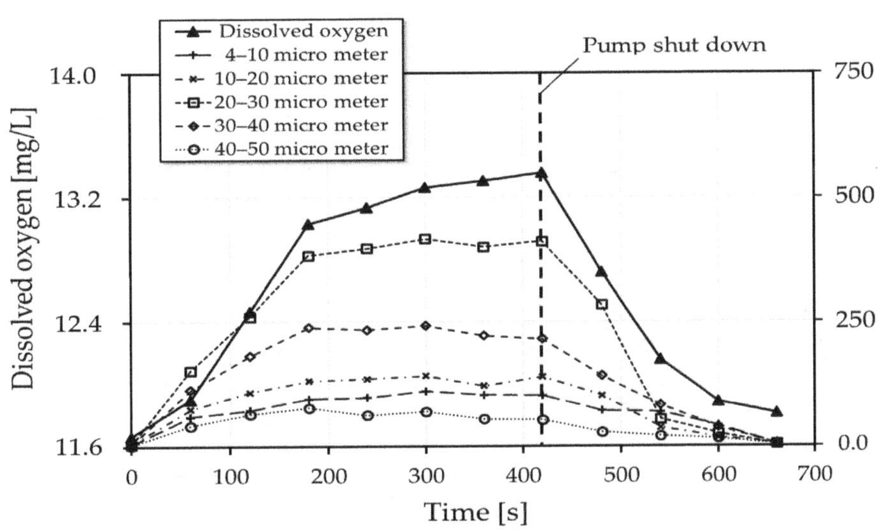

Figure 11. Dissolved oxygen and microbubble concentration in respect to time at the pump pressure of 5 bar and the air flowrate ratio of 4.0 percent.

3.5. Correlation between Microbubble Diameter and Terminal Rise Velocity

The terminal rise velocity estimated at the lower reservoir where the microbubbles generated from the pump are stored as shown in Figure 3 is compared to theoretical Equation for spherical fluid body derived by Hadamard and Rybczynski as follows [23,24].

$$U_{t(H-R)} = \frac{2\Delta\rho g r^2}{9\mu}\frac{\mu+\mu'}{2\mu+3\mu'} \tag{4}$$

where $U_{t(H-R)}$, $\Delta\rho$, g, r, μ and μ' are the terminal rise velocity, the difference in density between air and water, gravitational acceleration, the radius of the microbubble, viscosity of water and internal viscosity of air, respectively. In case of negligible internal viscosity like air, Equation (4) and diameter of microbubble become following Equations:

$$U_{t(H-R)} = \frac{\Delta\rho g r^2}{3\mu} \tag{5}$$

$$d = \left[\frac{12\mu U_{t(H-R)}}{\Delta\rho g}\right]^{1/2} \tag{6}$$

where d is the diameter of microbubble.

As shown in Equation (6), the bubble diameter can be predicted from the terminal velocity. In the methodology to determine it, Parkinson et al. have introduced visualization and image analysis by attaching a high-speed camera [25]. In this study, it was estimated from the time required for the concentration change of microbubbles at the conditions under which the microbubble was generated at the maximum for pump pressure of 4–6 bar.

Figure 12 shows the terminal velocity with respect to the microbubble diameter. In the figure, Hadamard and Rybczynski indicate the theoretical relation between the microbubble diameter and the terminal velocity. The microbubble diameters for three different pump pressures shows a similar tendency to the theoretical formula of Hadamard and Rybczynski. In particular, the microbubble diameter of about 20 μm determined at the pump pressure of 5 bar shows a relatively good agreement to the value acquired from the theoretical formula. It is noted that the highest concentration of microbubbles measured by the particle counter is at the pump pressure of 5 bar. Throughout the theoretical comparison of the microbubble size based on the terminal velocity, it can be said that the experimental study on the characteristics of microbubbles generated from the pump with respect to operating conditions has been carried out properly.

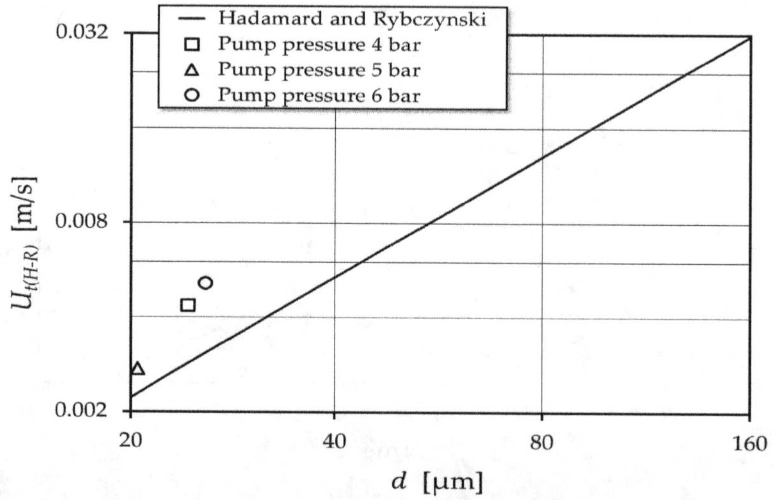

Figure 12. Terminal rise velocity with respect to microbubble diameter.

4. Conclusions

Bubble size and the concentration of a microbubble pump have been investigated by controlling intake air flowrates and pump pressure of 4, 5 and 6 bar, while the rotational frequency of the pump impeller was kept constant. The results are summarized as follows:

(1) Microbubble pumps with a regenerative impeller induce higher pressure differences from the inlet to the outlet and local recirculation flow at each blade passage. Considering the microbubble generation by mixing the water and suction air at the blade passage with the rotation of the impeller, optimal pump pressure is important to generate a higher concentration of microbubbles. It is noted that the amount of air suction is also an important factor for optimal bubble generation.

(2) Among the three different pump pressures, the bubble concentration of the test microbubble pump is highest when the pump pressure is 5 bar. The highest bubble concentration is distributed between the 20 and 30 micrometer bubble diameters, while the air flowrate ratio (*Ra*) is 4.0 percent.

(3) In the condition of higher bubble concentrations, it can be seen that pressure and efficiency of the microbubble pump have a similar tendency, regardless of the air supply. In the present pump, the maximum difference of the pump efficiency is less than 1.2 percent. So, the performance differences of the microbubble pump can be neglected, depending on the air flowrate ratio.

(4) Throughout the measurement of the dissolved oxygen concentration during microbubble generation, the tendency of the amount of oxygen dissolved to change with respect to time is the same as the generated amount of microbubbles. It is noted that the time variation of the oxygen concentration and the concentration of microbubbles have a close relationship from generation to extinction.

(5) Using a theoretical comparison of the terminal rise velocity with respect to the bubble size derived by Hadamard and Rybczynski, the microbubble diameters obtained from present measurements show a similar tendency to the theoretical formula.

Author Contributions: C.-M.J. designed the experiments; J.-Y.Y. advised throughout the research; S.-Y.J. performed the experiments, analyzed the data, and wrote the paper.

Nomenclature

c	concentration of microbubbles
d	diameter of microbubble
g	gravitational acceleration
r	radius of microbubble
A	cross-sectional area of side channel of microbubble pump
ΔP	pressure rise between inlet and outlet of microbubble pump
Q	volumetric flowrate of water
Ra	air flowrate ratio
T	torque acting on rotational axis of microbubble pump impeller
U_t	rotational velocity of microbubble pump impeller
$U_{t(H-R)}$	terminal rise velocity
Greek Letters	
η	machinical efficiency of microbubble pump
μ	viscosity of water
$\mu\prime$	internal viscosity of air
$\Delta\rho$	difference in density between air and water
ρ_w	density of water
ϕ	flow coefficient
ψ	pressure coefficient
ω	angular velocity of microbubble pump impeller

References

1. Khuntia, S.; Majumder, S.K.; Ghosh, P. Microbubble-aided water and wastewater purification: A review. *Rev. Chem. Eng.* **2012**, *28*, 191–221. [CrossRef]
2. Takahashi, M. Effect of shrinking microbubble on gas hydrate formation. *J. Phys. Chem. B* **2003**, *107*, 2171–2173. [CrossRef]
3. Takahashi, M. Potential of Microbubbles in Aqueous Solutions: Electrical Properties of the Gas-Water Interface. *J. Phys. Chem. B* **2005**, *109*, 21858–21864. [CrossRef] [PubMed]
4. Agarwal, A.; Ng, W.J.; Liu, Y. Principle and application of microbubble and nanobubble technology for water treatment. *Chemosphere* **2011**, *84*, 1175–1180. [CrossRef] [PubMed]
5. Kim, H.S.; Lim, J.Y.; Park, S.Y.; Kim, H.J. Effects on swirling chamber and breaker d3isk in pressurized-dissolution type micro-bubble generator. *KSCE J. Civ. Eng.* **2017**, *21*, 1102–1106. [CrossRef]
6. Takahashi, M. Base and technological application of micro-bubble and nanobubble. *Mater. Integration* **2009**, *22*, 2–19.
7. Ikeura, H.; Kobayashi, F.; Tamaki, M. Removal of residual pesticide, fenitrothion, in vegetables by using ozone microbubbles generated by different methods. *J. Food Eng.* **2011**, *103*, 345–349. [CrossRef]
8. Yoshida, A.; Takahashi, O.; Ishii, Y.; Sekimoto, Y.; Kurata, Y. Water purification using the adsorption characteristics of microbubbles. *Jpn. J. Appl. Phys.* **2008**, *47*, 6574–6577. [CrossRef]
9. Yin, J.; Li, J.; Li, H.; Liu, W.; Wang, D. Experimental study on the bubble generation characteristics for an venturi type bubble generator. *Int. J. Heat Mass Transf.* **2015**, *91*, 218–224. [CrossRef]
10. Uesawa, S.; Kaneko, A.; Nomura, Y.; Abe, Y. Study on bubble Beakup behavior in a venturi tube. *Multiphane Sci. Technol.* **2012**, *24*, 257–277. [CrossRef]
11. Gordiychuk, A.; Svanera, M.; Benini, S.; Poesio, P. Size distribution and Sauter mean diameter of micro bubbles for a Venturi type bubble generator. *Exp. Therm. Fluid Sci.* **2016**, *70*, 51–60. [CrossRef]
12. Zhao, L.; Mo, Z.; Sun, L.; Xie, G.; Liu, H.; Du, M.; Tang, J. A visualized study of the motion of individual bubbles in a venture-type bubble generator. *Prog. Nuclear Energy* **2017**, *97*, 74–89. [CrossRef]
13. Rodríguez-Rodríguez, J.; Sevilla, A.; Martínez-Bazán, C.; Gordillo, J.M. Generation of Microbubbles with Applications to Industry and Medicine. *Ann. Rev. Fluid Mech.* **2015**, *47*, 405–429. [CrossRef]
14. Nesbitt, B. *Handbook of Pumps and Pumping*; Elsevier: London, UK, 2005.
15. Karlsen-Davies, N.D.; Aggidis, G.A. Regenerative liquid ring pumps review and advances on design and performance. *Appl. Energy* **2016**, *164*, 815–825. [CrossRef]
16. Quail, F.J.; Scanlon, T.; Baumgartner, A. Design study of a regenerative pump using one-dimensional three-dimensional numerical techniques. *Eur. J. Mech. B/Fluids* **2012**, *31*, 181–187. [CrossRef]
17. Jeon, S.Y.; Kim, C.K.; Lee, S.M.; Yoon, J.Y.; Jang, C.M. Performance enhancement of a pump impeller using optimal design method. *J. Them. Sci.* **2017**, *26*, 119–124. [CrossRef]
18. Shimizu, T.; Yamamoto, Y.; Tenma, N. Experimental analysis of two-phase flows and turbine pump performance. *Int. J. Offshore Polar Eng.* **2016**, *26*, 371–377. [CrossRef]
19. Ahmadun, F.-R.; Pendashedh, A.; Abdullah, L.C.; Biak, D.R.A.; Madaeni, S.S.; Abidin, Z.Z. Review of technologies for oil and gas produced water treatment. *J. Hazard. Mater.* **2009**, *170*, 530–551.
20. Lundh, M.; Jonsson, L.; Dahlquist, J. The influence of contact zone configuration on the flow structure in a dissolved air flotation pilot plant. *Water Res.* **2002**, *36*, 1585–1595. [CrossRef]
21. Jang, C.M.; Han, G.Y. Enhancement of Performance by Blade Optimization in Two-Stage Ring Blower. *J. Them. Sci.* **2010**, *19*, 383–389. [CrossRef]
22. Jang, C.M.; Lee, J.S. Shape Optimization of a Regenerative Blower Used for Building Fuel Cell System. *Open J. Fluid Dyn.* **2012**, *2*, 208–214. [CrossRef]
23. Hadamard, J.S. Hadamard, Mouvement permanent lent d'une sphère liquide et visqueuse dans un liquide visqueux. *Acad. C. R. Sci.* **1911**, *152*, 1735–1752.
24. Rybczynski, W. Über die fortschreitende Bewegung einer flüssigen Kugel in einem zähen medium. *Bull. Acad. Sci. Cracovie Ser. A* **1911**, *1911*, 40–46.
25. Parkinson, L.; Sedev, R.; Fornasiero, D.; Ralston, J. The terminal rise velocity of 10–100 μm diameter bubbles in water. *J. Colloid Interface Sci.* **2008**, *322*, 168–172. [CrossRef] [PubMed]

10

Development and Assessment of Two-Stage Thermoacoustic Electricity Generator

Ahmed Hamood [1], Artur J. Jaworski [1,*] and Xiaoan Mao [2]

[1] School of Computing and Engineering, University of Huddersfield, Queensgate, Huddersfield HD1 3DH, UK; A.Hamood@hud.ac.uk
[2] Faculty of Engineering, University of Leeds, Woodhouse Lane, Leeds LS2 9JT, UK; X.Mao@leeds.ac.uk
* Correspondence: a.jaworski@hud.ac.uk

Abstract: This paper presents the development and assessment of a two-stage thermoacoustic electricity generator that aims to mimic the conversion of waste heat from the internal combustion engine exhaust gases into useful electricity. The one wavelength configuration consists of two identical stages which allow coupling a linear alternator in a "push-pull" mode because of the 180° out of phase acoustic excitation on two sides of the piston. This type of coupling is a possible solution for the low acoustic impedance of looped-tube traveling-wave thermoacoustic engines. The experimental set-up is 16.1 m long and runs at 54.7 Hz. The working medium is helium at maximum pressure of 28 bar. In practice, the maximum generated electric power was 73.3 W at 5.64% thermal-to-electric efficiency. The working parameters, namely load resistance, mean pressure and heating power, were investigated. System debugging illustrates the effect of local acoustic impedance of the regenerator on the start-up process of the thermoacoustic engine. The additional modelling showed that the feedback loop length can be reduced by using a combination of acoustic inertance and compliance components.

Keywords: thermoacoustic electricity generator; multi-stage; traveling-wave heat engine; push-pull; inertance-compliance; acoustic streaming

---

1. Introduction

Due to the widespread utilisation of high-grade heat sources in industry and transportation, there has been an increase in waste heat rejected to the environment. Therefore, many technologies have been developed for waste heat recovery applications for a range of scales and heat-grades. Thermoacoustic traveling-wave engines (TATWE) have drawn attention because of their advantages which include no mechanical moving parts, longevity and the use of environmentally friendly gases as working media [1]. Thermoacoustic engines are capable of converting heat into acoustic power. The acoustic power generated by the TATWE can be used to generate electricity by driving an electromechanical linear alternator or generate cooling by driving a thermoacoustic refrigerator [2] or a pulse tube cryocooler [3,4]. In general, TATWE take the form of an acoustic resonator filled with a gas and containing a thermoacoustic core consisting of a porous medium (stack or regenerator) with heat source and heat sink (i.e., heat exchangers) adjacent to it. The gas in the vicinity of the solid surface of the porous medium undergoes a thermodynamic cycle somewhat similar to the Stirling cycle.

The first TATWE in a looped-tube configuration was presented by Yazaki et al. [5]. It can be likened in certain respects to a standing-wave thermoacoustic engine because of the two-wavelength loop containing the thermoacoustic core at a specific location. Yazaki et al. [5] designed and built their air-filled engine to study the spontaneous gas oscillations in a travelling wave setting. The experimental results showed that the travelling wave outperformed the standing wave engines at the same frequency and wavelength. The low efficiency of this engine was discovered to be caused by low acoustic impedance (acoustic impedance is the ratio of oscillating pressure to volume flow rate). Looped-tube TATWE can have more than one thermoacoustic core in the same engine. These so-called multi-stage TATWEs are a solution to a low acoustic impedance and work at low onset temperature [6,7]. The multi-stage TATWE can be built either having identical stages each having a power extraction point or a number of geometrically non-identical stages in series and a single alternator in the loop. Although the cross section of the thermoacoustic cores in series increases in the wave propagation direction to reduce acoustic loses [8,9], identical stages run at lower acoustic losses as the staged power extraction points allow the device to run without high acoustic power "spots".

de Blok [8,10] built four engines demonstrating the possibility of cascading stages in looped tube TATWE. The engines consisted of four identical stages each having a power extraction point, and they were generating electricity in a range of 18 W to 1.64 kW. Zhang and Chang [7] studied the onset temperature, mean pressure, type of working medium (gas), hydraulic radius as well as the number of stages in a four-stage engine configuration similar to de Blok's [8]. They also investigated replacing one of the engine stages with a refrigerator stage [11]. The results showed that the device can reach a relative Carnot COP (coefficient of performance) of 28.5% at a cooling temperature of 5 °C. Yang et al. [12] adopted the four-stage configuration and adapted it by replacing one stage with acoustic compliance acting as acoustic impedance control. The engine design aimed at low-grade temperature. A thermal-to-electric efficiency of 1.51% was achieved at the hot temperature of 120 °C. Yang et al. [13] developed the engine by adding five acoustic loads. A maximum thermal efficiency achieved was 9.6% at the hot temperature of 195 °C. Li et al. [14] suggested an upgrade for the system with four identical stages by installing a branch containing a refrigerator stage and a dual linear alternator.

Another multi-stage configuration proposed by Li et al. [15] was built as three identical stages with linear alternators connected in-line to shift the phase difference. At a mean pressure of 40 bar of helium, each stage of the engine generated 1080 W of acoustic power at 36% total efficiency. Wu et al. [16] developed the system to generate useful electricity. At 50 bar mean pressure and 650 °C heating temperature, the engine generated 1.57 kW of electricity at thermal-to-electric efficiency of 16.8%. Bi et al. [17] optimized the acoustic network of the engines and ran it at a higher mean pressure of 60 bar; the engine generated 4.69 kW at 15.6% of thermal-to-electric efficiency.

Linear alternators convert the acoustic power into electricity at high transduction efficiency. The alternators used in thermoacoustic technology must have flexure bearings and gas clearance seals, which require no lubrication, to make them compatible with the low-maintenance requirements of thermoacoustic devices. The main disadvantage is that they are expensive. Hamood et al. [18] showed that the linear alternator performance improves with increasing the acoustic impedance. It is then advantageous to utilize multi-stage TATWE where high acoustic impedance is favourable for the linear alternator to generate higher electrical power. However, for "self-matched" identical stages they require many expensive linear alternators to run at high acoustic impedance and relatively low acoustic losses—hence finding a way to limit the number of alternators is important. The aim of the current research is to demonstrate a configuration that enables coupling one linear alternator to two power extraction points of two stages that work in pressure antiphase, thus reducing the number of expensive alternators in the system. Here, the effective acoustic field driving the linear alternator is the algebraic sum of the two fields belonging to the two stages.

2. Experimental Apparatus

The experimental apparatus consists of two identical engine stages each having a power extraction point, and a linear alternator connected to these two power extraction points. The conceptual design of the electricity generator is shown in Figure 1a, while Figure 1b shows a photograph of the actual device. The identical stages generate acoustic waves having similar pressure and volume flow rate amplitudes, but which are out of phase by 180° between the two stages. When these out of phase acoustic fields act upon the alternator, one is "pushing" while the other is "pulling" the piston. Hence, the active acoustic impedance running the alternator is the sum of the two push-pull acoustic fields, and this will increase the power output at a specific acoustic impedance [18].

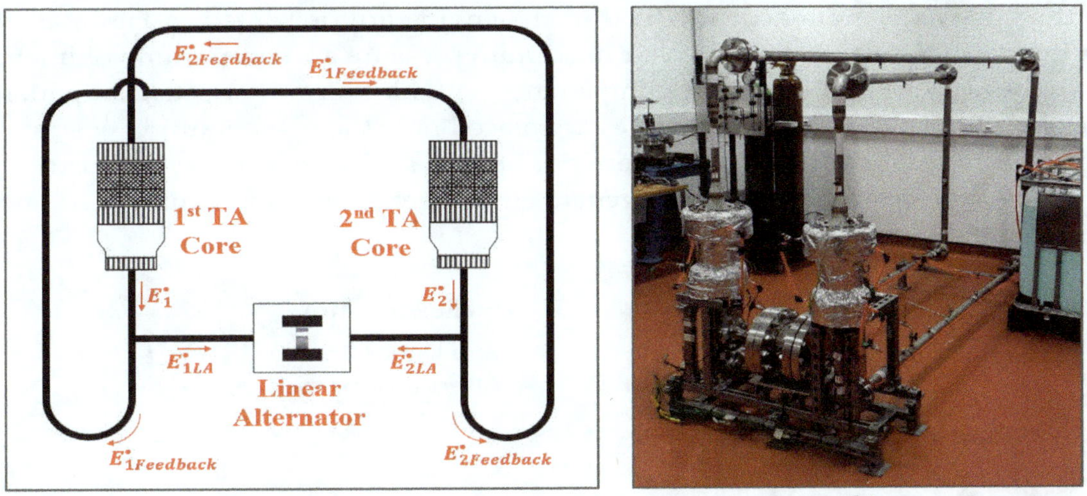

(a) (b)

Figure 1. (**a**) Conceptual drawing of the engine; (**b**) Photo of the thermoacoustic engine reported here.

The DeltaEC (design environment for low-amplitude thermoacoustic energy conversion) simulation tool was used to simulate the acoustic field and optimize the dimensions of the device components. DeltaEC is a design tool for thermoacoustic applications developed based on the linear thermoacoustic theory [19]. It enables a continuous optimization process to investigate the dimensions that offer the best performance. After a complex trade-off optimization process which considers the performance and parts' availability, a successful model was generated.

In addition to the system optimisation, DeltaEC was also used to study the favourable acoustic field for the linear alternator to generate high electrical power out of the acoustic input with high efficiency. Specifications of the linear alternator (Q-Drive 1S132M) were used as input values for the model. Figure 2 shows the generated power (black curves) and efficiency (blue curves) at a frequency of 56 Hz. They are plotted as a function of phase difference between velocity and pressure. The local acoustic impedance (which is a parameter here) is shown in the legend. The simulation results confirm that the linear alternator generates higher electrical power and runs more efficiently at higher acoustic impedance. In addition, the phase difference of around 15° is favourable for the linear alternator to generate the highest power, while between 40–60° to run at highest possible efficiency.

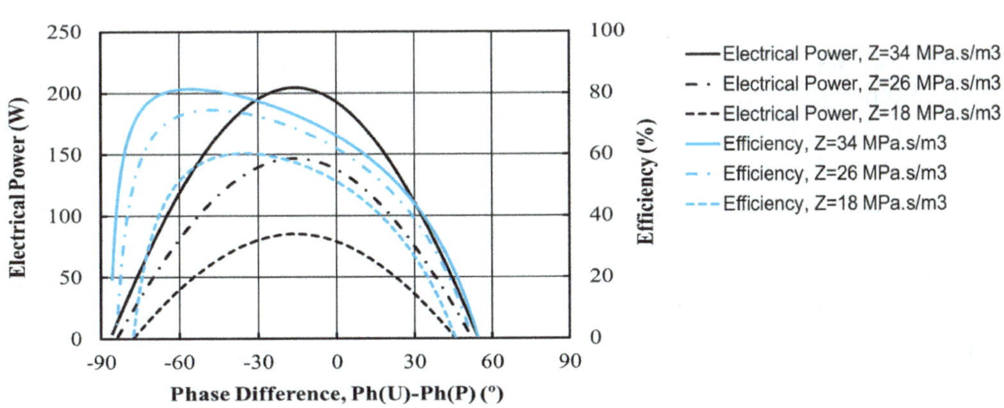

Figure 2. LA electrical power generated and efficiency at push-pull coupling for different acoustic impedance and phase difference.

This paper describes the experimental rig only in brief, since the full details can be found in [18]. The thermoacoustic generator is a 16.1 m long, looped tube two-stage thermoacoustic engine, and uses pressurized helium at 28 bar as the working gas. It runs at a frequency of 54.7 Hz. The simulation results for the one-wavelength acoustic field in the engine are presented in Figure 3. The figure shows that the regenerators are located near maximum pressure and minimum volumetric velocity amplitudes to be near the highest acoustic impedance and minimize the viscus dissipation. The two branches leading to the linear alternator sides are placed near the regenerator to ensure the highest possible acoustic impedance at the linear alternator branches. The acoustic power distribution along the engine shows that the acoustic power is generated in regenerators and mainly dissipated through the linear alternator branches.

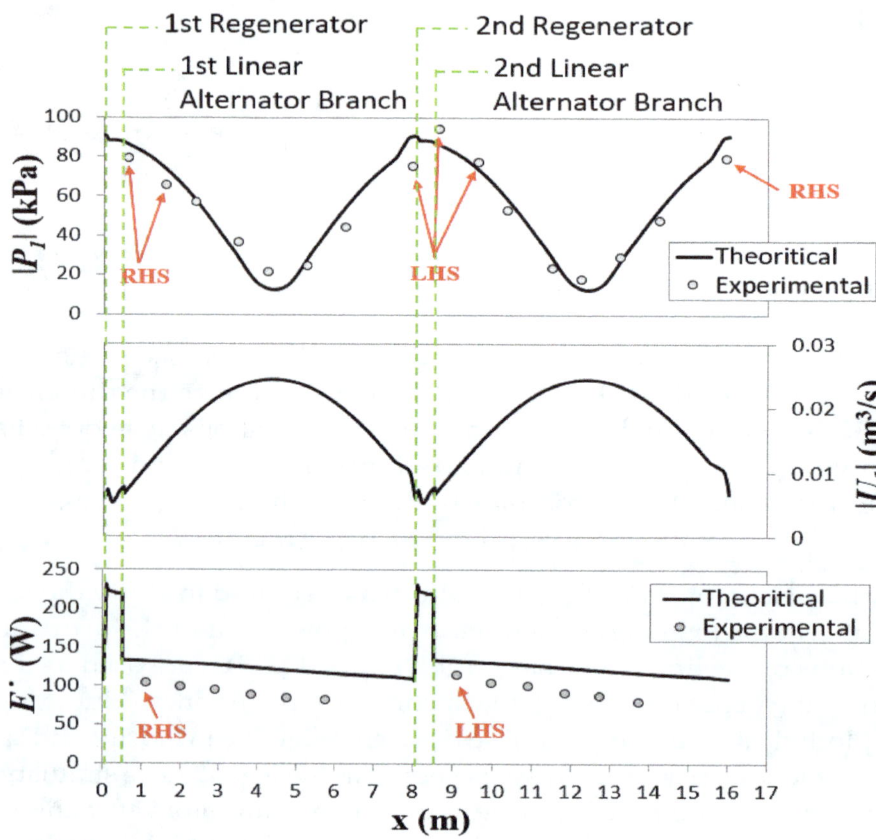

Figure 3. Distribution of pressure amplitude, volumetric velocity and acoustic power along the engine.

Figure 4 shows a cross-section of the thermoacoustic core. The regenerator holder, hot heat exchanger and thermal buffer tube have been manufactured as one piece to eliminate potential gas leakage problems which might appear at elevated temperatures at the seals of hot parts. The ambient heat exchanger is a cross flow heat exchanger having staggered fins at both water and helium flow directions. It is made out of a block of copper. The diameter of the heat exchanger on the helium side is 101.75 mm, and its thickness is 30 mm. The fins are 0.5 mm in width leaving 1 mm channels; on the helium side the fins are 8 mm long, while they are 5 mm long on the water side. At the design drive ratio, the peak-to-peak displacement is roughly half of the heat exchanger length. The porosity of the ambient heat exchanger is 31.2% on helium side.

The regenerator is made of 445 stainless steel mesh screen layers, piled up inside the regenerator holder. Regenerator length is 73 mm, and its diameter is 102 mm. The wire diameter in the mesh screen is 65 μm while the wire-to-wire aperture is 180 μm. On each end of the regenerator, there are coarse diamond mesh screens of 1.2 mm thickness, which act as spacers. The spacers allow the gas leaving the heat exchangers to mix and spread over the entire regenerator cross section. The regenerator hydraulic radius and the volume porosity have been calculated using the wire diameter, aperture and the amount of packed mesh per unit volume. The hydraulic radius is 60.5 μm and the volumetric porosity is 78.9%.

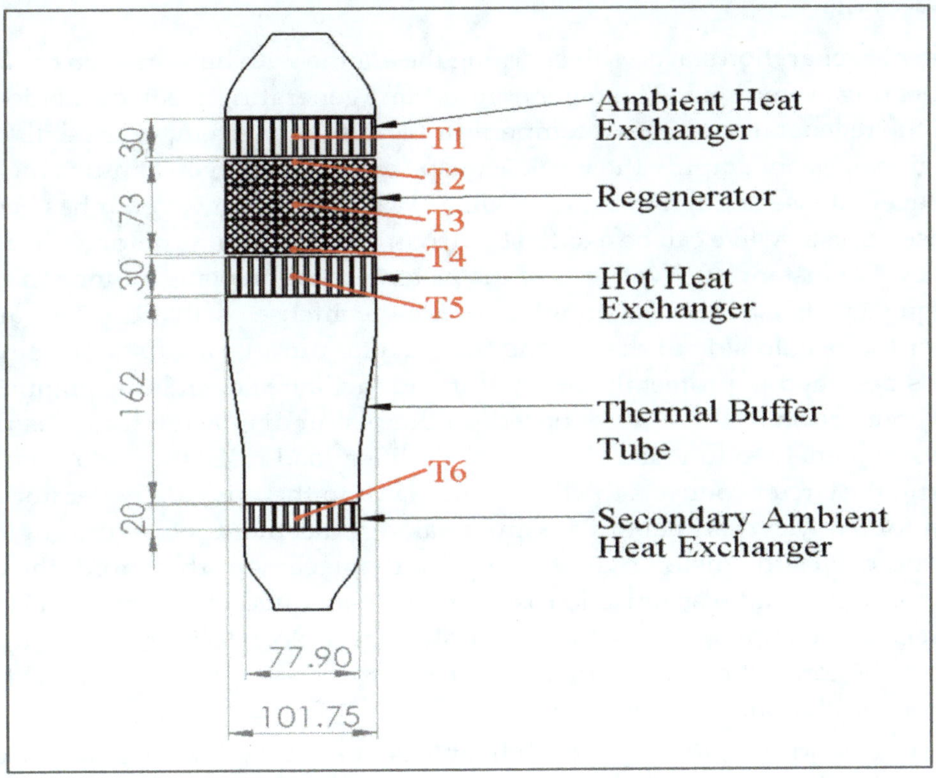

Figure 4. Cross-section of the thermoacoustic core consisting of ambient heat exchanger, regenerator packing, hot heat exchanger, thermal buffer tube and secondary ambient heat exchanger.

The hot heat exchanger has been manufactured from a low carbon steel. The choice of the material is based on a trade-off between the thermal conductivity and mechanical strength at elevated temperature. It has the face diameter of 102.2 mm (4 inch) and length of 40 mm along the flow direction. It is equipped with pairs of 100 W cartridge heaters. On the helium side, the comb-like structure creates channels of 1 mm width and fins that are 7 mm long and 0.5 mm thick. The porosity of the hot heat exchanger on the helium side is 34.4%. At the design amplitude, the peak-to-peak displacement is roughly one third of the heat exchanger length.

Below the hot heat exchanger is the thermal buffer tube providing thermal buffer between the hot and secondary ambient heat exchangers. It is 162 mm long having a conical middle section which reduces the internal diameter from 102.2 mm to 77.9 mm. The conical section is expected to reduce the Rayleigh streaming in the thermal buffer tube, as recommended by Swift [1].

The last part of the thermoacoustic core is the secondary ambient heat exchanger. The purpose of this part is to prevent heat from flowing beyond the core into the resonator. It is similar to the main ambient heat exchanger but with smaller dimensions. It is made of copper and has a porosity of 38%. The diameter of the heat exchanger on helium side is 77.5 mm, and its thickness is 20 mm. The fins are 0.5 mm in width; the fins are 9 mm long on the helium side and 5 mm long on the water side. At the design amplitude, the peak-to-peak displacement is roughly equal to the heat exchanger thickness.

The acoustic network delivers the acoustic power generated in the thermoacoustic core to the linear alternator branch and the rest is fed to the other thermoacoustic core. The network comprises of a straight standard $1\frac{1}{2}$ inch tube. The last 275 mm of the feedback loop is a standard 1-inch tube to adjust the phase difference at the linear alternator for a better performance. The linear alternator used in the rig is Q-Drive 1S132M. This alternator is asymmetric in that on one side the piston is exposed to the gas while the other side is connected to a shaft forming part of the electromagnetic armature.

Subsequently, the gas flow on two sides of the piston is not symmetrical—this feature is being corrected to some extent by introducing bespoke PVC inserts on the armature side, cf. [18].

3. Experimental Results

The experiment preparation starts with charging the engine with helium to 28 bar and then turning the cooling and heating systems on. The regenerators start generating weak oscillations when helium at the hot side of the regenerator reaches a temperature of 230 °C at a temperature difference of 185 °C. Normally, the engine does not amplify the weak acoustic oscillations (even at much higher temperature differences) to a level intense enough to drive the linear alternator. However, it has been found that in practice the intense acoustic wave can be excited by driving the linear alternator as an acoustic driver at a specific frequency. For instance, a few cycles of the piston excitation using a function generator and an amplifier at a frequency of 50.8 Hz was enough to excite the intense oscillation. This allows delivering an acoustic power to the cold side of the regenerator at a favourable acoustic phasing. An electrical control circuit was designed to protect the alternator and facilitate starting the engine. It switches the linear alternator connection in three ways based on the piston displacement measured by the laser displacement sensor: namely to function generator/amplifier, load resistance and a short circuit. At no oscillations present, the circuit connects the linear alternator to the function generator/amplifier which excites the piston for a few cycles at about 1.5 mm peak displacement. Once the engine amplifies the acoustic power and drives the piston over 2 mm peak displacement threshold, the circuit connects the linear alternator to the load resistance to dissipate the generated electricity and control the piston displacement. In case the engine drives the linear alternator close to its maximum stroke of 6 mm, the circuit switches the connection of the linear alternator to a short circuit to protect the alternator by stopping the piston oscillation.

At no oscillation condition, there is a high heat loss of about 450 W per stage from the hot heat exchanger (this value is deducted in performance calculation in this paper). As the hot heat exchanger is manufactured as one piece with the thermal buffer tube and the regenerator holder, the hot heat exchanger cannot be insulated from these two pieces. A possible way to reduce the conduction heat loss from the regenerator holder to the ambient heat exchanger is to place a low heat conductivity gasket between them. A gasket made out of thermiculite 715, Flexitallic model number SCRC04003T71515, was used. This gasket material has a low thermal conductivity of 0.3 W/m.K. The minimum available gasket thickness of 1.5 mm was selected. At this thickness, the gasket can seal up to 140 bar at a temperature of up to 540 °C.

The experiments showed that the insulating gasket improved the regenerator temperature difference and the performance of the engine. For example, the regenerator temperature difference increased from 297 °C to 308 °C and the generated electricity from 48.6 W (cf. previous work [18]) to 62.2 W at 900 W heating power, 28 bar mean pressure and 30.8 Ω load resistance.

There is an acceptable agreement between the measurements and the calculated results. The circular symbols in Figure 3 indicate the measured pressure amplitude and acoustic power (calculated using a two-microphone method, [20]), while the continuous line shows the calculated values along the engine. The measured values of pressure amplitude showed small differences between corresponding points for the two stages. This is caused by the construction of the asymmetrical linear alternator. All the left-hand side (LHS) points which are facing the armature of the linear alternator have slightly higher amplitudes than the right-hand side (RHS) points which are facing a flat side of the piston.

3.1. *Effect of Load Resistance*

In the experiments, a resistive load was connected to the linear alternator to measure and dissipate the generated electricity. The load value varied from 26.3 Ω to 92.5 Ω. Any value lower than 26.3 Ω damps the oscillations and the performance decreases at loads higher than 92.5 Ω, at similar heating power and mean pressure. For the linear alternator acting alone, at the nominal operating frequency, an increase in the load resistance will normally lead to a decreased acoustic load imposed by the linear alternator upon the oscillatory flow into the branch. As a result, the acoustic pressure in the branch, as well as the acoustic pressure difference across the alternator, increase in amplitude. The piston displacement also increases slightly, and so the linear alternator with the branch act more like a standing wave resonator which will draw less power from the engine loop. When such an alternator is coupled to the engine loop (as is the case here), the resulting acoustic pressure in the engine loop is of a higher amplitude.

The linear alternator piston applies an acoustic load to the acoustic field at each of the linear alternator branches. The value of the load resistance dominates the acoustic load which dominates the acoustic field and performance of the engine. Figure 5a shows the experimental results for the acoustic power generated by one engine stage, the acoustic power delivered to one side of the linear alternator and the piston displacement at different load resistances. Increasing the load resistance will decrease the linear alternator acoustic load which allows the piston to oscillate at higher displacement. Figure 5b shows the electricity output measured using the load resistance connected to the linear alternator against the predicted values obtained from the DeltaEC model. The experimental values are indicated by the symbols, while the continuous line shows the model prediction. The experimental values represent the average of four experimental readings, while the error bars correspond to their standard deviation. In experiments, the device generated 62.2 W of electricity at load resistance of 30.8 Ω (the best performance of the engine will be presented later in Section 4.1). The load resistance, the amplitude of acoustic pressure at the linear alternator and the temperature differential across the regenerator taken from experiments were applied as the boundary conditions for the DeltaEC model. A maximum electrical power of 85.02 W was predicted when the load resistance is 30.8 Ω.

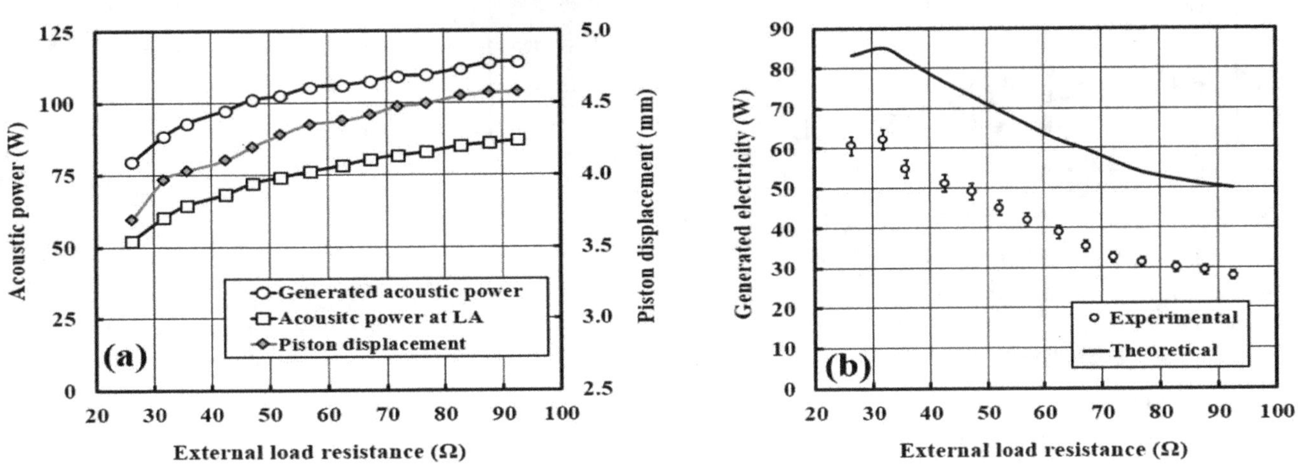

Figure 5. (**a**) Acoustic power generated by one engine stage, acoustic power delivered to one side of the linear alternator and piston displacement, (**b**) Electricity generated as a function of load resistance on the linear alternator. Heating power is 900 W.

The experimental and simulated electricity output profiles are comparable at all magnitudes of load resistance. However, significant discrepancies are observed. The main reason is that the phase difference between the volumetric flow rate and pressure at the linear alternator in the experiment is not the linear alternator's favourable acoustic condition set during modelling. For instance, the phase difference in the simulation is −30°, while in the experiment it is 10.5°. Figure 2 shows that the linear alternator does not favour the experimental phase difference value. Unfortunately, when fitting the DeltaEC model to the experimental results one can only take care of a limited number of the most important parameters, for instance the pressure amplitudes and temperature data will take precedence over phase relationships. However, there are additional reasons for discrepancies between modelling and experiments. For example, acoustic streaming which occurs in the experiment (explained in Section 4.2) and which is responsible for transferring heat from the hot to ambient heat exchanger is not included in the model. Similarly, the acoustic power dissipation through major and minor losses was calculated in the simulations using steady flow loss correlations for oscillating flow. In addition, DeltaEC performs calculations by integrating the one-dimensional wave and heat transfer equations, while the actual flow and heat transfer physics is three-dimensional in experiments. The accuracy of DeltaEC simulation results in predicting turbulence phenomena remains questionable, which may also be the underlying problem.

The power output of the electricity generator is a product of the acoustic power delivered to the alternator and its acoustic-to-electric transduction efficiency. The transduction efficiency should reach the maximum when the load resistance is equal to the coil resistance of the alternator [21], i.e., 2 Ω. The electrical power produced is also proportional to the piston displacement to the power of two. This increases continuously as seen in Figure 5a. Figure 6a shows that the acoustic-to-electric efficiency falls from 62.7% to nearly 28.4% by increasing the load resistance from 26.3 Ω to 92.5 Ω. The thermal-to-electric efficiency reaches the maximum of 6.91% at the highest electrical output when applying a load resistance of 30.8 Ω. Figure 6b shows the temperature difference measured across the regenerator (T2 and T4 shown in Figure 4) at various load resistances. At the same heating power, the temperature differential across the regenerator reduces gradually vs. the load resistance. Clearly, the heat transfer between hot and ambient heat exchangers increases due to a high-volume flow rate, but unfortunately this is not coupled with the increase in electricity production. This is because, while the acoustic power increases (cf. Figure 5a) the phasing between pressure and velocity (cf. Figure 2) becomes less favourable, and so the electrical power extraction decreases.

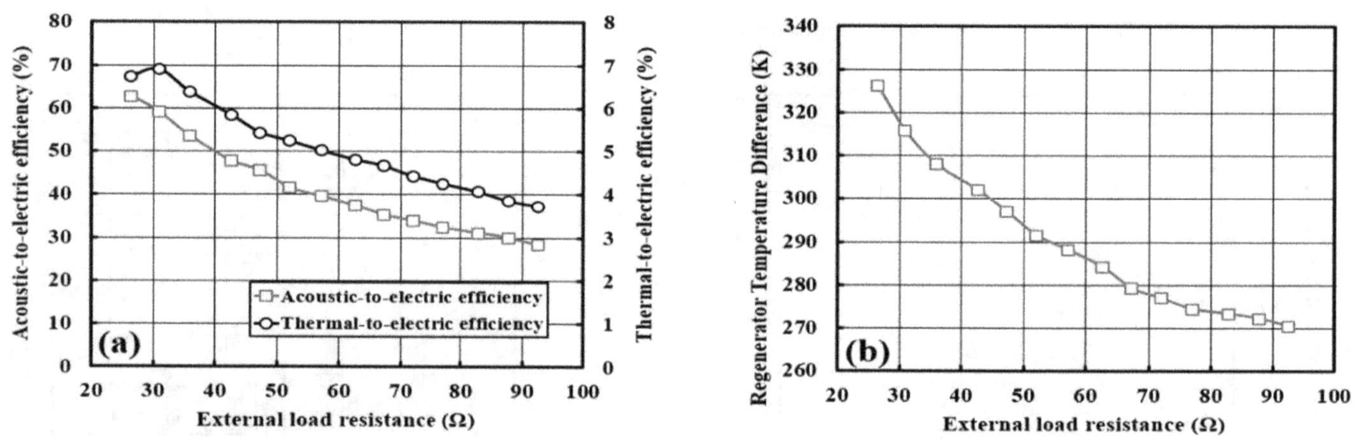

Figure 6. (**a**) Acoustic-to-electric and thermal-to-electric efficiencies vs. load resistance; (**b**) Temperature differential across the regenerator vs. load resistance. Heating power is 900 W.

3.2. *Effect of Mean Pressure*

The values of the mean pressure will affect both the power density of the acoustic field and the thermodynamic properties of the working gas. Swift et al. [1] determined the power density factor to be $p_m a A$, where a is the speed of sound, p_m is the mean pressure and A is the cross-sectional area. Higher power density will enable the thermoacoustic engine to run at higher acoustic impedance which in turn will allow higher acoustic to electric conversion at the linear alternator [18]. Varying the mean pressure changes the thermodynamic properties of the gas, including density and thermal and viscous penetration depths. These influence the processes of energy conversion in the thermoacoustic system. The mean pressure was varied in the range of 14–28 bar, at a load resistance of 30.8 Ω and heating power of 900 W. Any mean pressure less than 14 bar led to a non-harmonic oscillation which failed to maintain itself and was quickly damped.

Figure 7a shows the experimental values of the net acoustic power generated in one engine stage, the acoustic power delivered to one side of the linear alternator and the piston displacement vs. load resistance applied. It indicates that the engine performs better at higher mean pressure, as it provides higher power density and favourable phase difference to the linear alternator. Figure 7b shows the measured electrical power and the values predicted by the DeltaEC model. Symbols denote the experimental results, while the line shows the model prediction. In experiments, the engine generated 62.2 W of electricity at mean pressure of 28 bar (the best performance of the engine will be presented later in Section 4.1). The mean pressure, load resistance, the amplitude of acoustic pressure at the linear alternator and the measured temperature differential across the regenerator were applied as DeltaEC boundary conditions. A maximum electrical power of 85.02 W was predicted when the mean pressure is 28 bar. There is a clear trend of decreasing the generated electrical power with the decreasing mean pressure. The experimental and simulated electricity output profiles are comparable at all magnitudes of mean pressure. However, significant discrepancies are observed, which were explained in Section 3.1 in some detail and these explanations are applicable here too. Additional Figure 8 shows the effect of mean gas pressure on the acoustic-to-electric efficiency and thermal-to-electric efficiency (cf. Figure 8a) and the measured drive ratio (cf. Figure 8b).

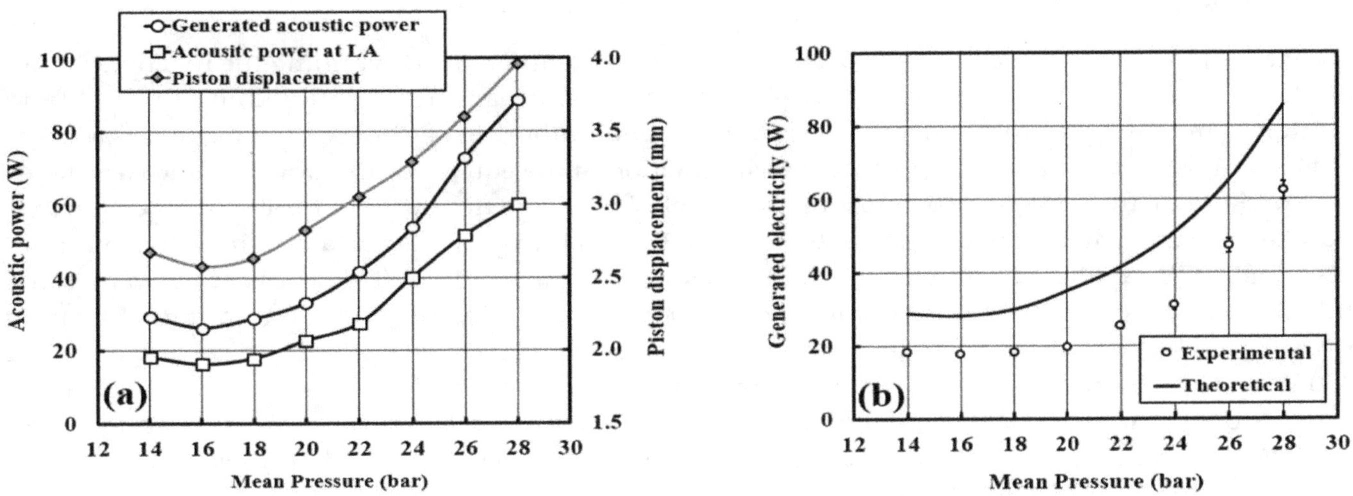

Figure 7. (**a**) Acoustic power produced in one engine stage, acoustic power on one side of the linear alternator and piston displacement vs. mean pressure; (**b**) Electricity generated by the device vs. mean pressure. Heating power is 900 W.

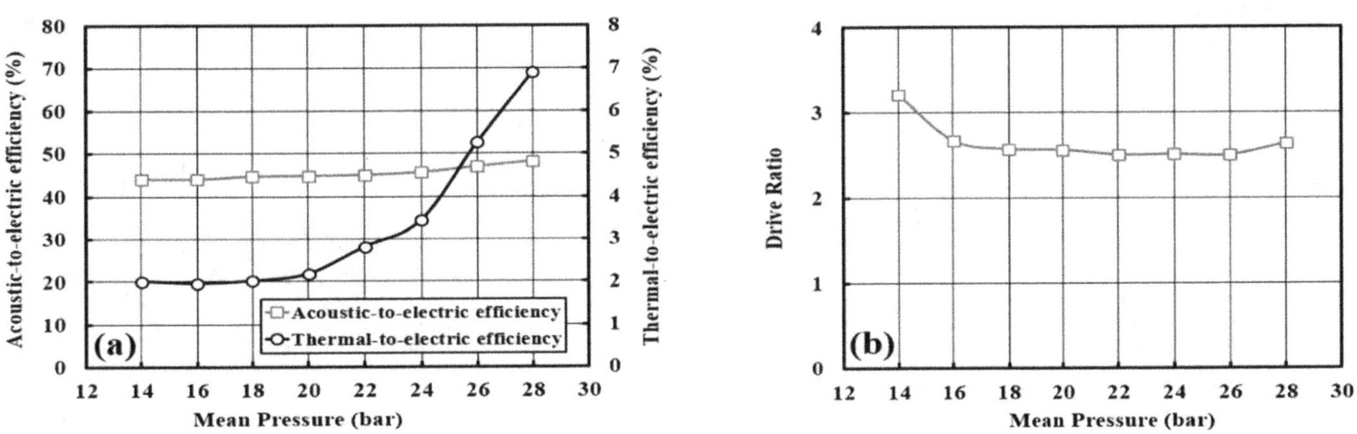

Figure 8. (**a**) Acoustic-to-electric and thermal-to-electric efficiencies vs. mean pressure; (**b**) Drive ratio vs. mean pressure. Heating power is 900 W.

Focusing on the low mean pressure range, it is not clear why the generated electrical power, generated acoustic power, drive ratio and piston displacement increase slightly when the mean pressure drops from 16 to 14 bar. Theoretically, these values should decrease based on the argument of power density being proportional to mean pressure. Most likely, this counterintuitive performance enhancement might be attributed to a phase difference at the regenerator being closer to the traveling wave for 14 bar (compared to 16 bar), which could lead to generating a higher acoustic power. Swift [1] pointed out that a resonator channel acts as an acoustic inertance and compliance simultaneously. Both contribute to the behaviour of the wave propagation in the channel. However, reducing the mean pressure increases the acoustic compliance effect which shifts the volumetric flow rate phase, while decreasing the acoustic inertance effect of a resonator which shifts the pressure phase. Unfortunately, the current setup does not allow the detailed measurements to validate this point. However, it is possible to inspect the DeltaEC modelling results in terms of the phase angle between oscillating pressure and volumetric velocity. When the mean pressure reduces from 28 to 16 bar the phase angle increases from 58.9° to 68.4°, i.e., the wave becomes "less travelling" and "more standing". On the other hand, a further decrease of mean pressure from 16 to 14 bar causes the phase angle to decrease from 68.4° to 68.0° to make the wave slightly "more travelling" again, which explains the apparent improvement of generator performance.

3.3. *Effect of Heating Power*

Heating power and oscillation intensity are the two parameters determining the regenerator hot side temperature. However, heating power is the dominant parameter determining the ability to maintain a high temperature difference across the regenerator during the oscillation. In this section, the value of the heating power represents the summation of the equal heating power of the two stages. At no oscillation, there is a high heat loss of about 450 W per stage from the hot heat exchanger which is deducted in performance calculations in this paper. The heating power was varied from the minimum power of 500 W capable of maintaining oscillations to a maximum of 1700 W, at 28 bar mean pressure and load resistance of 30.8 Ω. Figure 9a shows the generated electrical power at different heating power for both experiments and simulation. For both, the maximum is reached at a heating power of 1300 W. In the experiments, a maximum electrical power output of 72.5 W was obtained at 5.58% of thermal-to-electric efficiency, while the maximum efficiency of 7.3% was obtained at heating power of 700 W generating 51.1 W of electricity, as shown in Figure 10a. The thermal-to-electric efficiency decreases between 700 W and 1700 W.

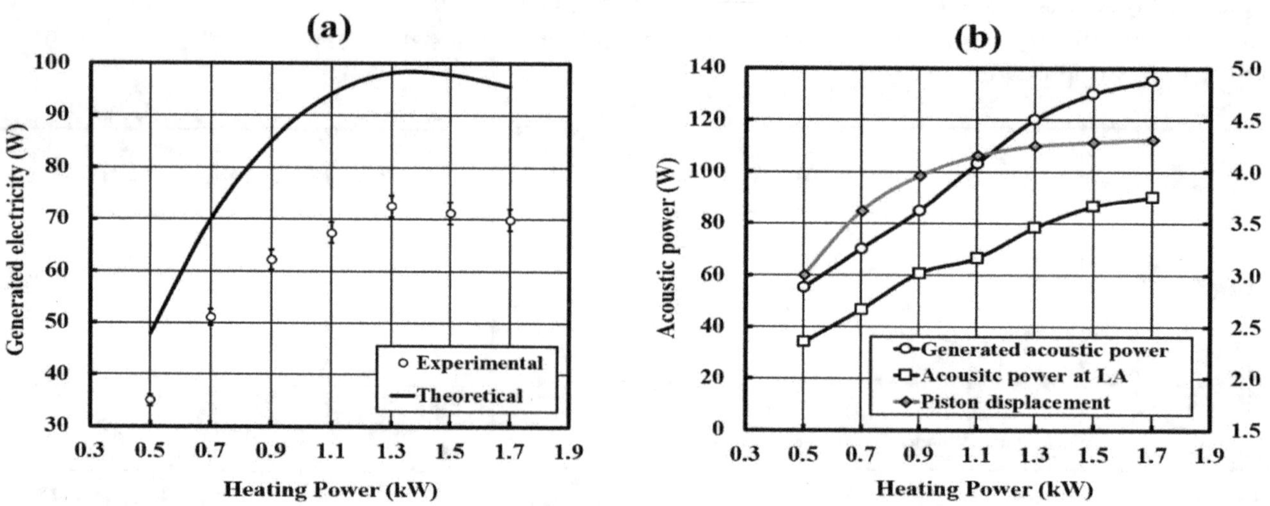

Figure 9. (**a**) Electricity generated by the device vs. heating power; (**b**) Acoustic power generated in one engine stage, acoustic power on one side of the alternator and piston displacement vs. heating power.

The existence of maximum generated electricity for heat input of 1300 W can be explained as a combination of two effects: On the one hand, increasing the heating power leads to the increase of the regenerator temperature difference (Figure 10b), generated acoustic power, acoustic power at the linear alternator and the piston displacement (Figure 9b). At the same time, the measured difference between volumetric flow phase and pressure phase increases towards the unfavourable values for the linear alternator which leads to the decrease in the alternator acoustic-to-electric efficiency, as shown in

Figure 10a. In experiments, the phase difference at 900 W heating power is 10.5° and it increases up to 34° at 1700 heating power. Figure 2 shows how shifting the phase difference affects the generated power and efficiency of the linear alternator. These competing effects lead to a maximum electricity production at 1300 W heating power rather than the highest heating power.

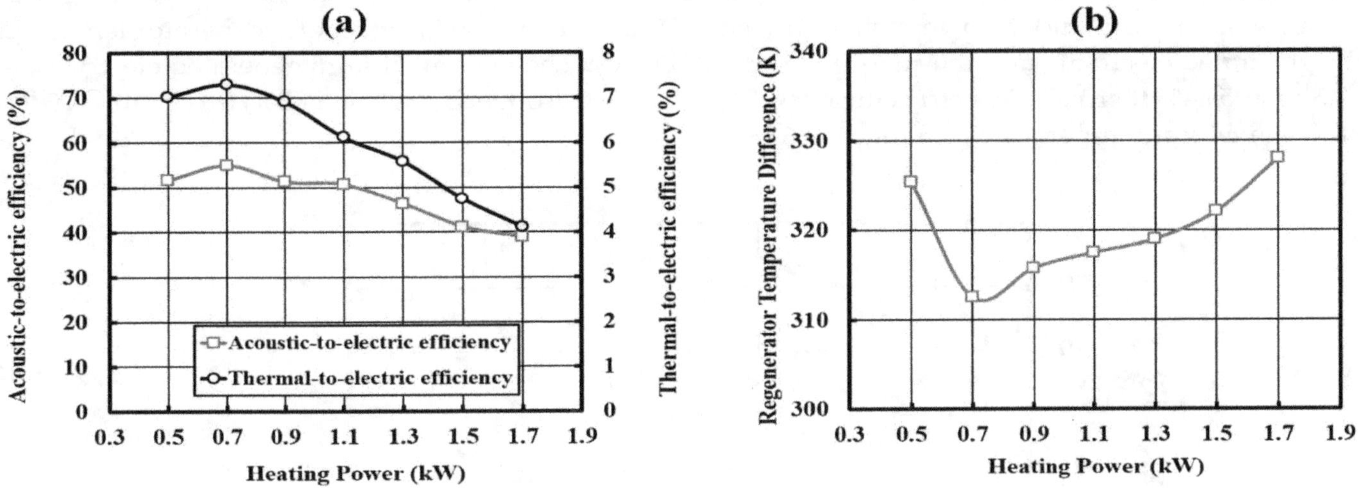

Figure 10. (**a**) Acoustic-to-electric and thermal-to-electric efficiencies vs. heating power; (**b**) Regenerator temperature difference vs. heating power.

The regenerator acts as an acoustic power amplifier. However, the flow resistance inside the regenerator plays a vital role in the power amplification as reported by Yu and Jaworski [22]. At a certain acoustic impedance, the flow resistance will dissipate most of the acoustic power fed through the regenerator cold end and this will decrease the acoustic power generation. Under such circumstances, the externally set temperature gradient will not have a significant effect. In fact, low conversion of heat into sound will lead to heating up of the regenerator hot side as shown at 500 W heating power in Figure 10b.

4. System Debugging

The aim of the debugging and optimization process was to solve and/or eliminate two problems: self-starting and streaming.

4.1. Start-Up Improvement

As mentioned in Section 3, the engine in its baseline configuration could not self-start and required "kick-starting" where a few cycles of initial excitation came from externally exciting the linear alternator. The successful solution to this problem turned out to be a slight reduction in the flow resistance. Yu and Jaworski [22] highlighted the relation between the flow resistance and local acoustic impedance and their effect on the net acoustic power and acoustic power input. It was concluded that the flow resistance plays a key role in determining the regenerator impedance as it determines the volumetric flow rate at a specific pressure amplitude. At a given pressure amplitude, the higher flow resistance increases the acoustic impedance by decreasing the volumetric flow rate.

Reducing the flow resistance in the regenerator was a possible solution to reduce the acoustic impedance, and hence increase the acoustic power leaving the regenerator, at specific acoustic power entering it, by reducing the acoustic power dissipated at the regenerator. The flow resistance could be reduced by decreasing the length of the regenerator or increasing the cross-sectional area. In the current research, the regenerator holder was welded to the hot heat exchanger and a heavy flange, therefore its length and diameter are fixed. The only way to reduce the regenerator length is to replace some of the regenerator mesh screens with coarse mesh (same as used for the spacers, cf. Section 2). The coarse mesh screens were applied on the cold side of the regenerator, for ease of replacement.

The effect of the regenerator length was investigated experimentally at 30.8 Ω load resistance, 1300 W heating power and 28 bar mean pressure. The regenerator length was increased once and reduced twice by a 1.2 mm step, which is the thickness of a single coarse mesh.

The engine self-starts at a regenerator length of 71.8 mm and 70.6 mm. The oscillation starts at a regenerator temperature difference of 280 °C. The reduction of the flow resistance was found to enhance the performance by a very small fraction. Figure 11 shows the effect of regenerator length on the generated electricity and thermal-to-electric efficiency. The new maximum generated electricity is 73.3 W at 5.64% thermal-to-electric efficiency. The relative Carnot efficiency is 11.3%, drive ratio is 3.4% at a regenerator temperature difference of 288.8 °C.

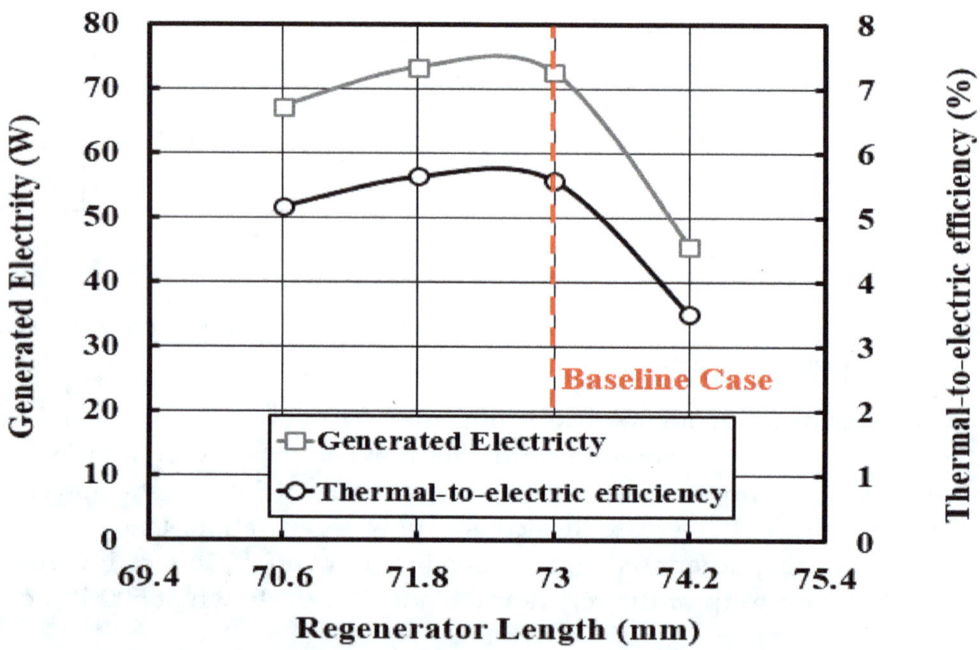

Figure 11. Generated electricity and thermal-to-electric efficiency versus the regenerator length.

4.2. *Efforts Towards Supressing Streaming*

Gedeon streaming exists in looped-tube or toroidal devices only. The reason is that a closed loop topology encourages a steady flow to circulate along such resonators. Gedeon [23] explained it as mass flow in the Stirling engines and pulse tube cryocoolers with a closed loop which leads to time-averaged convection enthalpy flux from the hot to the cold side. This phenomenon wastes heat in a thermoacoustic engine by removing heat from the hot side to the ambient of the regenerator without generating acoustic power. The devices suffering from a non-zero mass flow through the porous medium will show a non-linear temperature distribution within the porous medium. All the experimental tests showed a non-uniform temperature distribution, an example being shown in Figure 12. Many researchers [1,24,25] summarized that this kind of streaming can be suppressed either by placing a latex membrane or applying a non-symmetric flow resistance such as jet pump. The latex or elastic membrane will be transparent to the acoustic power while forming a barrier to the streaming flow, hence eliminating it.

The elastic membrane needs to be placed close to the minimum volumetric flow rate to suppress this streaming at the lowest possible acoustic power loss. Figure 3 shows that the best location is near the main ambient heat exchanger. Unfortunately, this location in the experimental rig was used to feed through the thermocouples, and hence, the membrane could not be placed there. Potential locations are between two flanges at three locations, as shown in Figure 13a. Figure 13b shows the locations of the membrane with reference to the theoretical volumetric flow rate.

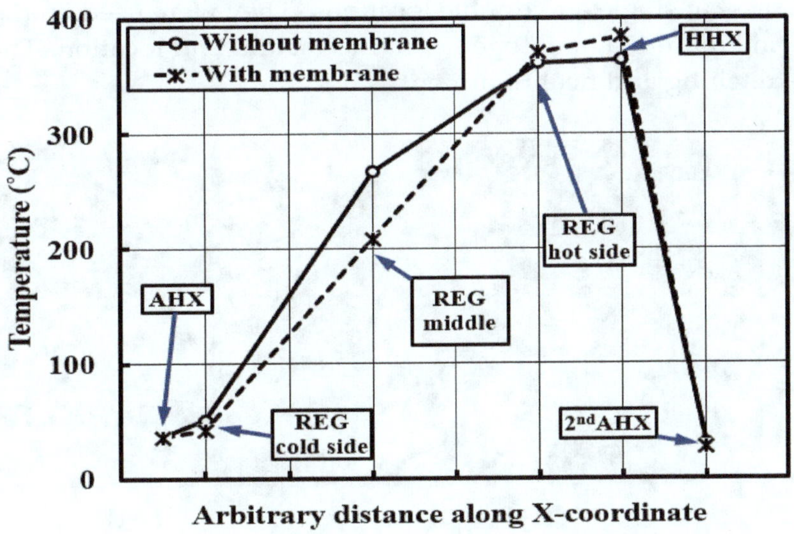

Figure 12. The effect of the presence of a concave membrane on the temperature profiles in the thermoacoustic core.

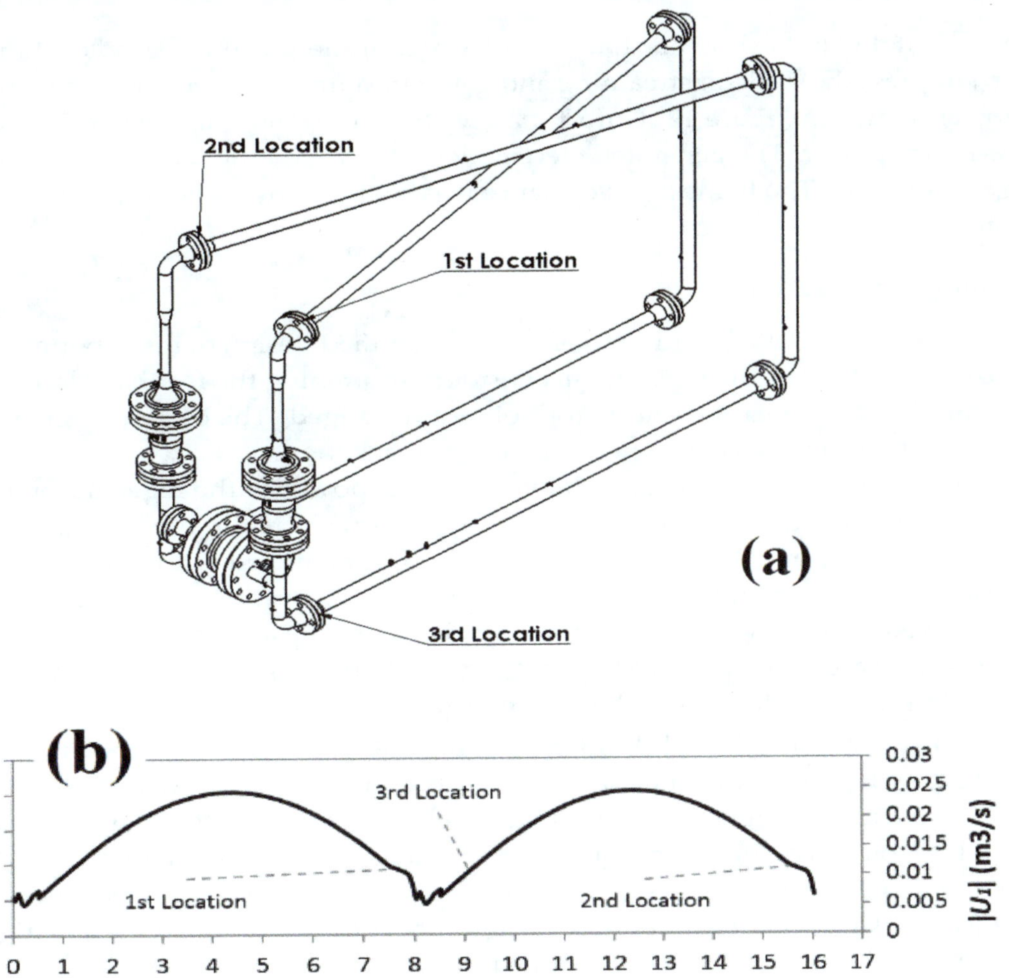

Figure 13. (**a**) Locations of the membrane along the engine loop; (**b**) locations of the membrane on the volumetric flow rate graph.

The membrane was selected based on its material elastic properties and thickness. A sheet of 100% genuine latex of 0.25 mm thickness, was used. Figure 14b–d show the three profiles of the membrane that were tested: flat, concave and loose. Figure 14a shows an example of an assembled membrane.

All three profiles were used in single and double locations. They were used on their own at the 1st and 3rd location, as shown in Figure 13a, and together at the 1st and 2nd location. The concave profile was made by continuous stretching and heat treatment.

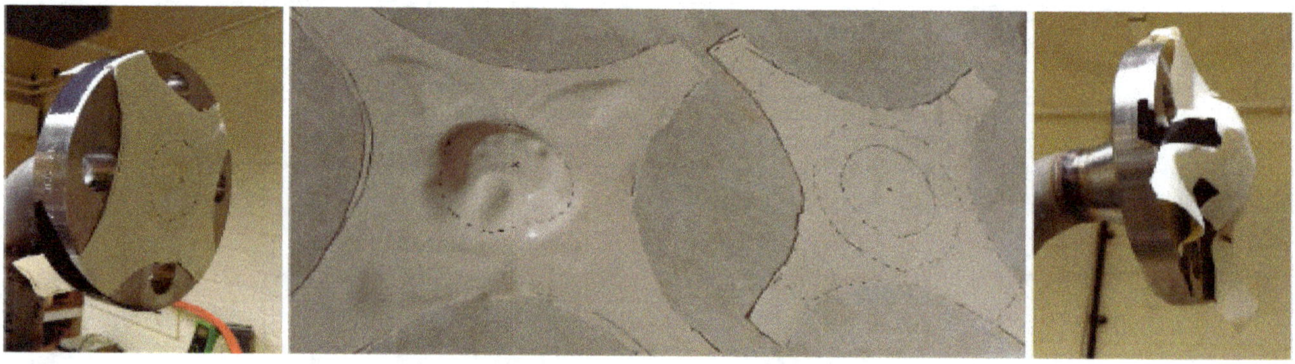

(a) (b) (c) (d)

Figure 14. (**a**) A membrane assembled on a flange; (**b**) concave membrane; (**c**) flat membrane; (**d**) loose membrane.

The experimental results showed that a single membrane placed at any location or a double membrane can suppress the Gedeon streaming and generate a uniform temperature distribution along the regenerator, as shown in Figure 12. Unfortunately, the membranes also act as flow resistance and dissipate the acoustic power. The generated electricity for the tested membrane locations and profiles varied from 0.5 W to 4.2 W. The highest performance was achieved by using one concave membrane at the 1st location.

5. Feedback Loop Optimization

This section presents a DeltaEC study to propose a modified design of the experimental apparatus to reduce its size. In particular, the new model considers shortening the feedback loop while keeping the current thermoacoustic cores and alternator holder unchanged. The current engine is 16.1 m long, of which approximately 15 m is a constant diameter feedback loop.

The function of the feedback loop is to deliver acoustic power to the regenerator at a favourable acoustic phasing. The current uniform section feedback loop shifts the pressure phase by 175° and volumetric flow rate phase by 50°. This phase shift could be achieved within a much shorter length by using a variable cross-section feedback loop. A wide cross-section pipe shifts the phase of the volumetric velocity since it acts as an acoustic compliance, while a narrow pipe shifts the pressure phase since it acts as an acoustic inertance [1]. The combination of compliance-inertance loop shifts the acoustic phasing in much shorter length than the constant diameter loop.

Many configurations of feedback loop combining inertances and compliances were studied, however, the paper will present the one that provides the shortest length without dissipating a high share of engine's generated acoustic power. Firstly, DeltaEC was used to simulate the acoustic field in the feedback loop only. The model considered the acoustic wave characteristics at the beginning and the end of the thermoacoustic core as boundary conditions of the compliance-inertance feedback loop. The new feedback loop reduced the engine length from 16.1 to 7.5 m. Subsequently, it was tested numerically on a full model and showed the same performance.

For a pipe of a certain length and diameter, the phase shifting capabilities strongly depend on the acoustic wave characteristics at the inlet. In this study, the phase shifting results for the local acoustic wave at the engine feedback inlet are shown in Figure 15. The selection of the pipe dimensions to act as an acoustic inertance is based on the pressure phase shifting and acoustic power dissipation. Figure 15 shows an example of the pressure phase shifting and acoustic power dissipation for different sizes of pipes at an acoustic impedance of 5.1 M Pa·s/m^3, 55° phase difference and 56.6 Hz frequency (which are the values at the inlet of the feedback loop to be replaced). A small diameter pipe can shift

power. For every pipe diameter there is a length range that appears to be very sensitive to the pressure phase shifting. This region needs to be avoided.

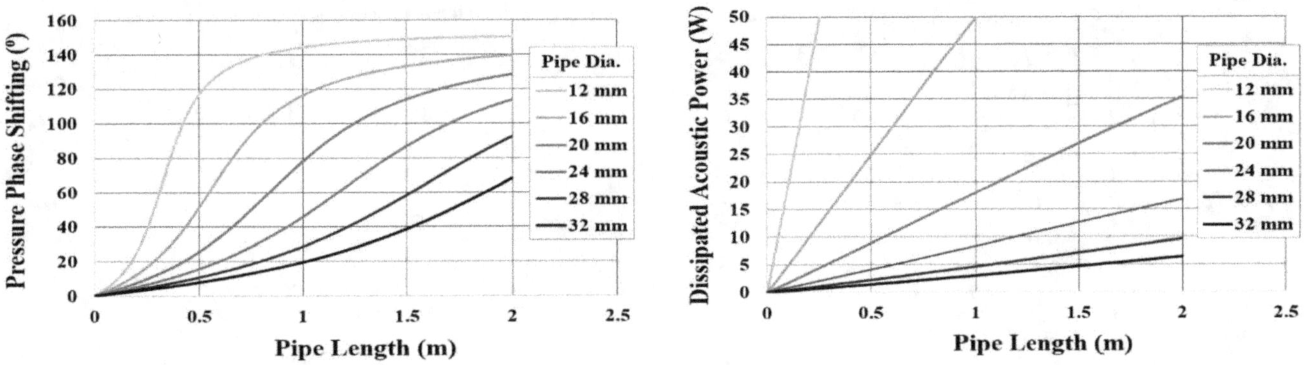

Figure 15. Effect of pipe length and diameter of the inertance pipe on (**a**) pressure phase shifting; (**b**) dissipated acoustic power.

Large diameter pipe shifts the velocity phase at low acoustic power dissipation. The phase shifting capabilities strongly depend on the acoustic wave characteristics at the inlet. The selection of the pipe diameter and length is based on the required phase shifting, at the acoustic wave characteristics at the inlet. Figure 16 shows an example of the effect of pipe length and diameter on the velocity phase shifting at an acoustic impedance of 3.8 M Pa·s/m^3, 39° phase difference and 56.6 Hz frequency (which are the values in the middle of the feedback loop where the compliance will be placed). Similar to the selection of inertance, the steep change regions of the phase shifting need to be avoided for better solution stability.

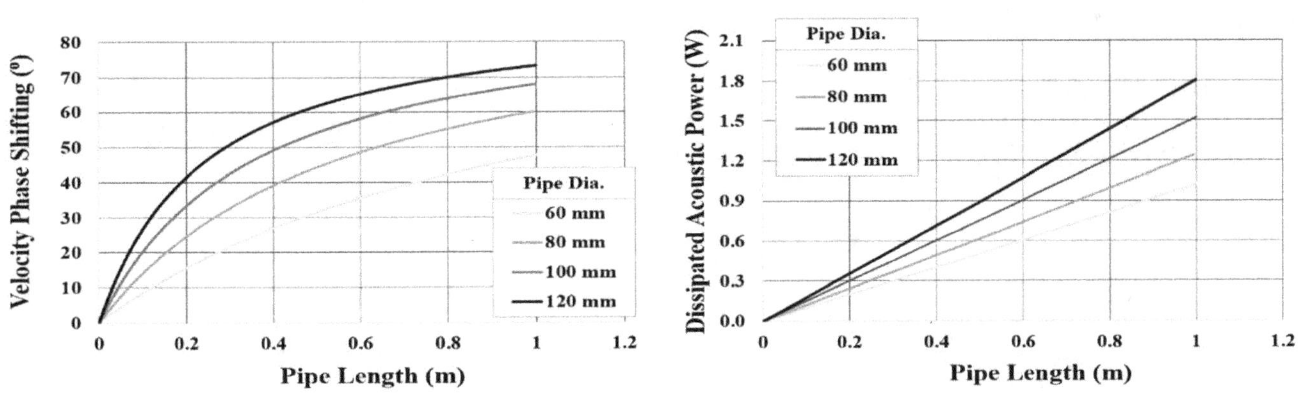

Figure 16. Effect of pipe length and diameter of the compliance pipe on (**a**) velocity phase shifting; (**b**) dissipated acoustic power.

The best feedback loop configuration studied consisted of inertance-compliance-inertance. The inertance was split into two parts with a compliance sandwiched between them so that the inertance after the compliance will be in lower acoustic impedance region. This will allow the use of a thinner pipe in the second inertance (after the compliance) which will lead to a shorter feedback loop without creating a high acoustic power loss. After continuous optimisation process, the ideal diameters of the inertances and compliance were replaced by those available for the commercially available pipes. This generated some discrepancy which was actually found to be advantageous as the new configuration allowed a reduction of the phase difference near the middle of the regenerator closer to the traveling wave phase difference (namely from 26° to 9°). This helped to increase the generated acoustic power, at a similar regenerator temperature difference, from 123.5 W to 159.5 W. However, the extra generated acoustic power is dissipated in the feedback loop. The original feedback loop dissipates 23.8 W while the new loop dissipates 59.5 W, as shown in Figure 17c. Figure 18 compares the engine configuration for both old and new feedback loops.

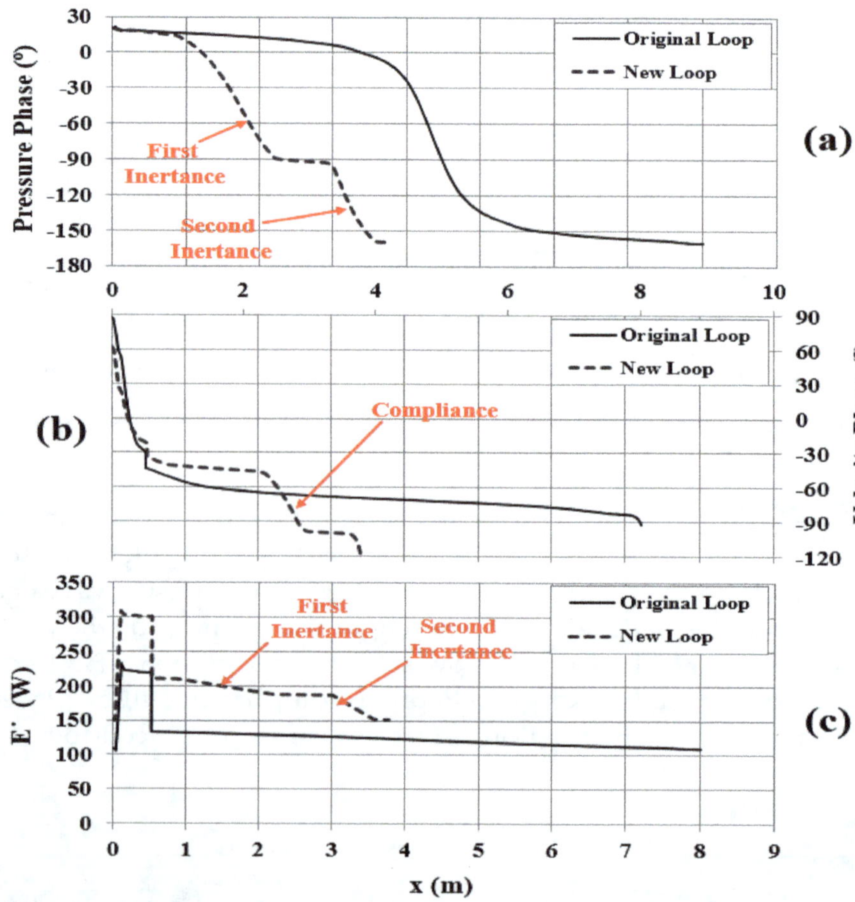

Figure 17. One-stage simulation results comparing the original and new feedback loop: (**a**) pressure phase, (**b**) velocity phase and (**c**) acoustic power flow along the engine.

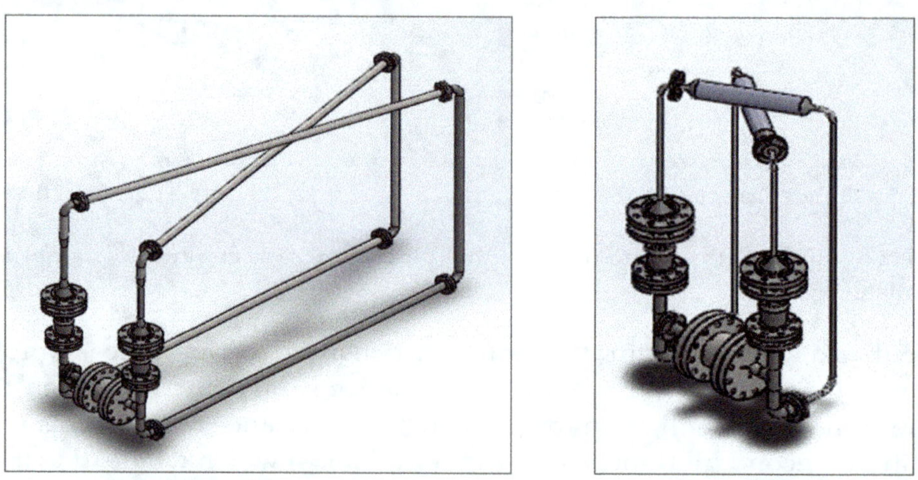

(a) (b)

Figure 18. Electricity generator with: (**a**) original feedback loop and (**b**) improved feedback loop.

Figure 17 compares the pressure phase, velocity phase and acoustic power in one stage for the original and new feedback loop. The first section of the feedback loop is a $1\frac{1}{2}$ inch diameter pipe with 300 mm length, which is part of the previous configuration. This is followed by a standard cone leading to the first inertance which is 1313 mm long and has a $\frac{3}{4}$ inch (20.9 mm) diameter. This shifts the pressure phase by approximately 100°, as shown in Figure 17a. The acoustic compliance is 420 mm long and 3 inches (77.9 mm) in diameter, which shifts the volumetric flow velocity phase by approximately 40°, as shown in Figure 17b. The second inertance is 584 mm long with dimeter of $\frac{1}{2}$ (12.2 mm) inch, and shifts the pressure phase by approximately 62°, as shown in Figure 17a. Both reducers connecting the compliance to the two inertances are non-standard, and of 50 mm length. The length and diameter of the two inertances and compliance were carefully optimized aiming to achieve the acoustic conditions at the shortest length possible.

6. Conclusions

Current work is focused on detailed studies and potential further improvements of a two-stage traveling-wave thermoacoustic engine. Here, the configuration of two identical half-wavelength stages allows the coupling of the linear alternator to two points with out-of-phase acoustic field, i.e., in so-called "push-pull" mode in an attempt to improve the impedance matching of the alternator to the engine, as well as reduce the ultimate cost by requiring only one alternator for two power extraction points. Modelling approaches are combined with experimental work in order to improve the overall performance of the prototype as well as improve the design to achieve more compact size.

In particular, the work presented here deals with system debugging, for instance improvements in electrical power output through limiting the axial heat leaks, and investigating the effects of regenerator length (i.e., regenerator impedance) on the start-up conditions in order to allow the engine self-excite without external power input and application of elastic membrane to eliminate Gedeon streaming. In addition, the paper presents a detailed account of the characterisation of the electricity generator system from the point of view of the mean pressure (range 14–28 bar) and heating power (500–1700 W) and load resistance (26.3 Ω to 92.5 Ω). It was found that the maximum electricity generated can reach 73.3 W at the heat input of 1300 W, load resistance of 30.8 Ω and mean pressure of 28 bar, with the overall thermal-to-electric efficiency of 5.64%. The maximum thermal-to-electric efficiency of 7.3% was obtained at heat input power of 700 W, while generating 51.1 W of electricity.

Finally, a design exercise was carried out aiming at reducing the size of the device while maintaining the same levels of performance. DeltaEC simulations have shown that introducing an inertance-compliance-inertance coupling instead of the constant diameter feedback pipe can reduce the resonator length from 16.1 m to 7.5 m, leading to a much smaller volume of the device.

Author Contributions: A.J.J. and X.M. guided the research. A.H. carried out the research by conceptualising the design, DeltaEC modelling, experimental data collection and producing the first draft. All authors critically evaluated the data and contributed to improving the manuscript through a number of iterations, in particular interpretation and discussion.

Acknowledgments: Artur J. Jaworski acknowledges funding from the Royal Society Industry Fellowship scheme (Ref. IF110094, 2012-2015) and EPSRC UK under $HARP^2$ programme (Ref. EP/R023328/1, 2018-2021). All authors acknowledge Innovate UK for financial support under TITAN project (ref no. 131497).

References

1. Swift, G.W. *Thermoacoustics: A Unifying Perspective for Some Engines and Refrigerators*; Acoustical Society of America (through the American Institute of Physics): Melville, NY, USA, 2002.
2. Luo, E.; Dai, W.; Zhang, Y.; Ling, H. Thermoacoustically driven refrigerator with double thermoacoustic-Stirling cycles. *Appl. Phys. Lett.* **2006**, *88*, 074102. [CrossRef]
3. Dai, W.; Luo, E.; Hu, J.; Ling, H. A heat-driven thermoacoustic cooler capable of reaching liquid nitrogen temperature. *Appl. Phys. Lett.* **2005**, *86*, 224103. [CrossRef]
4. Dai, W.; Yu, G.; Zhu, S.; Luo, E. 300 Hz thermoacoustically driven pulse tube cooler for temperature below 100 K. *Appl. Phys. Lett.* **2007**, *90*, 024104. [CrossRef]
5. Yazaki, T.; Iwata, A.; Maekawa, T.; Tominaga, A. Traveling wave thermoacoustic engine in a looped tube. *Phys. Rev. Lett.* **1998**, *81*, 3128. [CrossRef]

6. Chen, B.; Yousif, A.A.; Riley, P.H.; Hann, D.B. Development and assessment of thermoacoustic generators operating by waste heat from cooking stove. *Engineering* **2012**, *4*, 25859. [CrossRef]
7. Zhang, X.; Chang, J. Onset and steady-operation features of low temperature differential multi-stage travelling wave thermoacoustic engines for low grade energy utilization. *Energy Convers. Manag.* **2015**, *105*, 810–816. [CrossRef]
8. de Blok, K. Novel 4-stage traveling wave thermoacoustic power generator. In Proceedings of the ASME 2010 3rd Joint US-European Fluids Engineering Summer Meeting Collocated with 8th International Conference on Nanochannels, Microchannels, and Minichannels, Montreal, QC, Canada, 1–5 August 2010; pp. 73–79.
9. Hu, J.Y.; Luo, E.C.; Zhang, L.M.; Chen, Y.Y.; Wu, Z.H.; Gao, B. Analysis of a displacer-coupled multi-stage thermoacoustic-Stirling engine. *Energy* **2018**, *145*, 507–514. [CrossRef]
10. de Blok, K. Multi-stage traveling wave thermoacoustics in practice. In Proceedings of the 19th International Congress on Sound and Vibration, Vilnius, Lithuania, 8–12 July 2012.
11. Zhang, X.; Chang, J.; Cai, S.; Hu, J. A multi-stage travelling wave thermoacoustic engine driven refrigerator and operation features for utilizing low grade energy. *Energy Convers. Manag.* **2016**, *114*, 224–233. [CrossRef]
12. Yang, R.; Wang, Y.; Jin, T.; Feng, Y.; Tang, K. Development of a three-stage looped thermoacoustic electric generator capable of utilizing heat source below 120 C. *Energy Convers. Manag.* **2018**, *155*, 161–168. [CrossRef]
13. Yang, R.; Wang, Y.; Feng, Y.; Jin, T.; Tang, K. Performance of a looped thermoacoustic engine with multiple loads capable of utilizing heat source below 200 °C. *Appl. Therm. Eng.* **2019**, *148*, 516–523. [CrossRef]
14. Li, L.; Wu, Z.; Hu, J.; Yu, G.; Luo, E.; Dai, W. A novel heat-driven thermoacoustic natural gas liquefaction system. Part I: Coupling between refrigerator and linear motor. *Energy* **2016**, *117*, 523–529. [CrossRef]
15. Li, D.; Wu, Z.; Luo, E.; Zhang, L. Experimental Investigation on the Conversion between Heat and Power of the kW-class Thermoacoustic Engine. *Energy Procedia* **2014**, *61*, 1058–1062.
16. Wu, Z.; Yu, G.; Zhang, L.; Dai, W.; Luo, E. Development of a 3kW double-acting thermoacoustic Stirling electric generator. *Appl. Energy* **2014**, *136*, 866–872. [CrossRef]
17. Bi, T.; Wu, Z.; Zhang, L.; Yu, G.; Luo, E.; Dai, W. Development of a 5kW traveling-wave thermoacoustic electric generator. *Appl. Energy* **2017**, *185*, 1355–1361. [CrossRef]
18. Hamood, A.; Jaworski, A.J.; Mao, X.; Simpson, K. Design and construction of a two-stage thermoacoustic electricity generator with push-pull linear alternator. *Energy* **2018**, *144*, 61–72. [CrossRef]
19. Ward, B.; Clark, J.; Swift, G. *Design Environment for Low (Amplitude Thermoacoustic Energy Conversion DeltaEC Version 6.3 b11 Users Guide*; Los Alamos National laboratory: Santa Fe, NM, USA, 2012.
20. Fusco, A.M.; Ward, W.C.; Swift, G.W. Two-sensor power measurements in lossy ducts. *J. Acoust. Soc. Am.* **1992**, *91*, 2229–2235. [CrossRef] [PubMed]
21. Yu, Z.; Backhaus, S.; Jaworski, A.J. Design and Testing of a Travelling-Wave Looped-Tube Engine for Low-Cost Electricity Generators in Remote and Rural Areas. In Proceedings of the 7th International Energy Conversion Engineering Conference, Denver, CO, USA, 2–5 August 2009. [CrossRef]
22. Yu, Z.; Jaworski, A.J. Impact of acoustic impedance and flow resistance on the power output capacity of the regenerators in travelling-wave thermoacoustic engines. *Energy Convers. Manag.* **2010**, *51*, 350–359. [CrossRef]
23. Gedeon, D. DC gas flows in Stirling and pulse tube cryocoolers. In *Cryocoolers 9*; Springer: New York, NY, USA, 1997; pp. 385–392.
24. Wilcox, D.; Spoor, P. Gedeon streaming suppression in a small scale thermoacoustic-Stirling engine-generator. *J. Acoust. Soc. Am.* **2014**, *135*, 2408. [CrossRef]
25. Tijani, M.E.H.; Spoelstra, S. High temperature thermoacoustic heat pump. In Proceedings of the 19th International Congress on Sound and Vibration, Vilnius, Lithuania, 8–12 July 2012; pp. 8–12.

11

A Numerical Study on Influence of Temperature on Lubricant Film Characteristics of the Piston/Cylinder Interface in Axial Piston Pumps

Yueheng Song [1], Jiming Ma [2] and Shengkui Zeng [1,3,*]

1 School of Reliability and System Engineering, Beihang University, Beijing 100191, China; songyh125@buaa.edu.cn
2 Sino-French Engineering School, Beihang University, Beijing 100191, China; jiming.ma@buaa.edu.cn
3 Science and Technology on Reliability and Environmental Engineering Laboratory, Beihang University, Beijing 100191, China
* Correspondence: zengshengkui@buaa.edu.cn

Abstract: The loss of kinetic energy of moving parts due to viscous friction of lubricant causes the reduction of piston pump efficiency. The viscosity of lubricant film is mainly affected by the thermal effect. In order to improve energy efficiency of piston pump, this research presents a numerical method to analyze the lubricant film characteristics in axial piston pumps, considering the thermal effect by the coupled multi-disciplinary model, which includes the fluid flow field expressed by Reynolds equation, temperature field expressed by energy equation and heat transfer equation, kinematics expressed by the motion equation. The velocity and temperature distributions of the gap flow of piston/cylinder interface in steady state are firstly numerically computed. Then the distributions are validated by the experiment. Finally, by changing the thermal boundary condition, the influence of thermal effect on the lubricant film, the eccentricity and the contact time between the piston and cylinder are analyzed. Results show that with the increase of temperature, the contact time increases in the form of a hyperbolic tangent function, which will reduce the efficiency of the axial piston pump. There is a critical temperature beyond which the contact time will increase rapidly, thus this temperature is the considered as a key point for the temperature design.

Keywords: thermodynamic; numerical simulation; thermal effect; axial piston pumps

1. Introduction

As sliding pairs of axial piston pump move, both the metal parts and lubricants heat up, which causes temperature to increase significantly, decreases oil viscosity, reduces the bearing capacity of the oil, and intensifies the radial movement of the parts simultaneously [1]. The shape of the oil film changes, as does the lubrication mode, and the trajectory of piston could vary with the changing of oil film. The temperature impacts on oil film characteristics are important for analysis of the efficiency of the sliding pair.

Practical experience also indicates that the medium temperature significantly affects life. Cai et al. [2] tested the life of steel materials with a lubricating medium in the range of 23–175 °C and verified that the wear degree was directly related to the temperature. Specifically, the life of steel materials was closely related to the oil temperature distribution in the gap, and the distribution condition had a distinct influence on the oil characteristics. Unfortunately, the general test method could take temperature as the only variables; thus the results obtained by this way could only indicate the existence of a fuzzy relationship between temperature and life.

To indicate the relationship between the temperature and life more precisely, the behavior of lubricant film between the moving pair needs to be studied. Because the moving process is affected by fluid, solid, and thermal conditions together, it is a typical multiple-domain coupled problem. Many researchers have studied this problem around the bearing. McCallion et al. [3] solved the Reynolds and energy equations separately and neglected the effects of pressures during the calculation of temperature distribution. Ferron et al. [4] thoroughly studied the thermohydrodynamic (THD) performance of a plain journal bearing and determined its thermal characteristics. Rohde et al. [5] studied the elastic and thermal deformations in a slider bearing, noted that the variations in fluid viscosity with temperature were much more influential than solid deformation due to thermal and elastic effects. Researchers are still interested in the journal bearing where oil or gas is used as lubricants [6,7]. Lu et al. [8] conducted a thermal-fluid coupling study on characteristics of air–oil two phase flow and heat transfer in a micro unmanned aerial vehicle (UAV) bearing chamber. Some researchers took the cavitation into account [9] and others developed an efficient numerical method [10].

Axial piston pumps with constant pressure are the main type of engine-driven pumps (EDP), whose lifetime is affected by three sliding pairs, specifically the swash plate/slipper pair, cylinder/valve plate pair, and piston/cylinder pair, as shown in Figure 1.

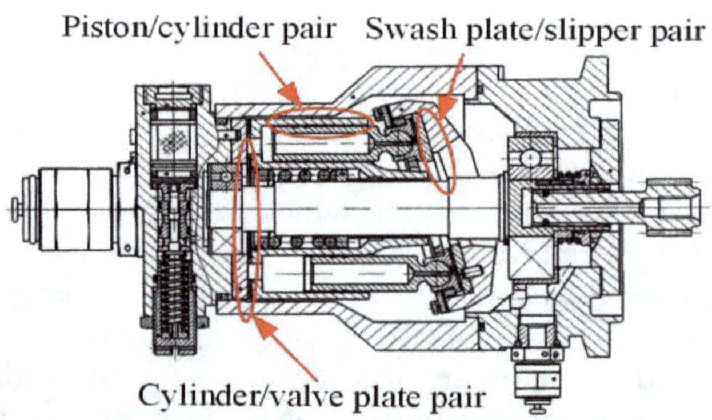

Figure 1. Sliding pairs in axial piston pump.

Several Elasto Hydrodynamic Lubrication (EHL) researches have focused on the piston pump. Gels et al. developed a simulation tool based on the Reynolds equation, which allowed to vary the geometry of the sliding parts. With the help of this tool, Gels have found an optimal compromise of piston/cylinder interface which reduces the losses of energy [11]. Ma et al. presented a method on the basis of the EHL model to analyze the behavior of the swash plate/slipper pair [12]. Olems presented an analytical model that integrates the multiphysics characteristic equations including Reynolds Equation and energy equation [13]. Given inlet medium temperature and operating conditions, the temperature distribution in the clearance of piston/cylinder interface was determined using the model. A series of studies of MAHA research center were carried out based on Olem's results. Ivantysynova et al. established a complete model of main sliding pairs for thermal analysis of oil film gap [14]. Kazama established a non-isothermal model of slipper/swash plate pair based on thermos-hydraulic dynamic theory [15]. Pelosi, Ivantysynova [16], and Zecchi [17] built an EHL model for the piston/cylinder pair and the swash plate/slipper pair. However, as shown in all these studies, few studies have modelled the interaction relationship among elastic, thermal and hydraulic behaviors only use numerical method without establishing the computer-aided design (CAD) model or using commercial software.

This study was designed to investigate the temperature distribution for the further accurate analysis on the thermal effect on the piston/cylinder interface. The flow velocity distribution in the piston/cylinder interface at a steady operating state of the piston pump (the output pressure of pump is steady) was calculated using the Reynolds equation. Then, the velocity distribution was imported into the model, considering the comprehensive effects of flow and temperature distributions in the gap, and the movement of moving components.

Only the numerical method is used during the whole simulation process. So, a parametric simulation model without the construction of a three-dimensional (3D) digital model was established, which is convenient for further application to the simulation of sliding pairs with different structures.

2. Description of Model

Under the assumption that the gap clearance (radial direction) was much smaller than the other two dimensions (circumferential and axial directions), an unwrapped Cartesian reference system was used for the description of the computation domain. The coordinates in this reference system are defined as Equation (1) and presented in Figure 2.

$$\begin{cases} x = \theta \\ y = r \\ z = z \end{cases} \tag{1}$$

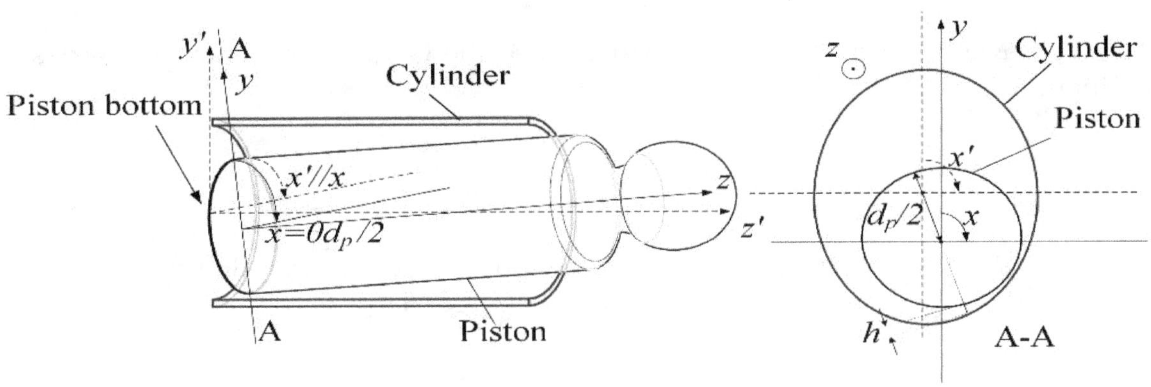

Figure 2. Definitions of coordinates and gap height.

2.1. Mathematical Model

The axial piston pump has an odd number of pistons arranged in a circular array within a housing that is commonly referred to as a cylinder block. This cylinder block is driven to rotate about its axis of symmetry by an integral shaft that is aligned with the pumping pistons. A piston/cylinder interface is an assembly with one piston, one cylinder, and a film of lubricant oil. When the piston pump is operating, the piston moves in a linear reciprocating pattern within the cylinder block, suctioning or discharging oil.

Analysis of thermodynamic behavior of piston/cylinder interface in piston pumps is a fluid-structure interaction problem. The motion for fluid flow, heat transfer, and the motion of the piston are expected to be determined simultaneously. Because it is a time-dependent problem, one shaft revolution is divided into a number of discrete time steps. For each time step, the problem was considered time-independent.

The momentum transport for the fluid flow in the gap is described by the Reynolds equation [18]:

$$\frac{\partial}{\partial x}\left(\frac{\partial p}{\partial x}\frac{h^3}{\mu}\right) + \frac{\partial}{\partial z}\left(\frac{\partial p}{\partial z}\frac{h^3}{\mu}\right) = 6\left((U_h - U_0)\frac{\partial h}{\partial x} + (V_h - V_0)\frac{\partial h}{\partial z} + 2\frac{\partial h}{\partial t}\right) \tag{2}$$

where h is the film thickness, μ is the oil viscosity, U_h is fluid velocity component along the circumferential direction on the boundary $y = h$, U_0 is fluid velocity component along the circumferential direction on the boundary $y = 0$, V_h is fluid velocity component along the axial direction on the boundary $y = h$, V_0 is fluid velocity component along the axial direction on boundary $y = 0$. $y = h$ indicates the interface between the cylinder and oil, and $y = 0$ indicates the interface between the piston and oil. We assumed that the velocity of oil at oil/solid interface was the same as the solid. So U_h, U_0, V_h, and V_0 were constant.

The temperature field of the fluid is governed by the convection-conduction equation [18]:

$$c_o \frac{\partial(\rho_o T_o)}{\partial t} + \rho_o c_o \mathrm{div}(\mathbf{V} \cdot T_o) = k_o \mathrm{div}(\mathrm{grad} T_o) + \phi_o \tag{3}$$

where, T_o is oil temperature, c_o is the specific heat capacity at constant pressure of oil, ρ_o is the density of oil, k_o is the thermal conductivity of oil, and ϕ_o is the viscous dissipation term.

The vector **V** in Equation (3) is the velocity of fluid composed by u and v which are determined by followed equations:

$$\begin{aligned} u &= \tfrac{1}{2\mu}\tfrac{\partial p}{\partial x}(y^2 - yh) + (U_h - U_0)\tfrac{y}{h} + U_0 \\ v &= \tfrac{1}{2\mu}\tfrac{\partial p}{\partial z}(y^2 - yh) + (V_h - V_0)\tfrac{y}{h} + V_0 \end{aligned} \tag{4}$$

The temperature depends on the fluid velocity, so the last term in Equation (3) describes the energy dissipation:

$$\phi_o = \mu\left[\left(\frac{\partial u}{\partial y}\right)^2 + \left(\frac{\partial v}{\partial y}\right)^2\right] \tag{5}$$

Then the local viscosity of grid point in the gap can be calculated using the temperature distribution with the Reynolds viscosity temperature equation [18]:

$$\mu = \mu_{\mathrm{ref}} e^{-\sigma(T_o - T_{\mathrm{ref}})} \tag{6}$$

where T_{ref} is the reference temperature, μ_{ref} is the viscosity of oil at reference temperature, and σ is the viscosity-temperature index.

The local viscosity is then used as new value in the following iteration cycle of algorithm.

For the temperature distribution inside the piston, the temperature distribution was determined using the heat conduction equation, which is written as Equation (7) [18]:

$$c_p \frac{\partial(\rho_p T_p)}{\partial t} = k_p \mathrm{div}(\mathrm{grad} T_p) + \phi_p \tag{7}$$

where T_p is the piston temperature, c_p is the specific heat capacity at constant pressure of piston, ρ_p is the density of piston, k_p is the thermal conductivity of piston, and ϕ_p is the viscous dissipation term.

Because the cylinder is geometrically much larger than the gap and the piston, the temperature is here considered a constant, T_c (case temperature).

2.2. Boundary Conditions

To simplify the model, only one piston/cylinder interface was considered. This piston/cylinder formed the system boundary of the thermal calculation, as shown in Figure 3. Because the nine piston/cylinder interfaces were symmetrical around the distribution circle, the behavior of these interfaces could be expected to be almost the same. For this reason, it was here considered reasonable to choose one interface as a representative. The numbers here represent the corresponding boundaries as labeled in Figure 3.

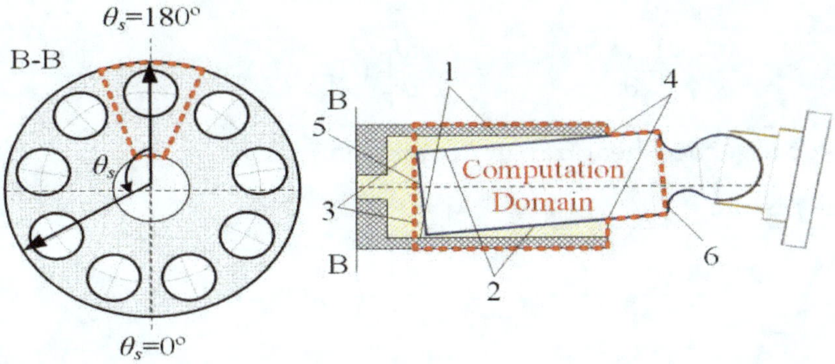

Figure 3. Computation domain and boundary.

The boundary conditions of fluid mechanics equations (Reynolds equation) are shown in Equation (8):

$$\begin{aligned} &\text{Interface 3}: P|_{\text{bottom edge}} = \begin{cases} P_{\text{suc}}(\theta_s = 0^\circ - 180^\circ) \\ P_{\text{dis}}(\theta_s = 180^\circ - 360^\circ) \end{cases} \\ &\text{Interface 4}: P|_{\text{top edge}} = P_{\text{case}} \end{aligned} \tag{8}$$

where P_{suc} is the output pressure of the pump when the piston is in the oil suction area (θ_s = 0°–180°), P_{dis} is the output pressure of the pump when the piston is in the oil discharge area (θ_s = 180°–360°), and P_{case} is the case pressure of the pump.

The boundary conditions of thermal analysis are the environmental temperature and temperatures of the medium at the inlet and outlet temperatures. For one piston/cylinder interface, solving the energy equation of fluid and heat conduction equation of solid covers the changes in heat in one phase and between two phases. For each specific boundary, the following boundary conditions are defined. The numbers here represent the corresponding boundaries as labeled in Figure 3.

(1) On the film-cylinder interface:

$$\left.\frac{\partial T_{\text{o}}(x,z)}{\partial y}\right|_{y=\frac{d_c}{2}} = -\frac{k_{\text{c}}}{k_{\text{o}} d_{\text{p}}} h(x,z) \left(\frac{T_{\text{o}}(x,z) - T_{\text{a}}}{\Delta y}\right) \tag{9}$$

(2) On the film-piston interface:

$$\left.\frac{\partial T_{\text{o}}(x,z)}{\partial y}\right|_{y=\frac{d_p}{2}} = -\frac{k_{\text{p}}}{k_{\text{o}} d_{\text{p}}} h(x,z) \left.\frac{\partial T_{\text{p}}(x,z)}{\partial y}\right|_{\text{piston-film}} \tag{10}$$

(3) Oil on the bottom edge of the film:

$$T_{\text{o}}|_{\text{bottom edge}} = T_{\text{mix}} \tag{11}$$

where T_{mix} is the mixing temperature, which is determined based on the energy balance [19].

(4) Oil on the top edge of the film:

$$T_{\text{o}}|_{\text{top edge}} = T_{\text{in}} \tag{12}$$

where T_{in} is inlet temperature of oil.

(5) On the bottom surfaces of the piston:

$$\left.\frac{\partial T_{\mathrm{P}}(x,y)}{\partial z}\right|_{z=0} = -\frac{\eta d_{\mathrm{P}}}{2k_{\mathrm{P}}}\left(T_{\mathrm{P}}(x,y)|_{z=0} - T_{\mathrm{mix}}\right) \tag{13}$$

where η is piston convection heat transfer coefficient.

(6) On the top of the piston:

$$\left.\frac{\partial T_{\mathrm{P}}(x,y)}{\partial z}\right|_{z=l_t} = -\frac{\eta d_{\mathrm{P}}}{2k_{\mathrm{P}}}\left(T_{\mathrm{P}}(x,y)|_{z=l_t} - T_{\mathrm{in}}\right) \tag{14}$$

Once the oil film temperature distribution was determined, mechanical analysis could progress.

2.3. Deformation Equation

For an accurate description of thermodynamic behavior of piston/cylinder interface, it is not sufficient to just solve the energy and Reynolds equation, but one also needs to consider deformation due to temperature and pressure. In this paper, the deformation matrix method was used to calculate the elastic deformation and thermal deformation of metal parts. The deformation equations are shown in Equation (15) [20]:

$$\begin{aligned} h_{\mathrm{E}}(i,j) &= \sum_f\sum_g [p_{f,g} + p_c|_{f,g}]\,C_P|_{f,g}^{i,j}\Delta A \\ h_{\mathrm{T}}(i,j) &= \sum_f\sum_g [T_P|_{f,g} - T_{\mathrm{ref}}]\,C_T|_{f,g}^{i,j} \end{aligned} \tag{15}$$

where h_{E} is the elastic deformation, $p_{f,g}$ is the oil pressure at node (f, g), $p_c|_{f,g}$ is the contact pressure, which is zero in this case, because the piston bears no force in the initial free case, $C_P|_{f,g}^{i,j}$ is the elastic deformation matrix, which represents the displacement of node (i, j) caused by the unit load acting on the node (f, g).

Similarly, h_{T} is the thermal deformation, $C_T|_{f,g}^{i,j}$ is the thermal deformation matrix, $T_P|_{f,g}$ is the temperature at node (f, g), and T_{ref} is the reference temperature and set to 25 °C.

2.4. Force Analysis

The force must be analyzed to solve motion equations. To clarify the description of force analysis, some definitions are presented in Figure 4.

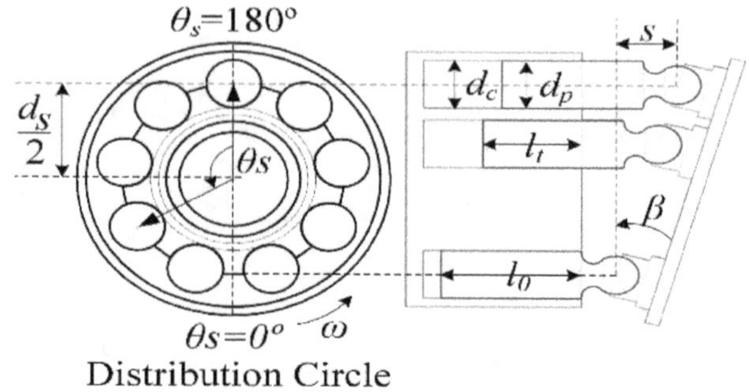

Figure 4. Definitions used for mechanical analysis.

When the pump is operating, the cylinder block slides over the valve plate. With the change in position within the distribution circle (θ_s), the displacement chambers of each piston are connected to the low-pressure port (θ_s = 0°–180°) or the high-pressure port (θ_s = 180°–360°). The pistons were joined to the slippers that glided over the swash plate. By changing the swash plate angle β, the piston stroke s could be adjusted. θ_s, l_t, and $\mathbf{s}$ were subject to Equation (16).

$$\begin{aligned} \theta_s &= \omega t \\ l_t &= \tfrac{d_s}{2}(1-\cos\theta_s)\tan\beta \\ \mathbf{s} = l_0 - l_t &= l_0 - \tfrac{d_s}{2}(1-\cos\theta_s)\tan\beta\cdot\mathbf{k} \end{aligned} \tag{16}$$

The loads acted on the piston in the cavity of piston/cylinder interface in the specific coordinates system is shown in Figure 5.

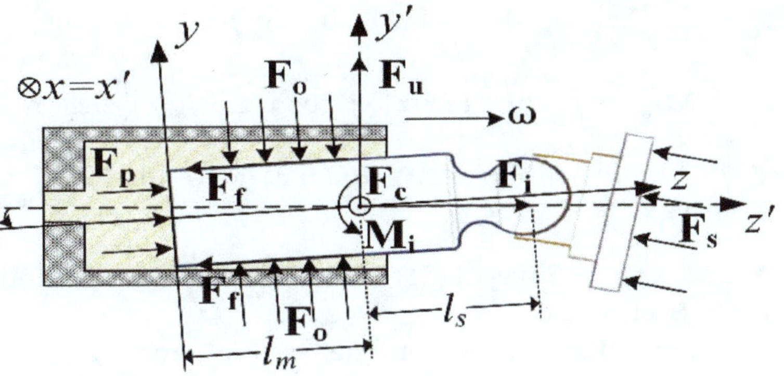

Figure 5. Graph of mechanical analysis.

Because of the overturning angle α, the velocity of piston $\mathbf{s}$ had a component normal to the $\boldsymbol{\omega}$, so the Coriolis force $\mathbf{F_c}$ took place. With the movement of cylinder block around the shaft, the centrifugal force $\mathbf{F_u}$ acted on the center of gravity of the piston. Because the piston movement took place within six degrees of freedom, there was also the inertia force in z-direction $\mathbf{F_i}$ and the inertia moment $\mathbf{M_i}$. All these forces acted on the center of gravity and were defined using Equation (17):

$$\begin{aligned} \mathbf{F_c} &= -2m\boldsymbol{\omega}\times\dot{\mathbf{s}} \\ \mathbf{F_u} &= m\omega^2\tfrac{d_s}{2}\mathbf{j}' \\ &\begin{cases} \mathbf{F_i} = -m\dot{\mathbf{s}} \\ \mathbf{M_i} = -J\dot{\boldsymbol{\omega}} \end{cases} \end{aligned} \tag{17}$$

The swash plate was removed for the calculation of acting force of slipper $\mathbf{F_s}$. $\mathbf{F_s}$ was resolved through the force equilibrium equation in the x-direction. $\mathbf{F_s}$ was decomposed into the y-direction and z-direction as presented in Equation (18):

$$\begin{aligned} \mathbf{F_{sz}} &= \mathrm{F_s}\cos(\alpha+\beta)\cdot\mathbf{k} \\ \mathbf{F_{sy}} &= \mathrm{F_s}\sin(\alpha+\beta)\cdot\mathbf{j} \end{aligned} \tag{18}$$

The pressure force $\mathbf{F_p}$ was calculated using the pressure of displacement chamber P_c, which differed between the low-pressure port (θ_s = 0°–180°) and the high-pressure port (θ_s = 180°–360°):

$$\mathbf{F_p} = \frac{\pi d_p{}^2}{4} P_c\cdot\mathbf{k} \tag{19}$$

The acting force of oil was normal to the piston surface and is defined as follows:

$$\mathbf{F_o} = \int_0^{2\pi}\int_0^{l_t} p(x,z)\cdot\mathbf{j}\cdot\cos x\frac{d_p}{2}dzdx \tag{20}$$

Due to the movement of piston, the friction of oil $\mathbf{F_f}$ acted on the piston. $\mathbf{F_f}$ was calculated using Equation (21):

$$\mathbf{F_f} = \left(\int_0^{2\pi} \int_0^{l_t} \left(-h(x,z) \left(\frac{\partial p(x,z)}{\frac{d_p}{2}\partial x} + \frac{\partial p(x,z)}{\partial z} \right) + \frac{\mu \dot{s}}{h(x,z)} \right) \frac{d_p}{2} dz dx \right) \cdot \mathbf{k} \tag{21}$$

Because the forces defined by Equations (17) and (19) acted through the center of gravity of the piston, their resultant moments were all zero. This left the moments produced by $\mathbf{F_s}$, $\mathbf{F_o}$, and $\mathbf{F_f}$, which can overturn the piston. These moments were calculated using Equations (22)–(24):

$$\mathbf{M_{F_s}} = \mathbf{F_s} \cdot l_s \sin(\alpha + \beta) \tag{22}$$

$$\mathbf{M_{F_o}} = \int_0^{2\pi} \int_0^{l_t} p(x,z) \cdot \mathbf{j} \cdot \cos x (z - l_m) \frac{d_p}{2} dz dx \tag{23}$$

$$\mathbf{M_{F_f}} = \left(\int_0^{2\pi} \int_0^{l_t} \left(-h(x,z) \left(\frac{\partial p(x,z)}{\frac{d_p}{2}\partial x} + \frac{\partial p(x,z)}{\partial z} \right) + \frac{\mu \dot{s}}{h(x,z)} \right) \cos x \frac{d_p{}^2}{2} dz dx \right) \cdot \mathbf{k} \tag{24}$$

where $p(x,z)$ in Equations (20), (21), (23) and (24) is the pressure distribution obtained by Section 2.1. $h(x,z)$ will be determined in Section 2.6.

Assuming the vector sum of forces and moments are $\mathbf{F_{sum}}$ and $\mathbf{M_{sum}}$.

$$\begin{aligned} \mathbf{F_{sum}} &= \mathbf{F_c} + \mathbf{F_u} + \mathbf{F_i} + \mathbf{F_s} + \mathbf{F_p} + \mathbf{F_o} + \mathbf{F_f} \\ \mathbf{M_{sum}} &= \mathbf{M_i} + \mathbf{M_{F_s}} + \mathbf{M_{F_o}} + \mathbf{M_{F_f}} \end{aligned} \tag{25}$$

2.5. Movement Analysis

The movement of the piston was analyzed during force analysis. According to Newton's second law, several types of force were identified. They are shown in Equation (26).

$$\begin{aligned} \mathbf{v_e}|_{t+dt} &= \mathbf{v_e}|_t + \frac{\mathbf{F_{sum}}}{m} dt \\ \mathbf{e}|_{t+dt} &= \mathbf{e}|_t + \mathbf{v_e}|_t dt \\ \mathbf{v_\alpha}|_{t+dt} &= \mathbf{v_\alpha}|_t + \frac{\mathbf{M_{sum}}}{J} dt \\ \boldsymbol{\alpha}|_{t+dt} &= \boldsymbol{\alpha}|_t + \mathbf{v_\alpha}|_t dt \end{aligned} \tag{26}$$

2.6. Calculation of the Thickness of the Oil Film

The thickness of the oil film $h(x,z)$ was determined as described below.

As shown in Figure 6, supposing that the space between the piston and cavity was filled with an oil film, the thickness of that oil film distribution could be determined according to the geometry characteristics of piston and cavity. In section C-C, O_1 was the center of the cavity, O_1' was the projection of O_1 on the y axis and O_2 was the center of piston, assuming that the piston overturning angle was α and projection of eccentric distance at bottom of piston $\mathbf{e}$ on y axis was e_0. Piston shaft and cavity shaft intersected in the position $(0, 0, l_1)|_{(x,y,z)}$. The following geometric relationships were established:

$$e_0 = l_1 \tan \alpha \tag{27}$$

For an arbitrary cross section C-C of piston, O_1 and O_2 were used to define a straight line that crossed the surface of the piston at P_1 and crossed the cavity surface at P_2. Any position on surface of piston P_1 could correspond to the circumferential angle θ. $|P_1P_2|$ was the thickness of the oil film, whose value is here written as $\overline{h}(\theta, z)$. In order to simplify the calculation, there was a transformation of system of coordinates. As shown in Figure 6b, O_2 was located at (0, 0, z) and O_1 at ($-e_2$, $-e_1$, z). e_1 and e_2 were subject to Equation (28):

$$\begin{aligned} e_1 &= |O_1'O_2| = (z - l_1)\tan\alpha \\ e_2 &= \frac{e_1}{\tan\theta} \end{aligned} \tag{28}$$

In this way, the algebraic expression of straight line determined by O_1O_2 was as follows:

$$y = x\tan\theta \tag{29}$$

On the cross section C-C, cavity was an ellipse, and the piston was a circle. Their algebraic expression was as shown in Equation (30):

$$\begin{aligned} \frac{4(x+e_2)^2}{d_c{}^2} + \frac{4(y+e_1)^2}{d_c{}^2\sec^2\alpha} &= 1 \\ x^2 + y^2 &= \frac{d_p{}^2}{4} \end{aligned} \tag{30}$$

The coordinates of P_1 and P_2 were determined by solving Equations (29) and (30) simultaneously. The results are presented in Equations (31) and (32):

$$\begin{cases} x_{P_1} = \frac{d_p}{2}\cos\theta \\ y_{P_1} = \frac{d_p}{2}\sin\theta \end{cases} \tag{31}$$

$$\begin{cases} x_{P_2} = \sqrt{\frac{d_c{}^2\sec^2\alpha}{4(\sec^2\alpha+\tan^2\theta)}} - e_2 \\ y_{P_2} = \sqrt{\frac{d_c{}^2\sec^2\alpha}{4(\sec^2\alpha+\tan^2\theta)}}\tan\theta - e_1 \end{cases} \tag{32}$$

Oil film thickness $\overline{h}(\theta, z)$ under specific overturning angle α was determined using the coordinates of P_1 and P_2:

$$\overline{h}(x,z) = \sqrt{(x_{P_1} - x_{P_2})^2 + (y_{P_1} - y_{P_2})^2} \tag{33}$$

Along with the elastic and thermal deformations, oil film thickness should be written as:

$$h(x,z) = \overline{h}(x,z) + h_E(x,z) + h_T(x,z) \tag{34}$$

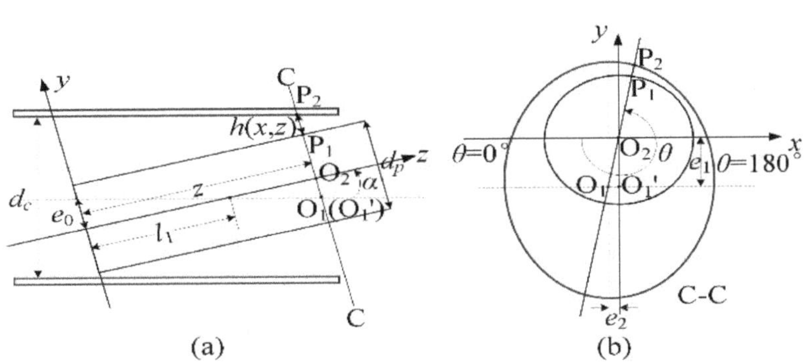

Figure 6. Graph of calculation of oil film thickness: (**a**) Axial section view; (**b**) Bottom view.

3. Simulation Algorithm and Conditions

Simulation of the thermodynamic behavior of the piston requires knowledge of the eccentric position, velocity of the piston, and all forces acting on it. Among these, eccentric position and forces are profoundly influenced by temperature distribution within the gap. An iterative method for the simultaneous calculation of all the requirements was here developed. Figure 7 shows a diagram of these calculations.

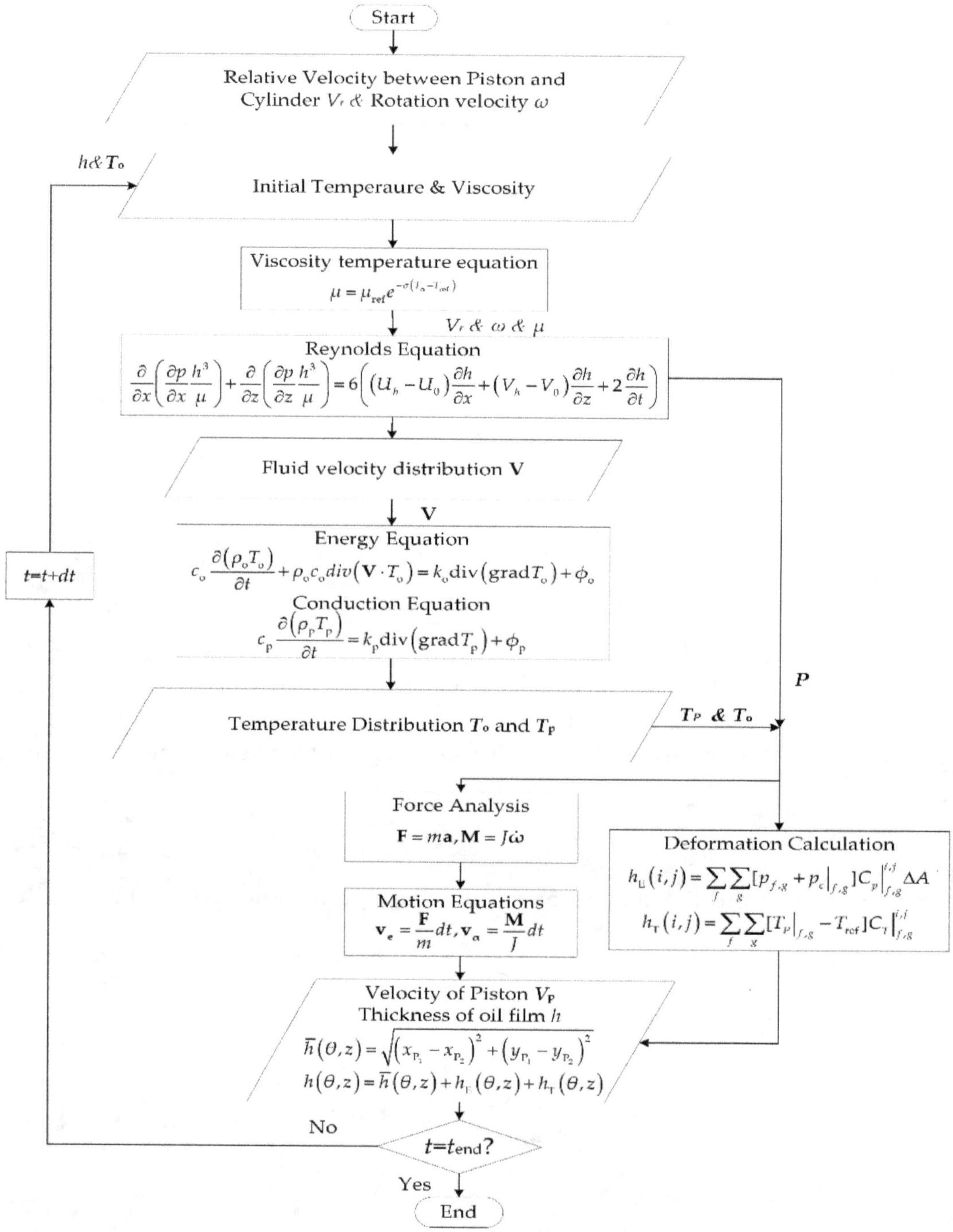

Figure 7. Flow chart of calculation.

With constant inlet and outlet pressure, the velocity distribution is considered constant and a basic condition of the Reynolds equation. When the velocity results were considered as the input of energy equation, the temperature distribution indicated by resolving of energy equation was the temperature distribution of oil film in steady state. The mechanical analysis was completed considering hydrodynamic forces and temperature distribution as inputs of motion equations. Using the overrelaxation iterative formula of the finite difference method, the Reynolds equation and the energy equation are solved by programming in MATLAB. If the absolute error of the results on all interior points obtained by two adjacent iterations is not greater than that of the prescribed error, then the iterative solution is considered to be convergent.

An axial piston pump with nine pistons was here selected for study. To start the simulation, some fundamental conditions were defined, which were the structure and working parameters of the pump, the material properties and thermal coefficients of the piston/cylinder interface, and the pressure of displacement chamber.

The structure and working parameters of a specific type of pump are presented in Table 1.

Table 1. Structure and working parameters.

Parameter	Value
Number of pistons	9
Rated working pressure	225 bar
Rated rotation speed	1000 rpm
Distribution circle diameter	80 mm
Piston diameter	30.14 mm
Cylinder cavity diameter	31.2 mm

The material properties and thermal coefficients of the pump are shown in Table 2 [21].

Table 2. Material properties.

	Density (kg/m^3)	Young Modulus (Gpa)	Specific Heat Capacity (J/(kg·K))	Heat Conductivity (W/(m·K))	Viscosity (20°C) (Pa·s)	Heat Convection Coefficient (W/(m^2·K))
Cylinder	8100	130	390	41.9	/	/
Piston	7850	210	490	50.28	/	50
Oil	862.1	/	1880	0.143	0.15	/

4. Results and Discussion

A simulation of the chosen piston pump was conducted with the model and numerical algorithm introduced in this paper.

In order to validate the developed model in this article, the simulated velocity and temperature distribution were compared with the experiment performed by Pelosi [22]. Constant inlet and outlet pressures on both sides of the piston when the pump was in normal operation were 16 bar (θ_s = 0°–180°) and 225 bar (θ_s = 180°–360°) (boundary $z = 0$ in Figure 8) and 5 bar (boundary $z = 1$ in Figure 8) respectively. For the temperature simulation, the input parameters were same as Pelosi [22], which are 57.5 °C for the case temperature, 48 °C for temperature at high pressure port, and 43 °C for temperature at low pressure port.

We selected the state where $l_t = l_0$, where the piston was in the position of $\theta_s = 0^\circ$ on the distribution circle, and the whole piston was in the cylinder, The corresponding α = 0.0067°. Figure 8 shows the velocity distribution of oil film at the interface of the oil and the piston. Figure 9a shows the result of thermal simulation, and Figure 9b shows the measurement result of the oil film at the interface of the oil and the piston carried by Pelosi [22]. Thus, we can conclude that the model could describe the oil temperature distribution, with reference to previous research.

Results indicated that the velocity distribution of oil film in piston/cylinder interface was affected not only by the relative velocity between piston and cylinder in the form of sinusoidal wave (z direction) but also the rotation of piston (x direction). The motion of the piston would lead the oil to follow its motion, and the change of the velocity direction of the oil would have hysteresis because of the viscous effect, so the vortex would be produced.

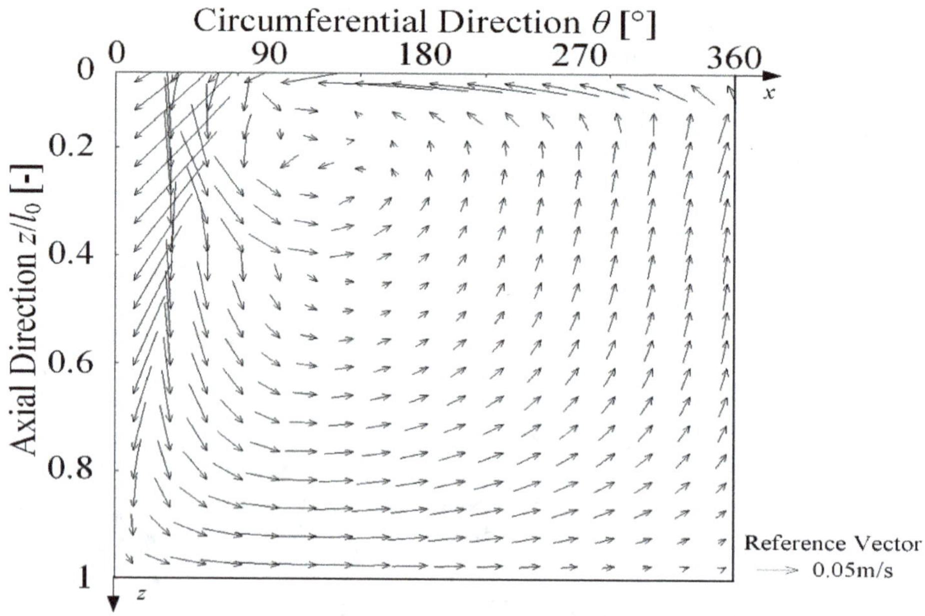

Figure 8. Velocity distribution.

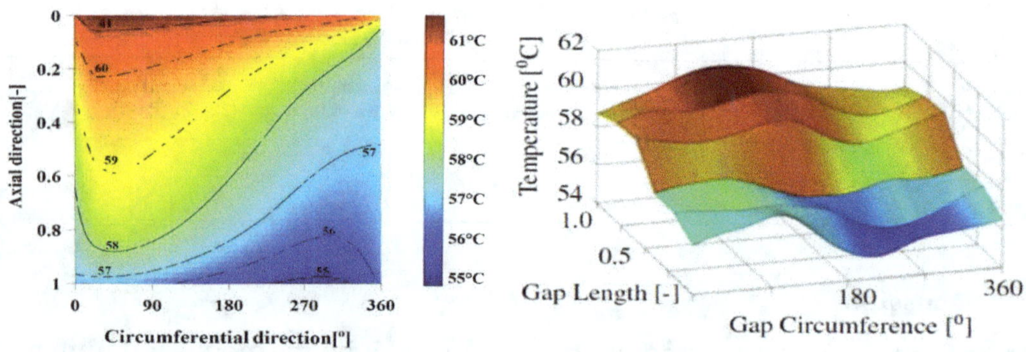

(a) (b)

Figure 9. (**a**) Simulated temperature distribution; (**b**) Measured temperature distribution by experiment [22].

Analysis of the Figure 8 also showed that the velocity of flow near the boundaries ($z = 0$ and $z = 1$) was faster than in the middle.

Comprehensive analysis of the contents of Figures 8 and 9 indicated that heat transfer was faster where the velocity was faster. Temperature diffused along the velocity, whether the temperature was high or low. This indicated that the accurate calculation of the velocity distribution was very important to the calculation of temperature distribution, considering the velocity has a highly significant effect on temperature diffusion.

The higher temperature region corresponded to a region where the fluid film was thinner, where more heat was generated. In these areas, the oil film tended to become thinner in next time step because high temperature reduced the carrying capacity of oil under the condition of lower viscosity. The eccentricity may be increased, but it still depended on the structure and the pressure in the chamber.

5. Analysis of Thermal Effect

With the thermal results, it was possible to calculate the fluid pressure and piston motion in the piston/cylinder interface. In order to investigate the effect of inlet oil temperature on the lubricating characteristics of oil film, we used the model proposed in this paper to simulate the movement of the bottom of piston in the chamber at different inlet temperatures. To facilitate the simulation, we assumed that the temperatures at high pressure port and low pressure port were the same as the inlet temperature T_{in}, and the T_{in} of the six simulations was 20 °C, 40 °C, 60 °C, 80 °C, 100 °C, and 140 °C respectively, and the case, temperature T_c was always 50 °C.

Figure 10 presents simulated trajectory of the piston bottom in the cavity over one shaft revolution with different inlet temperatures. The following conclusions can be summarized:

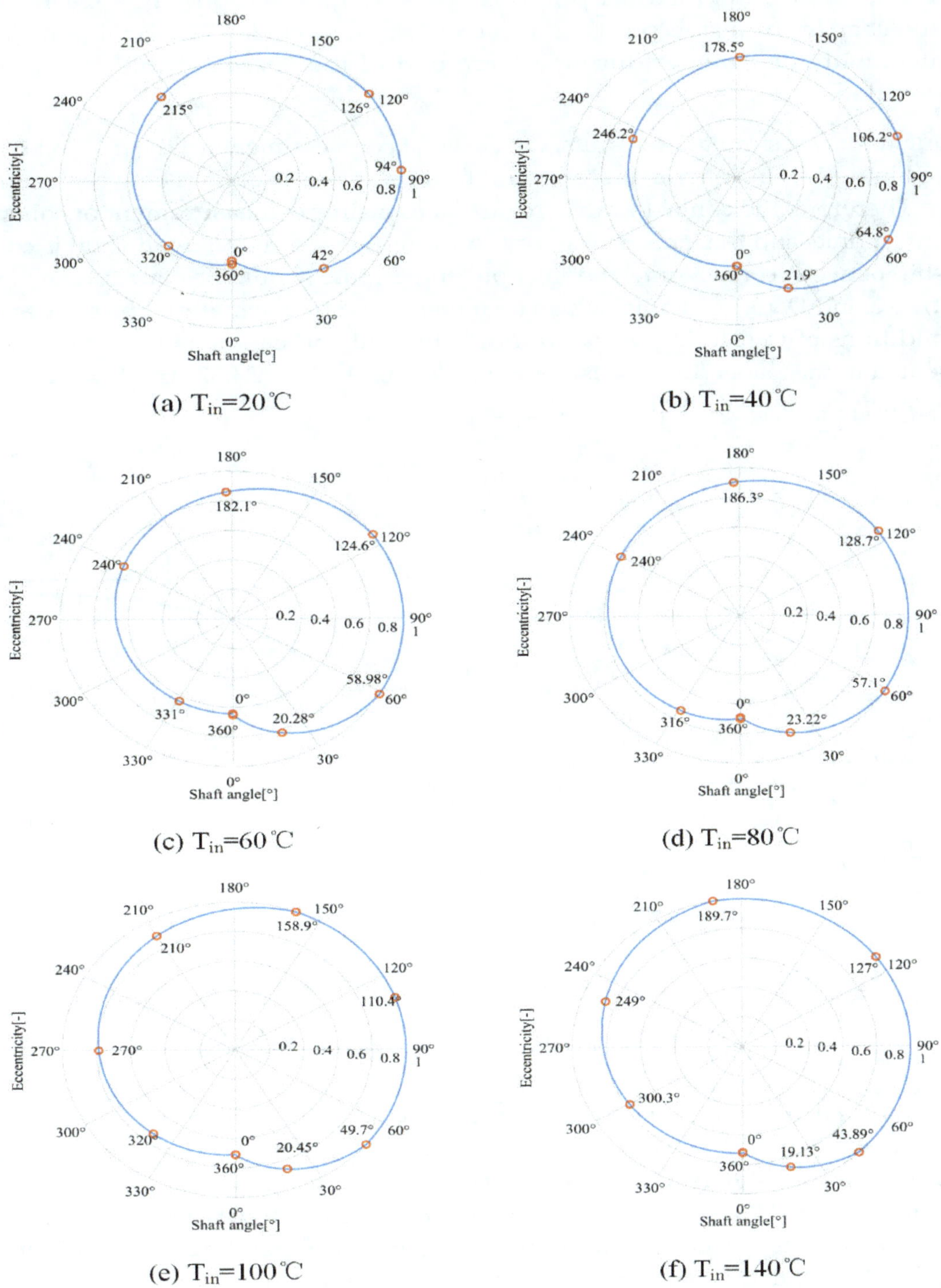

Figure 10. Simulated trajectory of the piston bottom in the cavity over one shaft revolution with different inlet temperatures.

(1). When the piston rotates within the suction zone (θ_s = 0°–180°), the oil pressure of the cylinder cavity is smaller, therefore the oil film supporting force is smaller, and the radial movement of the piston intensifies under the same inertial force compared with the condition when the pistons rotates into the outlet zone (θ_s = 180°–360°). The reduction of the eccentricity during the discharge area is caused by the very high pressure difference between the cylinder and the ambient pressure.

(2). With the increase of oil inlet temperature, the contact time between piston and cavity becomes longer and longer.

These results show that when the pump is operating, the eccentricity appears to be up and down with the changes of piston position in the distribution circle. The bottom of piston makes physical contact with the cylinder during a large part of the operating time, there will be more energy consumption.

According to the six simulation results, the curve of contact time with the inlet oil temperature is fitted out (see Figure 11). In general, there should be a positive correlation between contact time and solid friction. Therefore, the contact time is regarded as a simple measurement of solid friction, the longer the contact time, and the more the energy consumption. From our analysis, it is concluded that this relationship between contact time and inlet oil temperature should be expressed by Equation (35). The values *A*, *B*, *C* and *D* are constants, which are determined by the material properties, physics and operating conditions of the sliding pair and lubricating oil and can be obtained by the simulation method presented in this paper. In this case, $A = 0.15$, $B = 0.04$, $C = -84.37$, and $D = 0.25$. The R-square for the fitting, analyzed by MATLAB, is 0.97.

$$\frac{t_c}{t_{\text{total}}} = A \cdot \tanh(B(T_{\text{in}} + C)) + D \tag{35}$$

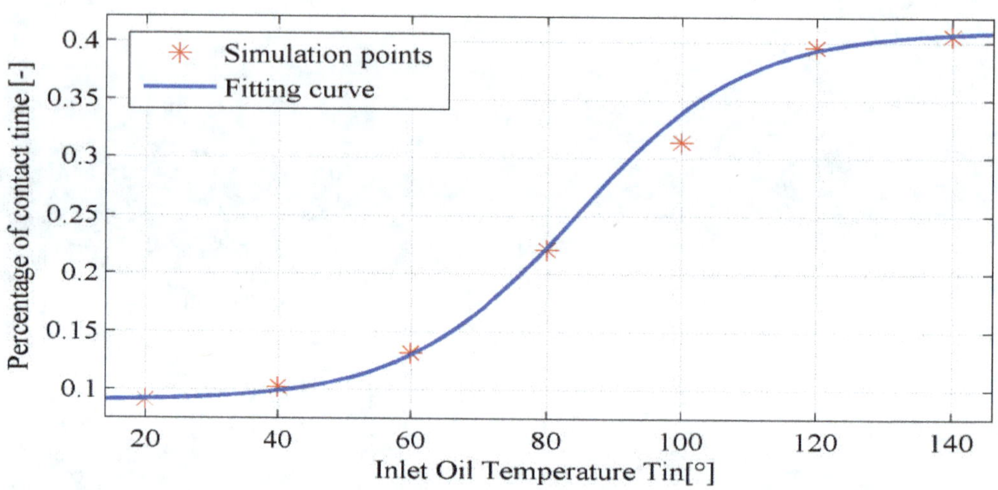

Figure 11. Curve of contact time with oil inlet temperature.

Figure 11 shows that the contact time increased with the increase of inlet oil temperature, and the growth rate increased first and then decreased, but would not increase all the time. From 20°C to 60°C, the contact time increased slowly and the temperature did not have much effect on the movement of the piston. From 60 °C to 100 °C, the contact time increased by a large margin, which may be due to the decrease in viscosity as the temperature continues to increase, resulting in a significant decrease in the bearing capacity of the lubricant film. The piston is in continuous contact with the cylinder under the action of centrifugal force. After 100 °C, the contact time between the piston and the cylinder will finally reach a balance due to the constraints of the structure of piston/cylinder interface and the working conditions (pressure).

6. Conclusions

A numerical multi-discipline modeling method suitable for describing the lubricant oil film characteristics in piston pump has been developed. The results of velocity field and temperature field are accurate compared with the experiment. Considering the thermal effect, the quantitative relationship between oil inlet temperature and contact time between piston and cylinder is obtained in this paper, and it is shown that the relationship is a hyperbolic tangent. Thus, it can be predicted when and where the contact will appear under different temperature conditions. This method can be helpful when it refers to the design and analysis of a piston pump.

In this paper, we consider the thermal effect on the oil film characteristics. In the future, we would include the oil pressure, axis rotate speed and the cavitation in the model and analyze all these factors' effects on the pump operation performance, such as oil leakage and pressure fluctuation. We will also establish a test rig to verify the simulation results.

Author Contributions: Conceptualization, Yueheng Song and Jiming Ma; Methodology, Formal Analysis, Investigation and Writing-Original Draft Preparation, Yueheng Song; Writing-Review & Editing, Supervision and Project Administration, Shengkui Zeng.

References

1. Yang, H.Y.; Ding, F.; Yang, X.P.; Lu, Q. Hydraulic power systems for trunk line aircrafts. *China Mech. Eng.* **2009**, *18*, 2152–2159.
2. Cai, Z.B.; Zhou, Y.; Qu, J. Effect of oil temperature on tribological behavior of a lubricated steel–steel contact. *Wear* **2015**, *332–333*, 1158–1163. [CrossRef]
3. McCallon, H.; Yousif, F.; Lloyd, T. The analysis of thermal effects in a full journal bearing. *J. Lubr. Technol.* **1970**, *92*, 578–587. [CrossRef]
4. Ferron, J.; Frene, J.; Boncompain, R.A. Study of the thermohydrodynamic performance of a plain journal bearing comparison between theory and experiments. *J. Lubr. Technol.* **1983**, *105*, 422–428. [CrossRef]
5. Rohde, S.M.; Oh, K.P. A thermoelastohydrodynamic analysis of a finite slider bearing. *J. Lubr. Technol.* **1975**, *97*, 450–460. [CrossRef]
6. Mi, J.T.; Meng, Y.G. THD analysis of rolling piston and journal bearings in rotary compressors. *Tribol. Trans.* **2016**, *59*, 195–207. [CrossRef]
7. Cheng, F.; Ji, W. A velocity-slip model for analysis of the fluid film in the cavitation region of a journal bearing. *Tribol. Trans.* **2016**, *97*, 163–172. [CrossRef]
8. Lu, P.; Zheng, X.; Li, W.; Yang, P.; Huang, H.; Yu, Y. A thermal-fluid coupling numerical study on the characteristics of air-oil two phase flow and heat transfer in a micro UAV bearing chamber. *RSC Adv.* **2017**, *7*, 44598–44604. [CrossRef]
9. Braun, M.J.; Hannon, W.M. Cavitation formation and modelling for fluid film bearings: A review. *J. Eng. Tribol.* **2016**, *224*, 839–863. [CrossRef]
10. Chan, C.W.; Han, Y.F.; Wang, Z.; Wang, J.; Shi, F.; Wang, N.; Wang, Q.J. Exploration on a fast ehl computing technology for analyzing journal bearings with engineered surface textures. *Tribol. Trans.* **2014**, *57*, 206–215. [CrossRef]
11. Gels, S.; Murrenhoff, H. Simulation of the lubricating film between contoured piston and cylinder. *Int. J. Fluid Power* **2010**, *11*, 15–24. [CrossRef]
12. Ma, J.M.; Chen, J.; Li, J.; Li, Q.L.; Ren, C.Y. Wear analysis of swash plate/slipper pair of axis piston hydraulic pump. *Tribolo. Int.* **2015**, *90*, 467–472. [CrossRef]
13. Olems, L. Investigations of the temperature behaviour of the piston cylinder assembly in axial piston pumps. *Int. J. Fluid Power* **2000**, *1*, 27–39. [CrossRef]
14. Ivantysynova, M.; Huang, C.C.; Andreas, J.J. Determination of Gap Surface Temperature Distribution in Axial Piston Machines. In Proceedings of the ASME International Mechanical Engineering Congress and Exposition, Chicago, IL, USA, 5–10 November 2006.
15. Kazama, T. Thermohydrodynamic lubrication model applicable to a slipper of swashplate type axial piston pumps and motors (Effects of operating conditions). *Tribol. Online* **2010**, *5*, 250–254. [CrossRef]

16. Pelosi, M.; Ivantysynova, M. The impact of axial piston machines mechanical parts constraint conditions on the thermo-elastohydrodynamic lubrication analysis of the fluid film interfaces. *Int. J. Fluid Power* **2013**, *14*, 35–51. [CrossRef]
17. Zecchi, M. A Novel Fluid Structure Interaction and Thermal Model to Predict the Cylinder block—Valve Plate Interface Performance in Swash Plate Type Axial Piston Machines. Ph.D. Thesis, Purdue University, West Lafayette, IN, USA, December 2013.
18. Wen, S.Z.; Huang, P. *Principle of Tribology*, 4th ed.; Tsinghua University Press: Beijing, China, 2012.
19. Khonsari, M.M.; Wang, S.H. On the fluid-solid interaction in reference to thermoelastohydrodynamic analysis of journal bearings. *J. Tribol.* **1991**, *113*, 398–404. [CrossRef]
20. Zhang, Z. Analysis of the Thermohydrodynamic Performance of Dynamically Loaded Bearings in Consideration of Non-Netonian, Derformation and Surface Topography. Ph.D Thesis, Shanghai Jiao Tong University, Shanghai, China, 2014.
21. Yang, S.M.; Tao, W.S. *Heat Transfer*, 4th ed.; Higher Education Press: Beijing, China, 2006.
22. Pelosi, M.; Ivantysynova, M. A Novel Thermal Model for the Piston/Cylinder Interface of Piston Machines. In Proceedings of the ASME Dynamic Systems and Control Conference, Hollywood, CA, USA, 12–14 October 2009.

12

Investigation on the Handling Ability of Centrifugal Pumps under Air–Water Two-Phase Inflow: Model and Experimental Validation

Qiaorui Si [1,*], Gérard Bois [2], Qifeng Jiang [3], Wenting He [1], Asad Ali [1] and Shouqi Yuan [1]

1 National Research Center of Pumps, Jiangsu University, Zhenjiang 212013, China; 741070483@qq.com (W.H.); 3360719084@qq.com (A.A.); shouqiy@ujs.edu.cn (S.Y.)

2 LMFL, FRE CNRS 3723, ENSAM, 8 boulevard Louis XIV, 59046 Lille CEDEX, France; Gerard.BOIS@ENSAM.EU

3 Key Laboratory of Fluid and Power Machinery, Xi Hua University, Chengdu 610039, China; qifeng.jiang@mail.xhu.edu.cn

* Correspondence: siqiaorui@ujs.edu.cn

Abstract: The paper presents experimental and numerical investigations performed on a single stage, single-suction, horizontal-orientated centrifugal pump in air–water two-phase non-condensable flow conditions. Experimental measurements are performed in a centrifugal pump using pressure sensor devices in order to measure the wall static pressures at the inlet and outlet pump sections for different flow rates and rotational speeds combined with several air void fraction (a) values. Two different approaches are used in order to predict the pump performance degradations and perform comparisons with experiments for two-phase flow conditions: a one-dimensional two-phase bubbly flow model, and a full "Three-Dimensional Unsteady Reynolds Average Navier–Stokes" (3D-URANS) simulation using a modified *k-epsilon* turbulence model combined with the Euler–Euler inhomogeneous two-phase flow description. The overall and local flow features are presented and analyzed. Limitations concerning both approaches are pointed out according to some flow physical assumptions and measurement accuracies. Some additional suggestions are proposed in order to improve two-phase flow pump suction capabilities.

Keywords: two-phase flow; pump performance; computational fluid dynamics; centrifugal pump; flow behavior

---

1. Introduction

Pumps play an important role in many energy engineering applications. Their main advantages are their high operational efficiency and reliability. In case of inlet two-phase flow conditions such as irrigation, refrigeration, petroleum, nuclear power industries, sewage treatment, etc., pump performance always fails under air–water two-phase inflow compared to single-phase operations. Especially, the need for the safe operation and design of a nuclear power reactor is of paramount importance to assure the safety evaluation of hypothetical loss-of-coolant accidents (LOCA) in light water reactors. Therefore, the investigation of pump behavior during blowdown is very important.

The degree of head degradation depends on the geometrical, physical, and thermal conditions. Compared with an axial flow pump that is able to suck high inlet gas volume fraction values, centrifugal pumps are less powerful when handling two-phase flow conditions (Gülich [1]). When the inlet gas void fraction increases, the pump head can be totally lost, and this often results in huge system instabilities and pump failure. Such a situation should be controlled using adequate prediction methods. Some improvements on pump ability to suck two-phase flow can also be achieved by a better understanding of local flow features. Several models have been developed in order to predict the two-phase flow performance of both the axial and centrifugal types of pumps. A first correlation was proposed in 1978 by Mikielewicz et al. [2], based on the concept of apparent loss coefficient. Among several theoretical approaches, the one-dimensional two-phase flow models proposed by Murakami et al. [3] and Minemura et al. [4] are still considered sufficiently accurate for engineering purposes for the prediction of pump surge or shut-off conditions.

A non-exhaustive list of important published works is proposed in the paper of Si et al. [5]. Related proposed models can be considered to be valid for low values of inlet void fraction (maximum 6–7% in volume). They cannot accurately predict the surge operating conditions that correspond to a rapid decrease in the pump performance, just because they only give mean flow characteristics along one mean streamline with no local information inside the impeller whole passage. Just before such severe conditions, several investigations have also detected the presence of stationary bubbles at impeller entrance channels for high gas fractions, being responsible for the starting point of the strong performance degradation of the pump (Sekoguchi et al. [6], Estevam et al. [7] and Barrios [8]). Numerical simulations, using the so-called "Three-Dimensional Unsteady Reynolds Average Navier–Stokes" (3D-URANS) approach, have also been applied to describe local phenomena more precisely in such flow conditions. They generally show significant deviations between predicted and experimental overall results when inlet void fraction values are higher than 6%. Recently, Müller et al. [9] have performed numerical calculations using a monodispersed phase distribution model, the results of which are compared with the experimental results obtained by Suryawijaya and Kosyna [10]. The comparison of impeller blade static pressure distributions showed that some improvements should be done for a high inlet gas flow rate in order to get a better fit between the numerical and experimental results. This means that there is still a need for experimental results on low specific speed pump geometries in order to obtain more data for numerical assessments.

The present paper presents a compilation of new experimental and numerical analysis performed by the same research team, starting from three years ago by Si et al. [5,11]. In the first part, the measured overall performances of the pump (head and global efficiency versus flow coefficients) are presented for three different rotational speeds and different inlet void fraction values up to pump breakdown. The results are then compared with the one-dimensional two-phase flow model developed by Minemura et al. [4]. In the second part, a comparison is proposed between the experimental results and 3D-URANS simulations using inhomogeneous flow model assumption. Based on the local flow feature analysis obtained from 3D-URANS simulations, radial impeller design modification is proposed to improve pump ability with respect to multiphase flows.

2. Pump Geometry and Experimental Test Rig

A single stage, low specific speed ($\Omega s = 0.65$) centrifugal commercial pump is chosen for two-phase flow investigations. The impeller casing is combined with a vaneless spiral volute. The design parameters of the pump, given by the manufacturer, are as follows: design volume flow rate: $Q_d = 50.6\ m^3/h$, design head: $H_d = 20.2$ m, design speed $N_d = 2910$ r/min, pump horizontal inlet tube diameter: $D = 65$ mm. The impeller inlet and outlet diameters are equal to 79 mm and 140 mm, respectively, and the impeller outlet width is equal to 15.5 mm. The six blades' inlet geometry corresponds to a three-dimensional twisted shape. The volute's law evolution obeys the Archimedes' spiral shape. The test rig is shown in Figure 1. As shown in this open loop, the air injection system is able to provide 0.2 mm diameter bubbles using a specific compressor device (more details can be found in Si et al. [5,11]). The air flow rate is measured by a microelectromechanical flow sensors system that could supply the volume air flow rate value on standard conditions (25 °C, 101.326 kPa). A gas–liquid flow mixture is sucked by the pump, goes through the regulating valve, and finally goes into the downstream tank. This open tank allows air bubbles to be separated from the water up to a second tank (upstream tank). The water flow rate was measured by an electromagnetic flowmeter set between the upstream tank and the air supply device. Measurements were performed using the followed procedure: a constant inlet void fraction was set by changing the throttle vane position and the corresponding water flow rate was consequently obtained. Measurement uncertainties are evaluated as a 1.2% error on the pump head and a 2.4% error regarding pump global efficiency. Figure 2 shows a general view of the sensor locations close to the pumping environment. Pump head and global efficiency values were obtained following ISO 9906:2012. It has to be noticed that only the air volume flow rate was measured at this step, as the bubble diameter distribution, including bubble number per unit volume at pump inlet, was not available.

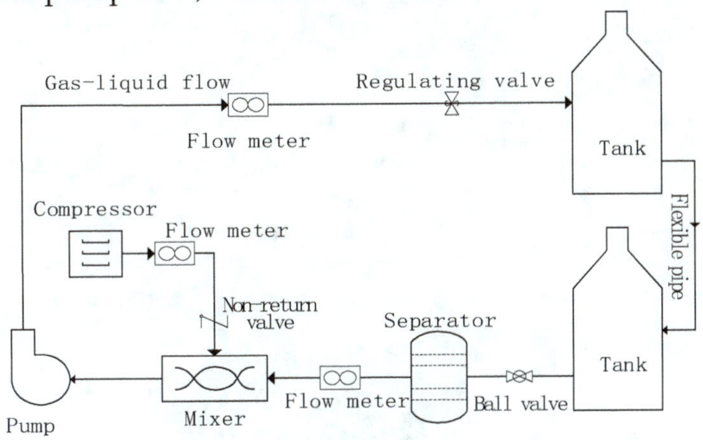

Figure 1. Test rig.

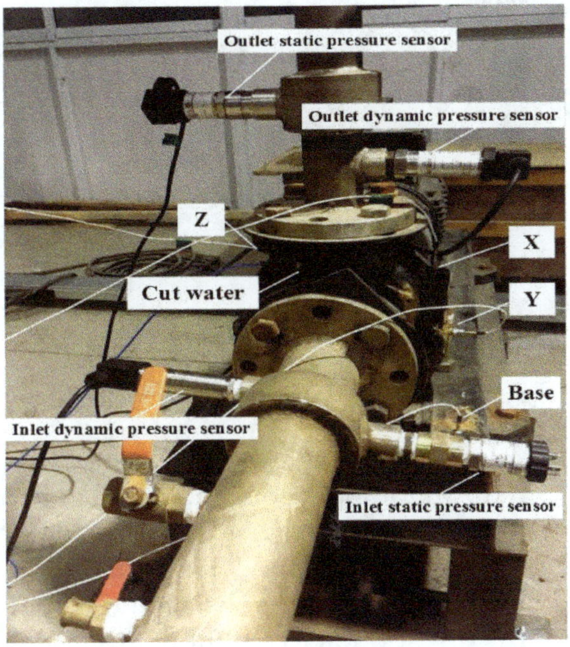

Figure 2. Sensor locations.

3. Numerical Model and Setups

3.1. Model and Mesh

Software Pro/E 5.0 was used for three-dimensional modeling and assembly. The computational domain included five separated domains: inlet pipe, ring, impeller, volute, and disc chambers, as illustrated in Figure 3. The hexahedron structured mesh was used to easily control the density of each node and properly adjust all of the nodes. In order to improve the accuracy of the simulation, the boundary layer grid was added on the volute shell wall after the local encryption of the volute tongue. The resulting pump model that consisted of 2,775,915 elements was chosen for rotating and stationary domains in total after grid independence analysis (see the paper of Si et al. [11]), the result of which is shown in Figure 4.

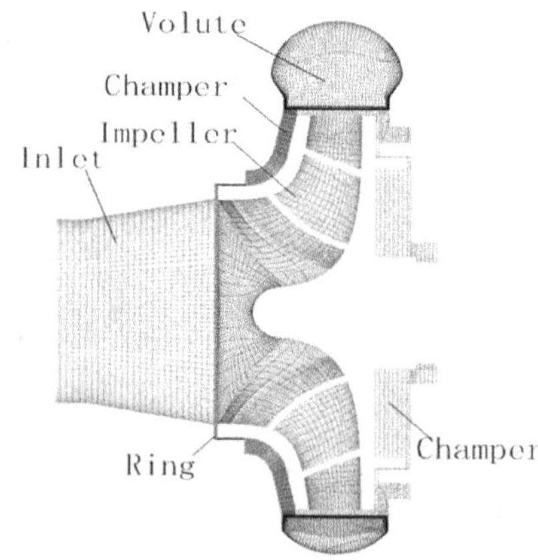

Figure 3. Three-dimensional (3D) numerical domain (partial) with local meshing.

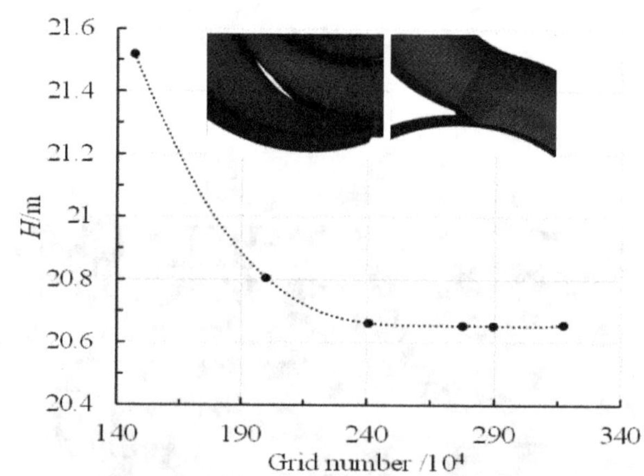

Figure 4. Grid independence results.

3.2. The Eulerian–Eulerian Inhomogeneous Two-Phase Flow Model

There are two kinds of multiphase flow models: homogeneous and inhomogeneous ones. The former assumes that the two-phase velocities are the same, not considering any velocity slip between the two phases. The latter considers not only the velocity slip, but also the interphase mass and momentum transfer terms. In the inhomogeneous model, each phase has its own fluid field and passes through the phase transfer unit. That is to say, each phase has its own velocity and temperature field, if any. This paper adopts the inhomogeneous model regardless of the temperature field, for which the liquid phase is continuous and the gas phase is discrete. The particle model assumes that the gas–liquid two-phase flow pattern corresponds to a bubbly flow, meeting the principle of the mass and momentum conservation:

$$\frac{\partial}{\partial t}(\alpha_k \rho_k) + \nabla \cdot (\alpha_k \rho_k \boldsymbol{w}_k) = 0 \tag{1}$$

$$\frac{\partial}{\partial t}(\alpha_k \rho_k \boldsymbol{w}_k) + \nabla \cdot (\alpha_k \rho_k \boldsymbol{w}_k \otimes \boldsymbol{w}_k) = -\alpha_k \nabla p_k + \nabla \cdot (\alpha_k \mu_k (\nabla \boldsymbol{w}_k + (\nabla \boldsymbol{w}_k)^{\mathrm{T}})) + \boldsymbol{M}_k + \boldsymbol{f}_k \tag{2}$$

where k—phase (l-liquid, g-gas); ρ_k—density of the k phase, kg/m^3; p_k—pressure of k phase, Pa; α_k—void fraction of k phase; μ_k—dynamic viscosity of k phase, Pa·s; $\boldsymbol{w}_k$—relative velocity of the k phase fluid, m/s; $\boldsymbol{M}_k$—due to interphase drag force; and $\boldsymbol{f}_k$—added mass force related to the contribution of the impeller rotation.

This also satisfied:

$$\alpha_g = 1 - \alpha_l = \frac{Q_g}{Q_g + Q_l} \quad (3)$$

where α_g—gas void fraction; α_l—liquid void fraction; Q_g—volume flow rate of the gas, m^3/h; and Q_l—volume flow rate of the liquid, m^3/h.

For this two-phase flow approach, the liquid phase is considered the continuous phase using a modified turbulence model that is based on the renormalization group (RNG) k-ε model [12]. Meanwhile, the gas phase is considered as the discrete phase using the zero-equation theoretical model, which means the action between the two phases only considers the so-called interfacial drag coefficient through the following relations [13]:

$$M_l = -M_g = \frac{3}{4} c_D \frac{\rho_l}{d_B} \alpha_g (w_g - w_l) |w_g - w_l| \quad (4)$$

with:

$$c_D = \begin{cases} \frac{24}{R_e}(1 + 0.15Re^{0.687}) & (R_e \leq 1000) \\ 0.44 & (R_e > 1000) \end{cases} \quad (5)$$

$$\text{And } Re = \rho_l \frac{|w_g - w_l|}{\mu_l} d_B \quad (6)$$

where d_B—diameter of the bubble, c_D—resistance coefficient.

3.3. The Modified Turbulence Model

As known, the turbulence model plays a significant role in the prediction of pumping two-phase flow. Since two-phase flow in centrifugal pumps is also an unsteady multiphase compressible flow concerning a liquid phase and a bubble phase, a corrected method was applied to solve that physical phenomenon. This method was proved effective on the prediction of cavitation flow in centrifugal pumps by Wang et al. [14]. The liquid density is defined as:

$$\rho_l = \rho_{ref} \sqrt[n]{\frac{p + C}{p_{ref} + C}} \quad (7)$$

where P_{ref} stands for the reference pressure at different void fraction, and ρ_{ref} is set as 998.2 kg/m^3. As for constant values of C and n, they are set to 300 MPa and seven for water, respectively, according to Dular and Coutier-Delgosha [15].

Moreover, the standard k–e model is deemed to overestimate the eddy viscosity in the mixture region. It cannot effectively resolve the detachment of the flow separation from the solid surface, and excessively attenuates the two-phase flow instability. Therefore, a modified RNG k–e model, as proposed by Coutier-Delgosha et al. [16], was employed in this work. It can successfully reduce the eddy viscosity by defining the turbulent viscosity, and based on RNG k-ε turbulence model, Johansen's filter-based model is used to correct the turbulent viscosity μ_t:

$$\mu_t = f(\rho_m) C_\mu \frac{k^2}{\varepsilon} \quad (8)$$

where $f(\rho_m)$ is the expression of the density correction which can be determined as:

$$f(\rho_m) = \begin{cases} 998.2 \text{ kg/m}^3 & (n = 1) \\ \rho_g + \frac{(\rho_m - \rho_g)^n}{(\rho_l - \rho_g)^{n-1}} & (n > 1) \end{cases} \quad (9)$$

where $f(\rho_m)$ changed with the density of the mixed flow following the rules, as shown in Figure 5. Coutier-Delgosha et al. [16] validated that n should be equal to 10 after comparing the experimental and numerical results.

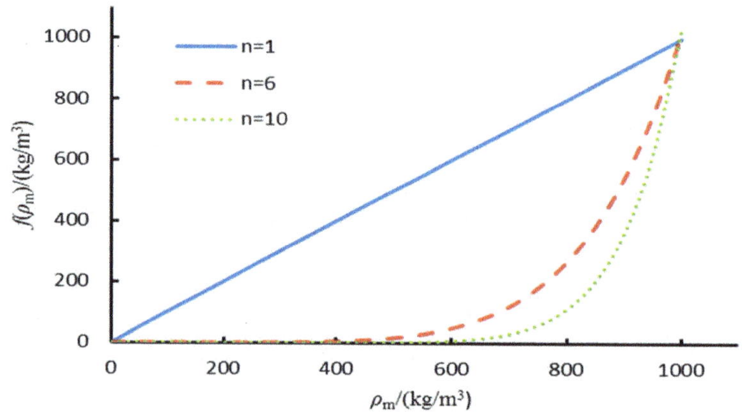

Figure 5. The law of the $f(\rho_m)$ with the density in the mixed phase.

Finally, the modified RNG k-ε turbulence model becomes as follows:

$$\frac{\partial(\rho_m k)}{\partial t} + \frac{\partial(\rho_m k \overline{u_j})}{\partial x_j} = \frac{\partial}{\partial x_j}\left[\left(\mu_m + \frac{\mu_t'}{\sigma_k}\right)\frac{\partial k}{\partial x_j}\right] + P_t - \rho_m \varepsilon \quad (10)$$

$$\frac{\partial(\rho_m \varepsilon)}{\partial t} + \frac{\partial(\rho_m \varepsilon u_i)}{\partial x_i} = \frac{\partial}{\partial x_j}\left[\left(\mu_m + \frac{\mu_t'}{\sigma_\varepsilon}\right)\frac{\partial \varepsilon}{\partial x_j}\right] + C_{\varepsilon 1}\frac{\varepsilon}{k} - C_{\varepsilon 2}\rho_m\frac{\varepsilon^2}{k} - R \quad (11)$$

Solutions of the simulations were processed in Ansys CFX 14.1 by adding the modified coefficients programs using CFX Expression Language (CEL).

3.4. Boundary Conditions

Considering the requirements of the test, the inlet boundary conditions are set according to the pressure measured in the experiment. A certain amount of gas is imported into the inlet, and the outlet boundary conditions are set as the mass flow rate. The liquid boundary adopts the non-slip solid wall condition, and the gas boundary adopts the free-slip solid wall condition. The smooth wall condition is used for the near-wall function. The initial inlet bubble diameter is set as 0.2 mm, according to the experimental setup device. The time step was set as 1.718×10^{-4} s, corresponding to a 3° impeller rotation for each step. A total time set was 0.206 s for simulated results storage in order to further process a data reduction of the unsteady flow field, corresponding to 10 impeller revolutions for which stable results were obtained.

4. Overall Pump Performances

In previous experimental work [5], as well as for pure water, pump head and global efficiency curves were found to be independent of rotational speed inside the range of 1800–2910 r/min. This result agrees with the similarity laws that have been applied for incompressible flow for impeller Reynolds number values (based on impeller diameter and angular speed) that are higher than the critical one that is usually adopted for turbomachinery applications. The rotational speed of 2910 r/min corresponds to the rated one. This result was achieved by considering that the maximum experimental error is 6% for the head at partial flow rates, and 2% around the maximum efficiency point (experimental error combines both measurement device accuracy and local unsteadiness due to two-phase flow stability problems).

Figure 6 shows the experimental pump performance head coefficient ψ_{tp} (defined in Equation (12)) versus flow coefficient φ for different inlet void fraction values for the rated rotational speed.

The corresponding theoretical head curve coefficient obtained with different rotational speeds is shown in Figure 7.

$$\psi_{tp} = \frac{gH}{u_2^2}\left(\frac{\rho_l}{\rho_g \times \alpha + \rho_l \times (1-\alpha)}\right) \tag{12}$$

$$\psi_{tp,th} = \psi_{tp}/\eta \tag{13}$$

As already pointed out by several previous researchers, pump performance starts to be significantly lower when the void fraction reaches 3% or more. The value of the head performance degradation also depends on the flow coefficient. For a given loop characteristic curve, both the head and volume flow rate decrease when *a* increases. A decrease in 20% in the head coefficient compared with single phase shut-off conditions is achieved for all of the water flow rates that are below nominal conditions when *a* reach 7%. A maximum *a* value, which is close to 10%, can be achieved without pump surge for a water flow coefficient around 0.065–0.077. These values correspond to flow rate values between 34–40 m^3/h, which are close to the pump-measured best efficiency (reached at $\varphi = 0.084$) for single-flow operations. Pump performance sharply drops when *a* is more than 8% for all of the flow rates.

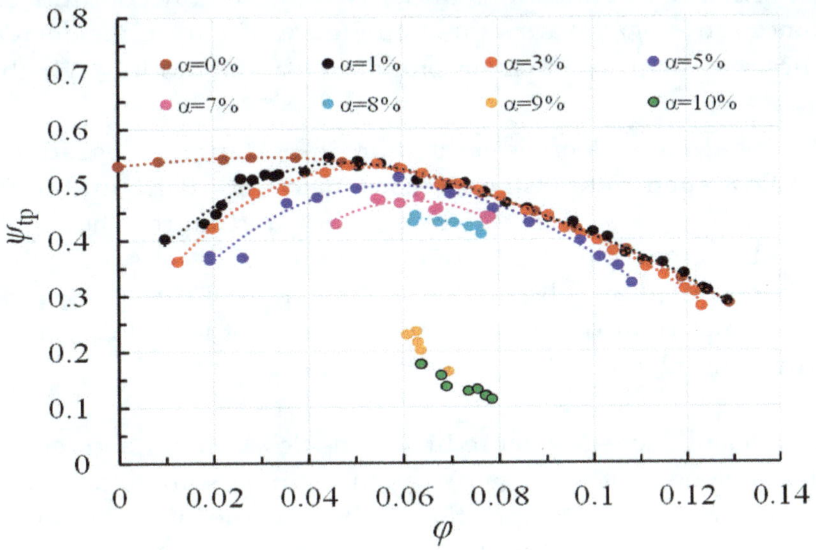

Figure 6. Head coefficient at n = 2910 r/min.

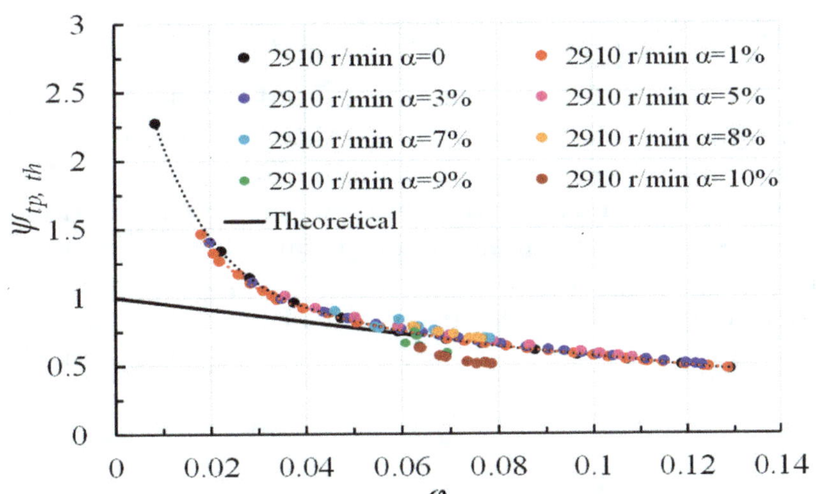

Figure 7. Theoretical head coefficient.

The concept of apparent losses is confirmed by the theoretical head coefficient curve always being the same whatever the inlet void fraction is (up to 8%), as shown in Figure 7. This is the reason why all of the following performance modifications due to two-phase flow conditions are defined by a head ratio ψ^* between the actual head coefficient ψ divided by the head coefficient ψ_0, which is related to the measurement results that are obtained only with water.

The starting point of the severe pump degradation rate is related to a specific flow coefficient corresponding to the change of the negative constant slope of the theoretical head coefficient, as shown in Figure 7. Below a flow coefficient approximately equal to 0.04, the theoretical head curve does not follow a straight line anymore, which corresponds to the hypothesis of a no swirl flow condition at the inlet and a constant outlet relative flow angle at the impeller exit plane. According to this point, an analysis on the influence of rotational speed is proposed in the following section for three different flow coefficients higher than $\varphi = 0.038$ in order to avoid strong swirling recirculation at the pump inlet.

5. Influence of Rotational Speed

Two additional rotational speeds have been selected (respectively: 2300 r/min and 1800 r/min) in order to get pump performance modifications under two-phase flow conditions. Pump experimental degradation performance ratios (ψ/ψ_0) are shown on Figure 8a–c for three decreasing flow coefficient values (solid black lines represent results from one-dimensional models, which are discussed later).

The more the rotational speed decreases, the more important the degradation ratio. This is more pronounced when the flow coefficient also decreases. Below $\varphi = 0.058$ and for the lowest rotational speed, the degradation ratio is close to 50% when $a = 8\%$ and more. The combination of low flow rates and high inlet void fraction is probably one of the possible causes for the high level of pump degradation, as already observed by Schäfer et al. [17] in another pump's geometry. One also has to consider two additional important parameters: the increase of local losses due to the bubble drag coefficient, and the Froude number effects.

When the flow coefficient decreases (by reducing the flow rate for a given rotational speed and/ or reducing the rotational speed and flow rate), the inlet pipe mean velocity decreases accordingly. In the present test of the pump inlet pipe, the limiting value of the inlet velocity, for which a transition from bubbly flow to plug or slug flow occurs, is equal to 2 m/s according to Baker's chart for horizontal pipes [18]. This means that such a transitional flow pattern may happen not only inside the impeller blade and at its blade passages, but also inside the incoming pump pipe, with non-uniform and non-homogeneous inlet flow conditions.

According to this analysis, it is not the rotational speed that is the effective parameter, but rather the mean velocity level at the pump inlet. In this respect, the experimental results obtained from the nominal rotational speed are considered to be in fair agreement with the bubbly flow regime assumption (up to a certain value of a around 7–8%), except for below a flow coefficient value $\varphi = 0.045$, which is evaluated as a limiting value. For 2300 r/min, the limiting value of φ is 0.06, and for 1800 r/min's set of results, only the ones corresponding to $\varphi = 0.07$ can be considered to fulfill the hypothesis of the bubbly flow regime. These experimental results will be compared with the one-dimensional approach proposed by Minemura and URANS calculations in a specific section of the present paper.

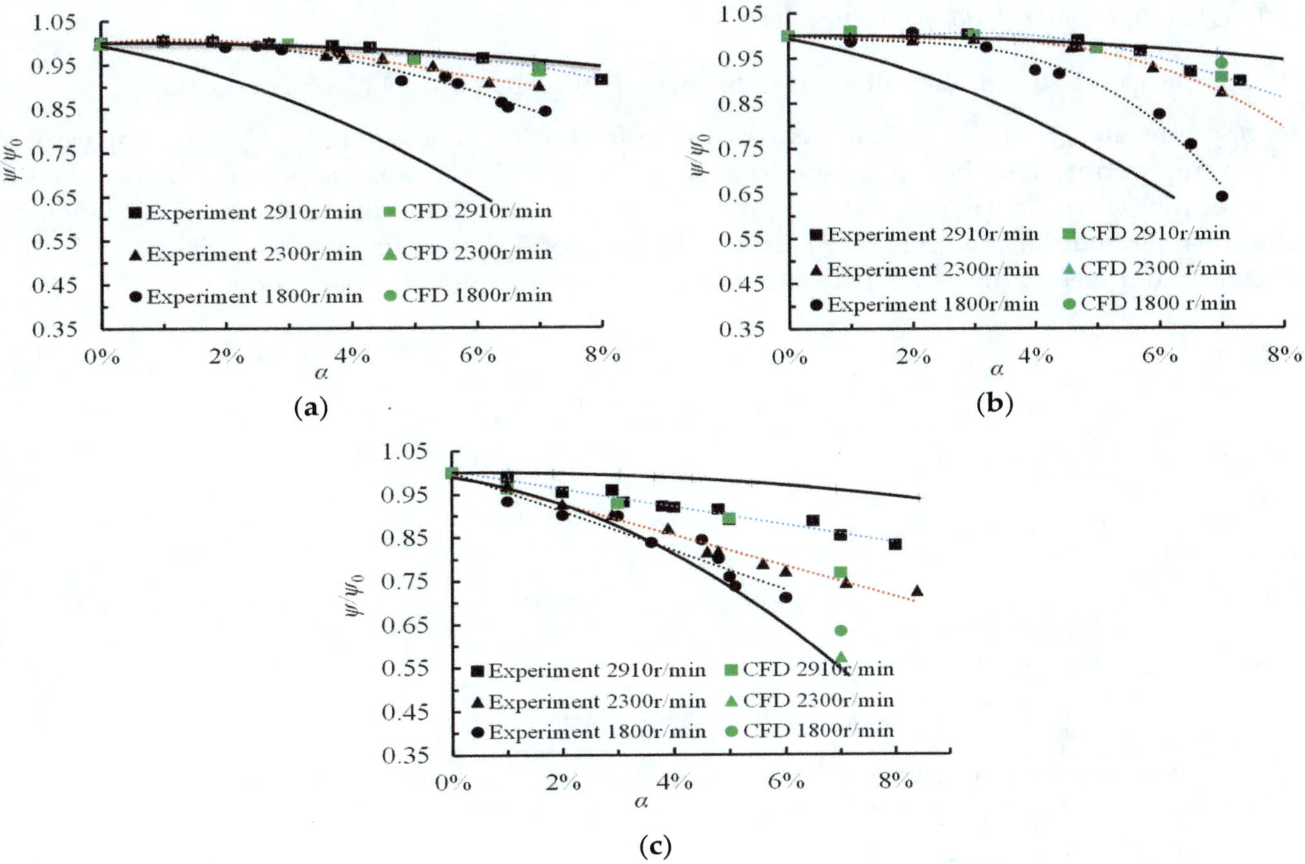

Figure 8. Pump degradation level for three flow coefficients. Comparison between experiments, one-dimensional (1D) model and "Three-Dimensional Unsteady Reynolds Average Navier–Stokes" (3D-URANS) calculations. Solid black lines represent the theoretical evolution from Minemura's approach using two different bubble drag coefficients (upper line for bubbly flow, lower line for churn flow): **(a)** $\varphi = 0.077$; **(b)** $\varphi = 0.058$; **(c)** $\varphi = 0.038$.

6. Numerical Results

6.1. Pump Head Deterioration Ratio ψ^*: Comparisons between Experiments and One-Dimensional Two-Phase Flow Model

Results deduced from Minemura's model [4] are first compared with experimental ones. They correspond to the solid lines plotted in Figure 8a–c. Up to $a = 6\%$, they all give the same performance deterioration ratio as the model used by Tomiyama et al. [19] (see upper solid black curve) and Hench [20] (see lower solid black curve) for all of the flow coefficients at nominal speed (2910 r/min), and the two highest ones for 2300 r/min. The drag coefficient in Equation (2) is also used in the 3D-URANS numerical simulation. Finally, for the smallest rotational speed and flow coefficient values ($\varphi = 0.038$), the evolution of ψ^* looks quite similar to the lower solid black curve (Figure 8c) corresponding to the churn flow regime used for the drag coefficient proposed by Zuber and Hench [21], the value of which depends on the bubble diameter and local void fraction. Since all of the experimental results regarding the pump deterioration ratio are located between these two curves, thus, it is believed that the complete flow structure should be considered not only inside the impeller, but also at the pump inlet section. Possible combinations of bubble and churn flow may arise due to the internal three-dimensional flow structure and void fraction inside the impeller [22]. In addition, it may arise when the flow rate and/or rotational speed are too low, including inlet recirculation. This was obviously not taken into consideration in previous studies when using a one-dimensional model. This is the reason why the URANS approach results are presented in the next section.

6.2. 3D-URANS Overall Performance Results

6.2.1. Comparison between the Initial and Modified Turbulence Model (See Section 3.3)

For pure water (Figure 9) and different inlet void fraction conditions (Figure 10a–d), comparisons of overall pump performance between experiments and 3D-URANS results are presented. For numerical and experimental comparisons on pump overall performance, the final mesh of 2,775,915 elements can be considered quite good (Figure 9) for a wide range of flow coefficients. A small shift in the flow coefficient can be seen probably due to some flow leakage during the experiments.

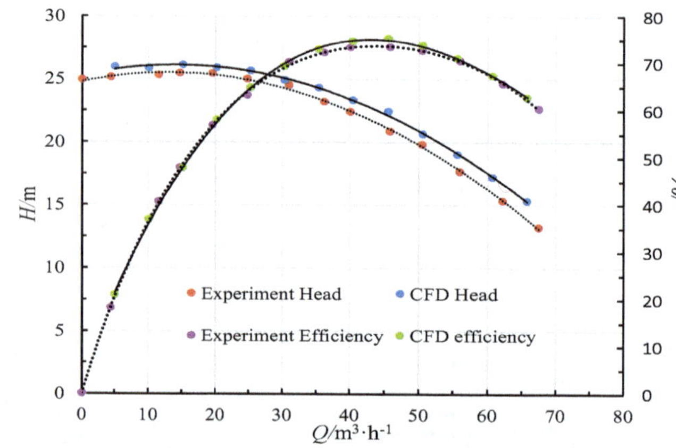

Figure 9. Comparison between experiment and numerical head and efficiency ($a = 0$).

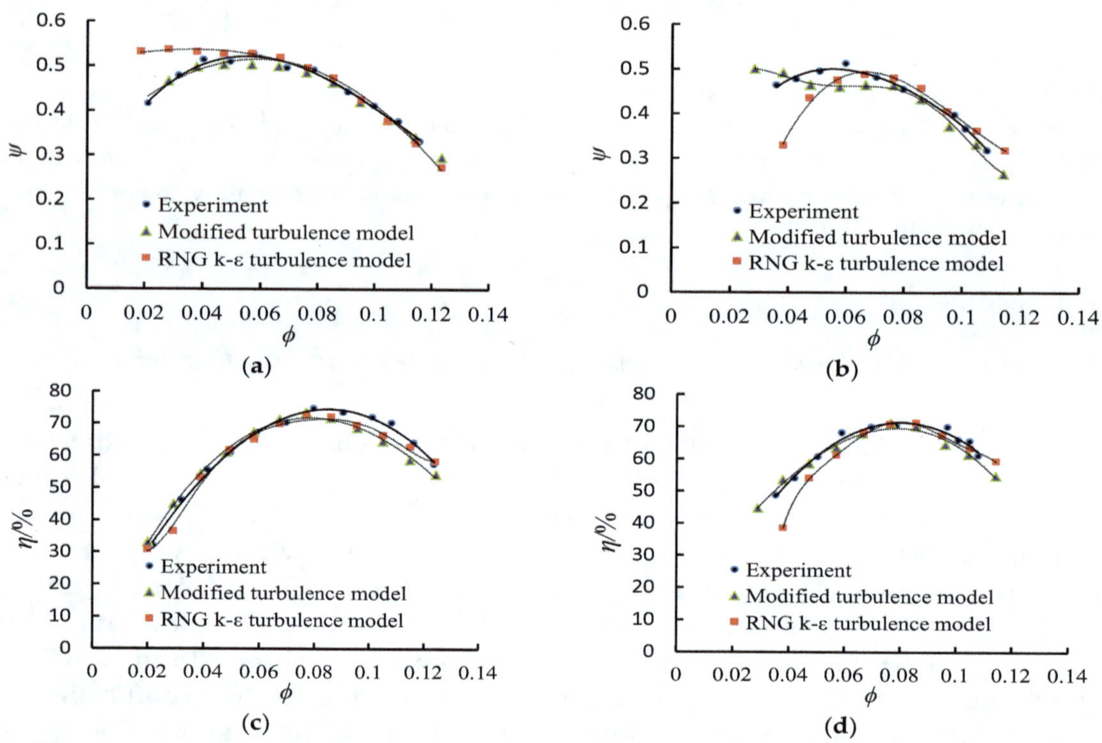

Figure 10. Comparison between experiment and numerical head coefficient and efficiency for four different a values: (**a**) Head coefficients when $a = 3\%$; (**b**) Head coefficients when $a = 5\%$; (**c**) Efficiency when $a = 3\%$; (**d**) Efficiency when $a = 5\%$.

When using the modified turbulence model, the numerical results show a better overall performance of the pump than the initial model, especially for low flow coefficients, as shown in Figure 10a,b with quite lower scattering between the experiments and the 3D-URANS calculation. Consequently, the efficiency curves are quite similar to the experimental curves (Figure 10c,d).

6.2.2. 3D-URANS Results Comparison for Different Inlet Void Fraction and Flow Coefficients

These results are also plotted in Figure 8a–c, using green symbols for comparisons with experimental ones. Compared with the one-dimensional models already presented in the previous section, the calculation results give slightly better results when the values of a increase up to 7% at the rated rotational speed. The effects of rotational speed are not well predicted, which are probably for the same reasons that have been highlighted previously.

Some additional effects must be taken into consideration when the flow coefficient is more reduced, such as for example pressure level effects on bubble diameter, which is not discussed in the present work. Concerning this particular point, one has shown that the main loss coming from the two-phase flow character appears in the first 20% of the blade passage, where the pressure increase is generally low. There is probably an effect when the rotational speed is decreasing, but it has not been evaluated in the present analysis.

6.3. Local Impeller Passage Flow Structures

6.3.1. Flow at Inlet Impeller Section

A particular focus on inlet flow conditions is shown in Figure 11 for low flow rates in order to illustrate the effects of a. Strong reverse inlet pre-rotation flow with high tangential absolute velocities can be seen for pure water. When $a = 7\%$, the flow pattern strongly changed at the pump inlet tube section; recirculation does not exist anymore, or is quite small. This is because air is going toward the pipe external radius due to the lower air velocities and the axial adverse pressure gradient along the streamline paths. Consequently, the reverse flow mainly takes place inside the first part of the impeller channel. This may explain what has been experimentally observed when a rapid breakdown of the pump occurs.

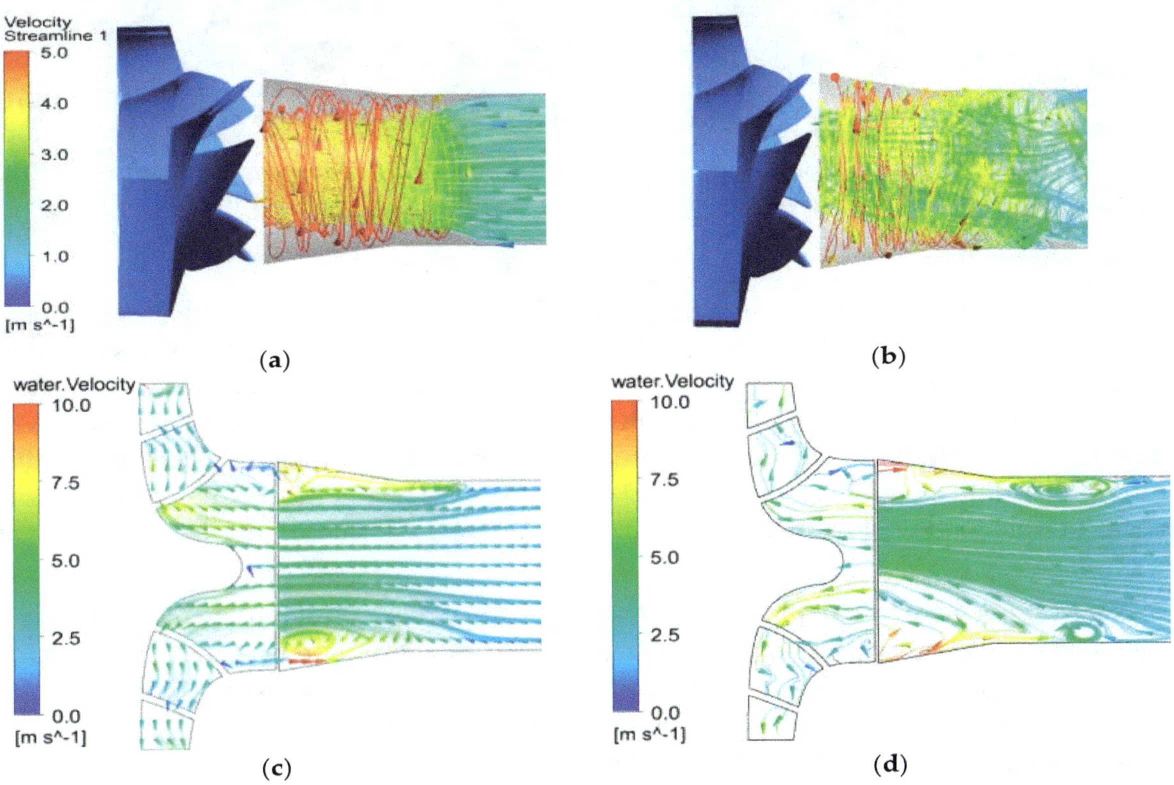

Figure 11. Local two-phase flow pattern inside the impeller passage (not including impeller wall regions), nominal rotational speed, $\varphi = 0.038$: (**a**) $a = 0\%$ (**b**) $a = 7\%$; (**c**) $a = 0\%$; (**d**) $a = 7\%$.

The following pictures in Figure 12 describe for a constant value of $a = 7\%$, the evolution of the surfaces of the constant local void fraction $\alpha = 15\%$ for two different water flow rates. Although interpretation difficulties arise in such complicated flow patterns, one can detect the air–water mixture locations at the upstream inlet bend of the impeller hub, and the shroud sections inside the regions indicated by red color areas. The more the flow rate decreases, the more it spreads and tends to block the inlet shroud region. This is the reason why the local void fraction appears partly on the suction side of the blade's leading edge, and not close to the shroud, but rather in the second part of the suction side toward the mid-height and the hub. When the flow rate decreases, because of a higher inlet incidence angle, the void fraction extends to the entire suction-side blade height. At the impeller inlet section, more flow blockage can be seen; this may explain why no water inlet recirculation occurs when the inlet void fraction is important, as already discussed earlier. Further downstream, this void fraction value extends toward the mid-impeller channel (green color circles). For $Q = 30\ \text{m}^3/\text{h}$, a separation zone can be seen starting from the pressure side of the hub region, which migrates toward the mid-impeller passage as well. As pointed by Gülich [1], such a complicated flow pattern results from the dominant body forces and inertia for the liquid behavior (due to the density ratio between water and gas), while gas distribution is mainly determined by the pressure field and secondary flows. In the next section, the void fraction on the pressure and suction sides of the impeller is going to be analyzed more specifically.

Figure 12. Constant void fraction iso-lines inside the impeller passage when $a = 7\%$: (**a**) $\varphi = 0.077$; (**b**) $\varphi = 0.058$.

6.3.2. Flow inside the Impeller Section

Void fraction distributions for decreasing flow rate values on the impeller blade and at the outlet radius of the impeller are shown in Figure 13a–d. Looking at the leading-edge regions of the impeller, a high local void fraction (between 20–30%) appears on the suction side mainly close to the shroud. High α values (up to 100%, corresponding to full air) appear on the pressure side of the shroud corner for $Q = 30\ \text{m}^3/\text{h}$, just after the inlet throat section of the impeller passage. The corresponding drag force, acting on a bubble, combined with the adverse pressure gradient inside the impeller passage, forces the gas bubble to move toward the pressure side where the local liquid relative velocity tends to be lower.

For a flow rate of 40 m^3/h, the local void fraction values remain relatively small at both sides of the impeller leading edge. This is perhaps the reason why the experimental results show that the pump is still working at a high inlet a (up to 10%; see Figure 2) without gas locking. Finally, for 20 m^3/h, a quite high local void fraction can be seen on the suction side of the leading edge due to incidence angle effects (see again Figure 11b for the meridional plane).

Further downstream, and for all of the flow rates, high void fractions are located near the pressure side, close to the second throat. This has been already mentioned by Mirakami and Minemura [3]. This is attributed to the increase in the velocity difference between the water and the gas when the void fraction increases. The corresponding drag force, acting on a bubble, combined with the adverse pressure gradient inside the impeller passage, forces the gas bubble to move toward the pressure side where the local liquid relative velocity tends to be lower. As an example, this is clearly obtained by the CFD results given in Figure 14 for the mid-section between the hub and the shroud, for $Q = 30$ m^3/h.

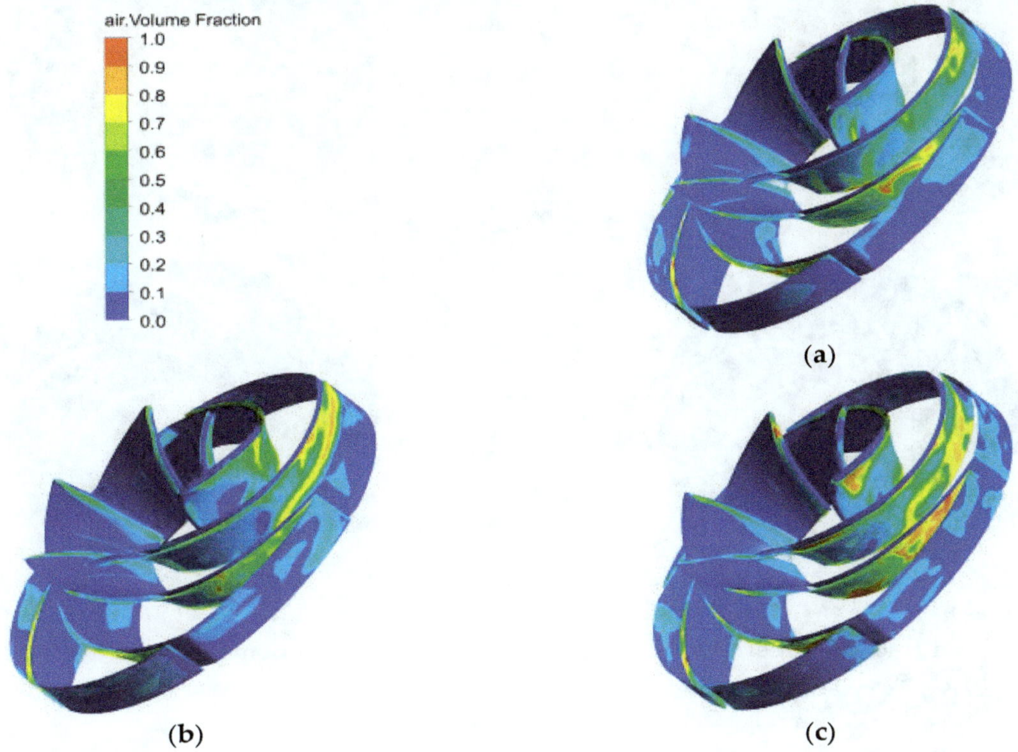

Figure 13. Void fraction pattern at the impeller outlet section, a = 7%: (**a**) $\varphi = 0.077$; (**b**) $\varphi = 0.058$; (**c**) $\varphi = 0.038$.

By looking at the additional maps for the specific local void distribution (Figure 15), we can have a better insight on the pressure and suction sides of the blade. For instance, at a given flow rate (Q = 30 m^3/h), it can be observed that on the pressure sides, the pattern of air bubbles differs between a = 5–7%. For a = 5%, high values of local void fraction are reached and spread along the blade height just after the leading edge. Then, due to the Coriolis force, the liquid is mainly present on this pressure side, and finally, due to the radial pressure gradient and slip velocity, air accumulates again close to the trailing edge. When a increases to 7%, air locations are shifted downstream in the hub region, and the liquid (and low local void fractions) area takes a larger surface and pushes air closer to the outlet blade section. As already mentioned, this last case exhibits strong 3D effects with flow separations and reverse flow regions.

One can observe that the more the flow rate decreases, the more of high void fractions are shifted toward the pressure side due to the decrease of the mean relative velocities. For $Q = 20$ m^3/h, all of the passages are filled with air. All of the blade passages do not have the same pattern due to the presence of the volute tongue.

At the impeller outlet plane, the void fraction does not exceed 30%, instead of reaching the lowest flow rate, for which the values may reach up to 60%. Just downstream of the impeller, due to strong mixing, the void fraction strongly decreases (not shown here). Close to nominal conditions, the void fraction values obtained from CFD are quite similar to the values predicted by the model developed by Minemura et al. [3]. On the pressure side, there is always a thin layer that develops with high void fraction values.

The next set of figures gives the pattern of the local void fraction in the mid-section plane between the hub and the shroud, for decreasing flow rates and a given a value of 7%. This corresponds to the maps given in Figure 16.

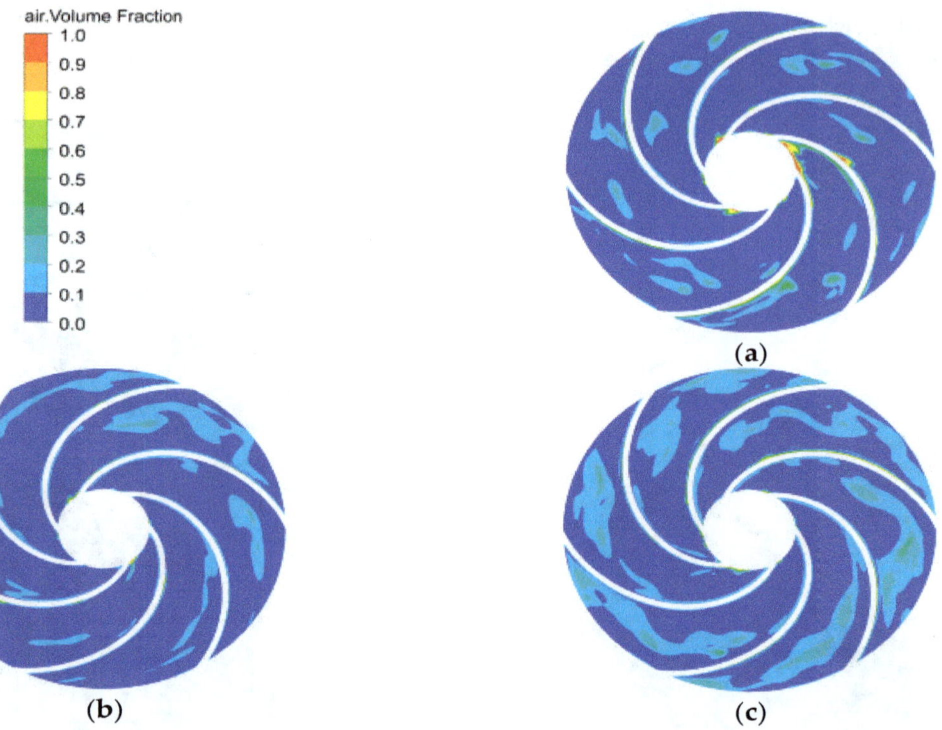

Figure 14. Void fraction pattern in the blade-to-blade mid-section plane, $Q = 30\ \mathrm{m^3/h}$, $\varphi = 0.058$: (**a**) $a = 3\%$; (**b**) $a = 5\%$; (**c**) $a = 7\%$.

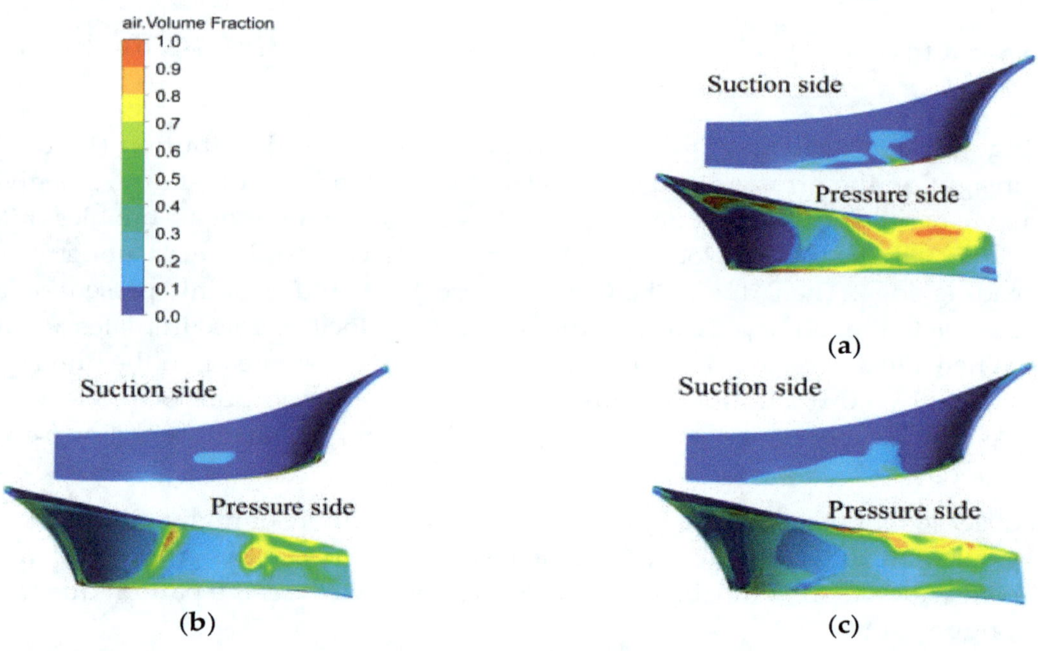

Figure 15. Void fraction pattern on the blade surface, $\varphi = 0.058$: (**a**) $a = 3\%$; (**b**) $a = 5\%$; (**c**) $a = 7\%$.

All of the flow patterns that have been found and described inside the impeller channels are the consequences of what happens at the impeller inlet section just before and after the blade leading-edge area. This means that one has to find the local design modifications forcing high two-phase flow values to be mixed with the main flow using additional local body forces. For this reason, it is expected that the blade clearance between the hub and the shroud can be an efficient way to create a local vortex for a better mixing, avoiding a local accumulation of the high void fraction values on the meridional plane inlet sections and in the blade-to-blade plane further downstream of the inlet passage throat. This will be carried out in a future work.

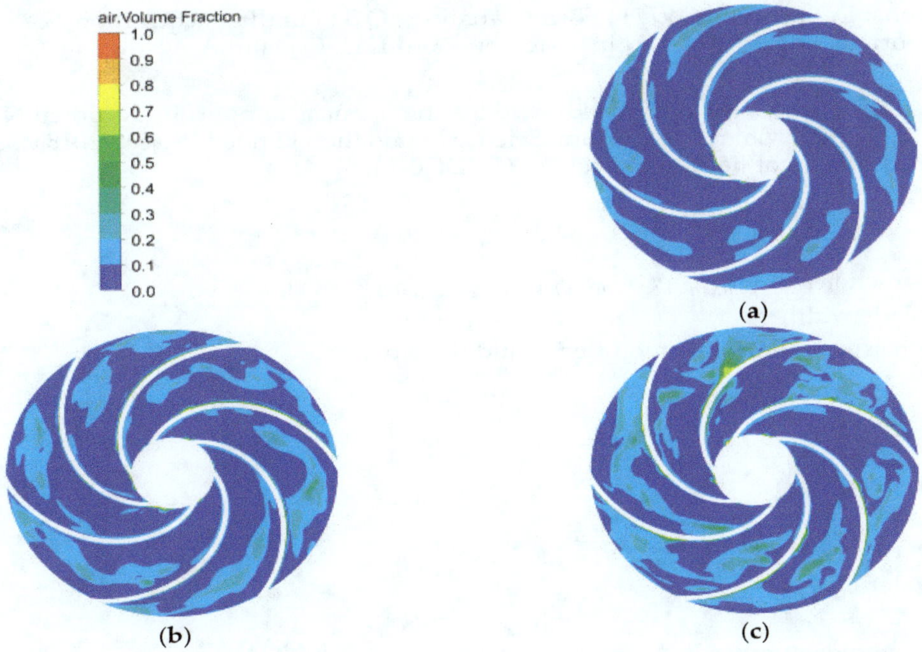

Figure 16. Void fraction pattern in the blade-to-blade mid-section plane for different flow rates, $a = 7\%$: (**a**) $\varphi = 0.077$; (**b**) $\varphi = 0.058$; (**c**) $\varphi = 0.038$.

7. Conclusions

Experimental and numerical analysis has been performed in a centrifugal pump for different flow coefficients and inlet void fractions up to 10%. The effect of rotational speed has also been investigated. The main conclusions are the following:

1. Pump performance degradation is more pronounced for low flow rates compared to high flow rates when the inlet void fraction increases. The starting point of a severe pump degradation rate is related to a specific flow coefficient, whose value corresponds to the change of the slope of the theoretical head curve. When increasing the inlet void fraction, the degradation slope curves increase (with a negative sign) with the decreasing flow coefficient. The more the rotational speed decreases, the more the experimental pump performance is affected at a given inlet void fraction value.
2. Existing one-dimensional models can be considered quite good tools for the first step of a two-phase flow analysis. They give good global indications based on the mean values along one streamline, but attention should be taken when using non-dimensional flow coefficients. The chosen particle fluid model with interface transfer terms looks quite suitable for evaluating pump performance degradation up to an a value of 7%.
3. For a higher a value, CFD can't always correctly predict the sudden breakdown of the pump performance as obtained by measurements. This is probably due to the flow regime inside the impeller, which does not correspond to a bubbly one anymore. However, for the lowest flow rate, the performance breakdown is well predicted with the modified turbulence model.
4. The difference between experimental and numerical results exist not because of rotational speed, but because of its consequence on the local velocity values that decrease according to the flow coefficient, and more specifically at the pump inlet tube and at the pump inlet section. Numerical simulations must take churn flow characteristics into account in order to get better results. Both bubbly and churn flow conditions may be present for inlet flow conditions depending on the experimental setup, rotational speed, and the pump flow coefficient, even at nominal conditions.
5. The numerical simulation gives interesting local flow information that would be taken into consideration for new design approaches at the pump inlet section for an improved two-phase flow suction capability of centrifugal pumps.

Author Contributions: Data curation, W.H.; Formal analysis, Q.S.; Funding acquisition, S.Y.; Investigation, Q.S. and B.G.; Writing—original draft, Q.S.; Writing—review & editing, Q.J. and A.A.

Acknowledgments: The authors gratefully acknowledge the financial support by National Natural Foundation of China (51609164, 51779107), China Postdoctoral Science Foundation Funded Project (2015M581735, 2016T90422) and Senior Talent Foundation of Jiangsu University (15JDG048).

Nomenclature

a	inlet void fraction $a = Q_{air,inlet}/(Q_{air,inlet} + Q_{water,inlet})$
b	impeller blade width
C	constant pressure value when modified liquid density
D	diameter
H	pump head
n	exponent
N	rotational speed
p	local pressure
P	shaft power
Q	volume water flow rate
R	radius
$R_{e,imp}$	Impeller Reynolds number $R_{e,imp} = u_2 \cdot R_2/v$
t	time
th	theoretical
tp	related to two-phase condition
u	circular velocity
v	water cinematic viscosity
z	impeller blade
Greek symbols	
η	global efficiency of the pump $\eta = \rho gQH/P$
ρ	density
ρ_m	density of fluid mixture $\rho_m = \rho_l (1 - \alpha) + \alpha\rho_g$
φ	flow coefficient $\varphi = Q/(2\pi \cdot R_2 \cdot b_2 \cdot u_2)$
ψ	head coefficient $\psi = gH/(u_2)^2$
Ωs	specific speed $\Omega s = \omega \cdot \frac{Q^{0.5}}{(gH)^{0.75}}$
ω	angular velocity
Subscripts	
α	local void fraction
B	bubble
d	design condition
g	gas
l	liquid
m	mixture fluid
ref	reference
0	related to $a = 0$
1	Impeller pump inlet
2	Impeller pump outlet

References

1. Gülich, J.F. *Centrifugal Pumps*; Springer: Berlin/Heidelberg, Germany; New York, NY, USA, 2010; ISBN 978-3-642-12823-3.
2. Mikielewicz, J.; Wilson, D.G.; Chan, T.C.; Goldfinch, A.L. A method for correlating the characteristics of centrifugal pumps in two-phase flow. *J. Fluids Eng.* **1978**, *100*, 395–409. [CrossRef]
3. Murakami, M.; Minemura, K. Effects of entrained air on the performance of a centrifugal pump First report-Performance and Flow Conditions. *Bull. JSME* **1974**, *17*, 1047–1055. [CrossRef]

4. Minemura, K.; Uchiyama, T.; Shoda, S.; Egashira, K. Prediction of air-water two-phase flow performance of a centrifugal pump based on one-dimensional two-fluid model. *J. Fluids Eng.* **1998**, *120*, 237–334. [CrossRef]
5. Si, Q.R.; Cui, Q.L.; Zhang, K.Y.; Yuan, J.P.; Bois, G. Investigation on centrifugal pump performance degradation under air-water inlet two-phase flow conditions. *La Houille Blanche* **2018**, *3*, 41–48. [CrossRef]
6. Secoguchi, N.; Takada, S.; Kanemori, Y. Study of air-water two-phase centrifugal pump by means of electric resistivity probe technique for void fraction measurement-First report-measurement of void fraction distribution in a radial flow impeller. *Bull. JSME* **1984**, *27*, 931–939. [CrossRef]
7. Estevam, V.; França, F.A.; Alhanati, F.J. Mapping the performance of centrifugal pumps under two-phase conditions. In Proceedings of the 17th International Congress of Mechanical Engineering, Sao Paulo, Brazil, 10–14 November 2003.
8. Barrios, L.; Prado, M.G. Experimental visualization of two-phase flow inside an electrical submersible pump stage. *J. Energy Resour. Technol.* **2011**, *133*. [CrossRef]
9. Müller, T.; Limbach, P.; Skoda, R. Numerical 3D RANS simulation of gas-liquid flow in a centrifugal pump with an Euler-Euler two-phase model and a dispersed phase distribution. In Proceedings of the 11th European Conference on Turbomachinery, Fuild Dynamics and Thermodynamics, Madrid, Spain, 23–25 March 2015.
10. Suryawijaya, P.; Kosyna, G. Unsteady measurement of static pressure on the impeller blade surfaces and optical observation on centrifugal pumps under varying liquid/gas two-phase flow condition. *J. Comput. Appl. Mech.* **2001**, *6759*, D9–D18.
11. Si, Q.R.; Bois, G.; Zhang, K.Y.; Yuan, J.P. Air-water two-phase flow experimental and numerical analysis in a centrifugal pump. In Proceedings of the 12th European Conference on Turbomachinery, Fluid Dynamics and Thermodynamics, Stockholm, Sweden, 3–7 April 2017.
12. Yakhot, V.; Orszag, S.; Thangam, S.; Gatski, T.; Speziale, C. Development of turbulence models for shear flows by a double expansion technique. *Phys. Fluids A Fluid Dyn.* **1992**, *4*, 1510–1520. [CrossRef]
13. Clift, R.; Grace, J.; Weber, M. *Bubbles, Drops and Particles*; Academic Press: New York, NY, USA, 1978.
14. Wang, J.; Wang, Y.; Liu, H.; Huang, H.; Jiang, L. An improved turbulence model for predicting unsteady cavitating flows in centrifugal pump. *Int. J. Numer. Methods Heat Fluid Flow* **2015**, *25*, 1198–1213. [CrossRef]
15. Dular, M.; Coutier-Delgosha, O. Numerical modelling of cavitation erosion. *Int. J. Numer. Methods Fluids* **2009**, *61*, 1388–1410. [CrossRef]
16. Coutier-Delgosha, O.; Fortes-Patella, R.; Reboud, J.L. Evaluation of the turbulence model influence on the numerical simulations of unsteady cavitation. *ASME J. Fluids Eng.* **2003**, *125*, 38–45. [CrossRef]
17. Schäfer, T.; Bieberle, A.; Neumann, M.; Hampel, U. Application of gamma-ray computed tomography for the analysis of gas holdup distributions in centrifugal pumps. *Flow Meas. Instrum.* **2015**, *46*, 262–267. [CrossRef]
18. Baker, O. Simultaneous Flow of Oil and Gas. *Oil Gas J.* **1954**, *53*, 185–190.
19. Tomiyama, A.; Kataoka, I.; Sakaguchi, T. Drag coefficients of bubbles. 1st report. drag coefficients of a single bubble in a stagnant liquid. *Nihon Kikai Gakkai Ronbunshu B Hen/Trans. Japan Soc. Mech. Eng. Part B* **1995**, *61*, 2357–2364.
20. Hench, J.E.; Johnston, J.P. Two-dimensional diffuser performance with subsonic, two-phase, air-water flow. *ASME J. Basic Eng.* **1972**, *107*, 105–121. [CrossRef]
21. Zuber, N.; Hench, J.E. *Steady State and Transient Void Fraction of Bubbling Systems and Their Operating Limits*; General Electric: Schenectady, NY, USA, 1962.
22. Zhu, J.J.; Zhang, H.Q. A review of experiments and modeling of gas-liquid flow in electrical submersible pumps. *Energies* **2018**, *11*, 180. [CrossRef]

13

Effect of Rotor Thrust on the Average Tower Drag of Downwind Turbines

Shigeo Yoshida [1,*], Kazuyuki Fujii [2], Masahiro Hamasaki [1] and Ao Takada [1]

1 Research Institute for Applied Mechanics, Kyushu University, 6-1 Kasugakoen, Kasuga, Fukuoka 816-8580, Japan; hamasaki@riam.kyushu-u.ac.jp (M.H.); takada@riam.kyushu-u.ac.jp (A.T.)
2 Interdisciplinary Graduate School of Engineering and Science, Kyushu University, 6-1 Kasugakoen, Kasuga, Fukuoka 816-8580, Japan; fujiik@riam.kyushu-u.ac.jp
* Correspondence: yoshidas@riam.kyushu-u.ac.jp

Abstract: A new analysis method to calculate the rotor-induced average tower drag of downwind turbines in the blade element momentum (BEM) method was developed in this study. The method involves two parts: calculation of the wind speed distribution using computational fluid dynamics, with the rotor modeled as a uniform loaded actuator disc, and calculation of the tower drag via the strip theory. The latter calculation considers two parameters, that is, the decrease in wind speed and the pressure gradient caused by the rotor thrust. The present method was validated by a wind tunnel test. Unlike the former BEM, which assumes the tower drag to be constant, the results obtained by the proposed method demonstrate much better agreement with the results of the wind tunnel test, with an accuracy of 30%.

Keywords: wind turbine; downwind; tower shadow; load; tower; BEM; actuator disc

1. Introduction

Horizontal axis wind turbines are categorized as upwind and downwind turbines according to the position of the rotor relative to the tower. Among these, upwind turbines have been predominant throughout the 30 year history of commercial wind turbines. The most essential reason for the unpopularity of the downwind turbines is the presence of a strong aerodynamic interaction between the rotor and the tower, which is commonly known as the "tower shadow effect". This phenomenon generates impulsive loads and infrasound when one of the blades passes through the wake of the tower [1]. The modeling of the tower shadow effect is a critical technical problem of downwind turbines.

Design loads are calculated on the basis of the international design standard IEC61400-1 [2] or the guidelines for certification bodies DNV GL [3]. In a large number of cases in which the design load is combined with the wind model, the wind turbines can experience failure conditions, as well as various types of wind and marine conditions. The flexibilities of the structure and the controls, in addition to the aerodynamics, hydrodynamics, aero-elastics, and control of the wind turbines, considerably influence the load. The blade element momentum (BEM) method is the most popular tool for determining these characteristics [4,5]. Therefore, the modeling of the tower shadow effect in the BEM is one of the most important technical challenges in the design and development of downwind wind turbines. However, most of the tower shadow models are too simple to express such phenomena, as they consider only the wind speed profile behind an isolated tower [5,6] and not the aerodynamic interaction between the rotor and the tower.

Several studies have been conducted until now that have focused on the variable loads of the downwind turbines caused by the tower shadow effect. Zhao et al. [7] studied the loads on a tower by performing a comparison between two- and three-bladed upwind and downwind rotors in two different rotor speed conditions using computational fluid dynamics (CFD). The peak to average values of the hub bending moment and the tower base bending moment of the downwind turbine were shown to be 5–6 times and 2.5 times larger, respectively, than those of the upwind turbine. Zahle et al. [8] conducted 2-dimensional CFD analysis for three typical tower concepts of downwind turbines as well as for an upwind turbine; a steep load variation was noted for the downwind turbine, and the implementation of streamlined and four-leg towers was shown to be an effective technique to reduce the load variation caused by the tower shadow effects. Matiz-Chicacausa and Lopez [9] investigated the possibilities of the application of the actuator line model (ALM) to the CFD analysis of the tower shadow effects. The ALM was shown to be useful for load estimation of a downwind turbine under specific conditions.

Although the abovementioned research contributed to the expansion of the understanding of the phenomenon, the authors did not aim to develop models for extensive load calculations based on the design standards and guidelines. Wang and Coton [10] developed a high-resolution tower shadow model for downwind turbines based on the prescribed wake vortex model and an efficient near wake dynamics model of the vorticity trailing the blades. The model demonstrated reasonable agreement with the experimental results, with the exception of the blades passing through the tower wake. However, the load of the tower was not discussed in this research. Yoshida and Kiyoki [11,12] developed the load equivalent tower shadow modeling method. The method considered a bell-shaped wind speed profile behind the tower, which helped obtain the equivalent rotor load variation with the wind turbine CFD analysis. This model was applied to the development of the first multimegawatt-scale commercial downwind turbines, including the SUBARU 80/20 [13] and, later, the Hitachi 2 MW, to realize high performance and safety in complex terrains [14,15]. The technology was extended to the later produced Hitachi 5 MW [16] offshore wind turbines. The model has been successfully applied in engineering applications so far; however, two major problems remained unsolved. The first concern was that the CFD modeling of the rotor–tower configuration required for each operating condition was time-consuming, which made the technique inconvenient for use in the design optimization process. The second issue was that no model was established for the tower shadow effects on the tower drag. Considering the situation, the tower load variation caused by the tower shadow effect was modeled by Yoshida by combining the BEM and the Lifting Line Theory [17]. However, the average tower drag was still not discussed in that research.

Considering these situations, this paper proposes a BEM model for the tower average drag. The formulation is presented in Section 2, and the validation of the method by a wind tunnel test is discussed in Sections 3–5.

2. Methodology

The formulation of the average loads of the towers of the downwind turbine by using the BEM is discussed in this chapter. The following two calculations are considered:

(1) Wind speed distribution by CFD
(2) Rotor thrust-induced tower average drag

2.1. Wind Speed Distribution by CFD

The wind speed distribution in front of the rotor is calculated using CFD. Figure 1 shows the top view of the rotor and the tower of a downwind turbine. The tower, which has the diameter D_T, is located in front of the rotor. The free stream wind speed U_0 decreases to $U_0(1 - a)$ at the rotor plane as a result of the rotor thrust. Here, a is the axial induction factor. Figure 2 is the schematic of the CFD involved in the present study. The tower is not included in the model, but the position is termed "virtual tower position" in this research. Further, u_T is the wind speed at the virtual tower position, which is between U_0 and $U_0(1 - a)$, depending on the condition. The rotor is modeled by an actuator disc model (ADM). The load distribution is assumed to be uniform to realize a convenient application to the BEM. The thrust per unit volume σ of the ADM is as shown in Equation (1):

$$\sigma = \frac{1}{2\pi r w}\frac{dT}{dr} \tag{1}$$

where w is the thickness of the ADM.

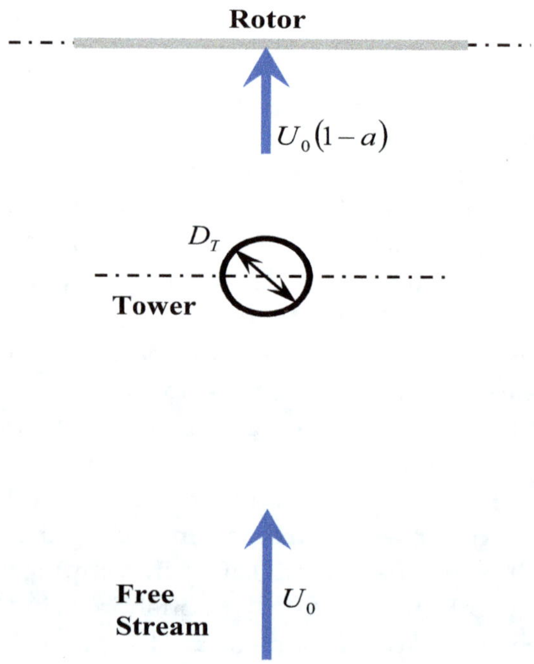

Figure 1. Top view of a downwind turbine. U_0, free stream wind speed, D_T, tower diameter.

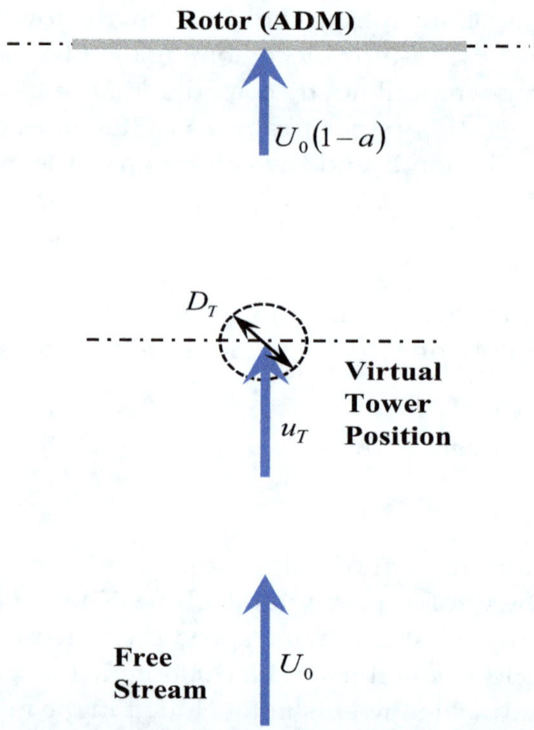

Figure 2. Computational fluid dynamics (CFD) with actuator disc model (ADM). u_T, wind speed at the virtual tower position.

2.2. Rotor Thrust-Induced Tower Average Drag

The wind speed at the tower and the tower drag decrease as the rotor thrust increases due to the aerodynamic interaction. Two factors, that is, the wind speed and the ambient pressure gradient caused by the rotor thrust, are considered in this research. The formulation is based on the assumption of the strip theory, according to which the flow between the sections normal to the tower axis is not taken into account.

(1) Wind Speed Effect

The tower drag decreases proportionally to the square of the wind speed when the tower section drag coefficient remains constant. This influence is modeled herein.

The section drag f_{XTO} of the isolated tower with no rotor interaction is as shown in Equation (2):

$$f_{XTO} = \frac{1}{2}\rho U_0{}^2 D_T C_{dTO} \tag{2}$$

where C_{dTO} is the drag coefficient at the tower section.

In case the effect of the rotor thrust comes into play, the wind speed at the tower position decreases. Furthermore, the tower section drag also decreases from f_{XTO} to f_{XT}. This is expressed in the following two ways, based on the wind speeds at the free stream and at the virtual tower position, as shown in Equation (3):

$$f_{XT} = \frac{1}{2}\rho U_0{}^2 D_T C_{dT} = \frac{1}{2}\rho u_T{}^2 D_T C_{dTO} \tag{3}$$

where C_{dT} is the tower section drag based on the free stream wind speed, and u_T is the wind speed at the virtual tower position, which is calculated by CFD without incorporating a tower model, as explained in the previous section. Here, the drag coefficient C_{dTO} is assumed to be constant as in Equation (2).

Therefore, the tower section drag deviation induced by the rotor thrust Δf_{XT} is calculated as in Equation (4):

$$\Delta f_{XTV} = f_{XT} - f_{XTO} = -\frac{1}{2}\rho U_0{}^2 D_T C_{dTO}\left(1 - \mu_T{}^2\right) \tag{2}$$

where the normalized wind speed at the virtual tower is

$$\mu_T = \frac{u_T}{U_0}$$

Therefore, the change in the relevant drag coefficient caused by the wind speed change ΔC_{dTV} is calculated by Equation (4), as given in Equation (5):

$$\Delta C_{dTV} = -C_{dTO}\left(1 - \mu_T{}^2\right) \tag{3}$$

This indicates that the term of the tower section drag decreases as the normalized wind speed at the virtual tower, induced by the rotor thrust, decreases.

(2) Effect of the Ambient Pressure Gradient

The tower drag is also dependent on the ambient pressure gradient around the tower. This influence is modeled herein.

The pressure at the virtual tower position p_T is calculated by Bernoulli's law, as in Equation (4):

$$p_0 + \frac{1}{2}\rho U_0{}^2 = p_T + \frac{1}{2}\rho u_T{}^2 \tag{4}$$

where p_0 and p_T are the pressures at the free stream and the virtual tower center, respectively. Therefore, the windward pressure gradient $\partial p_T/\partial x_T$ at the virtual tower center is calculated as shown in Equation (5) by the differential of Equation (4):

$$\frac{\partial p_T}{\partial x_T} = -\rho u_T \frac{\partial u_T}{\partial x_T} \tag{5}$$

The section drag by the pressure gradient at the tower section Δf_{XTP} is calculated as in Equation (6) by employing the pressure deviation from the front and back sides of the tower and the tower diameter. The pressure deviation is calculated by the pressure gradient and the windward reference distance Δx_T:

$$\Delta f_{XTP} = \frac{\partial p_T}{\partial x_T} \Delta x_T D_T \tag{6}$$

Assuming a uniform pressure gradient, the tower section drag is expressed as in Equation (7):

$$\Delta f_{XTP} = \int_{-\pi/2}^{\pi/2} \left(\frac{\partial p_T}{\partial x_T} D_T \cos \phi_T \right) \left(\frac{D_T}{2} \cos \phi_T \right) d\phi_T \tag{7}$$

The reference distance is calculated as in Equation (8) using Equations (6) and (7):

$$\Delta x_T = \frac{\pi}{4} D_T \tag{8}$$

Therefore, the change in the tower section drag owing to the pressure gradient is calculated as in Equation (9):

$$\Delta f_{XTP} = \frac{\pi D_T{}^2 \rho u_T}{4} \frac{\partial u_T}{\partial x_T} = \frac{\pi D_T \rho U_0{}^2 \mu_T}{4} \frac{\partial \mu_T}{\partial \xi_T} \tag{9}$$

where ξ_T is the normalized distance and can be written as

$$\xi_T = \frac{x_T}{D_T}$$

Therefore, the change in the relevant drag coefficient caused by the pressure gradient ΔC_{dTP} can be written as in Equation (10):

$$\Delta C_{dTP} = \frac{\pi}{2} \mu_T \frac{\partial \mu_T}{\partial \xi_T} \tag{10}$$

This indicates that the term of the tower section drag decreases as the negative pressure gradient caused by the rotor thrust increases.

(3) Total Average Tower Drag

From (1) and (2), the deviation of the drag Δf_{XT} and the drag coefficient ΔC_{dT} induced by the rotor thrust are derived as Equations (11) and (12), respectively:

$$\Delta f_{XT} = \Delta f_{XTV} + \Delta f_{XTP} = \frac{1}{2} \rho U_0{}^2 D_T \left[-C_{dT0} \left(1 - \mu_T{}^2 \right) + \frac{\pi}{2} \mu_T \frac{\partial \mu_T}{\partial \xi_T} \right] \tag{11}$$

$$\Delta C_{dT} = \Delta C_{dTV} + \Delta C_{dTP} = -C_{dT0} \left(1 - \mu_T{}^2 \right) + \frac{\pi}{2} \mu_T \frac{\partial \mu_T}{\partial \xi_T} \tag{12}$$

3. Wind Tunnel Test

3.1. Outline

A wind tunnel test for a wind turbine was conducted to validate the theory described in the previous chapter. The rotor–tower interaction was simulated by a dummy tower placed in front of the rotor of the upwind turbine.

3.2. Facility

The Boundary Layer Wind Tunnel at the Research Institute for Applied Mechanics, Kyushu University [18], was used for the test. The wind tunnel has a test section with a width of 3.6 m, a height of 2.0 m, and a length of 15 m. The maximum wind speed is 30 m/s.

3.3. Test Model

The general specifications and the outline of the model and the tower installed in the wind tunnel are presented in Table 1. The blockage ratio, which is the ratio of the rotor area to the cross section of the wind tunnel, is as small as 5.3%. The rotor speed is controlled by the variable dump load. Figure 3 shows the schematic of the test model and the dummy tower. Although the model is an upwind turbine supported by a tower with a diameter of 50.8 mm, a dummy tower is installed in front of the model to mimic the downwind turbine tower. The diameter of the dummy tower is 64 mm, which is 200% the value of the blade chord length. The dummy tower was placed at two positions; $-6D_T$ and $-4D_T$. Figure 4 shows an image of the test model. The dummy tower is shown as located in front of the turbine.

Table 1. Specifications of the test model and the dummy tower.

Item	Specification
Number of blades	2
Rotor diameter D	700 mm
Rotor radius R	350 mm
Tilt angle	0 deg
Coning angles	0 deg
Airfoil	NACA0013
Blade chord length	32 mm
Twist angle	0 deg
Tower diameter D_T	64 mm
Tower position	−338 mm ($-6D_T$) −225 mm ($-4D_T$)

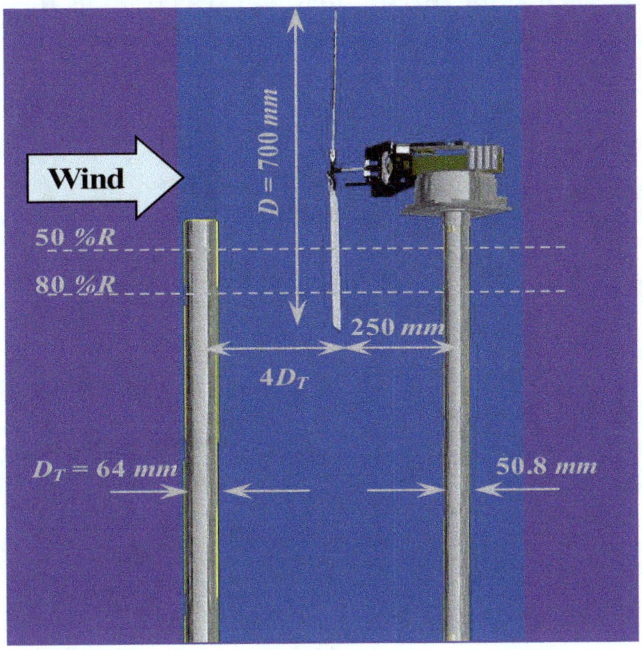

Figure 3. Schematic of the test model (dummy tower at $-4D_T$).

Figure 4. Test model and dummy tower installed in the wind tunnel.

3.4. Measurement

- Rotor speed: Magnetic encoder
- Load (top of the tower): 6-axis Load Cell (Nissho)
- Pressure on the dummy tower (50%R, 80%R, 8 points each): Pressure sensor (Otegiken)

3.5. Test Conditions

- Wind speed (U_0): 6 m/s (uniform, steady)
- Yaw angle: 0 deg
- Blade pitch angle (θ): 4, 6, 8 deg
- Tip speed ratio (λ): 6.6–9.2

3.6. Test Results

The measurements for the power coefficient C_P and the thrust coefficient C_T for the isolated rotor model are shown in Figure 5. The rotor torque was calculated from the rolling moment at the top of the tower. The value of thrust force at the top of the tower was used as the rotor thrust. The maximum value of C_P was noted at approximately $\lambda = 8$ and $\theta = 6$ deg. However, C_T tended to increase with an increase in the tip speed ratio or a decrease in the pitch angle.

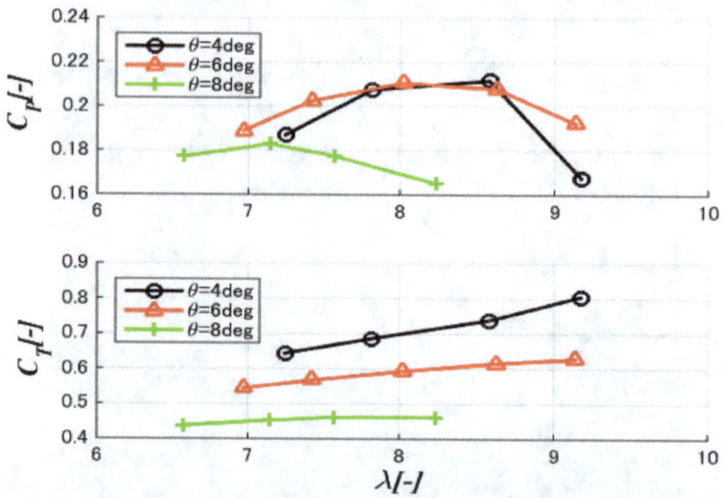

Figure 5. Relationship between the power (C_P) and thrust (C_T) coefficients and the tip speed ratio. θ, blade pitch angle.

The pressure distributions on the dummy tower in typical cases are shown in Figure 6. Here, 0 deg indicates the upwind side of the tower. The pressure on the 50%R was relatively higher than that at 80%R. In general, the pressures around the downwind side of the tower (180 deg) tended to be higher as the rotor thrust was larger.

The details of other characteristics are discussed in Section 5.

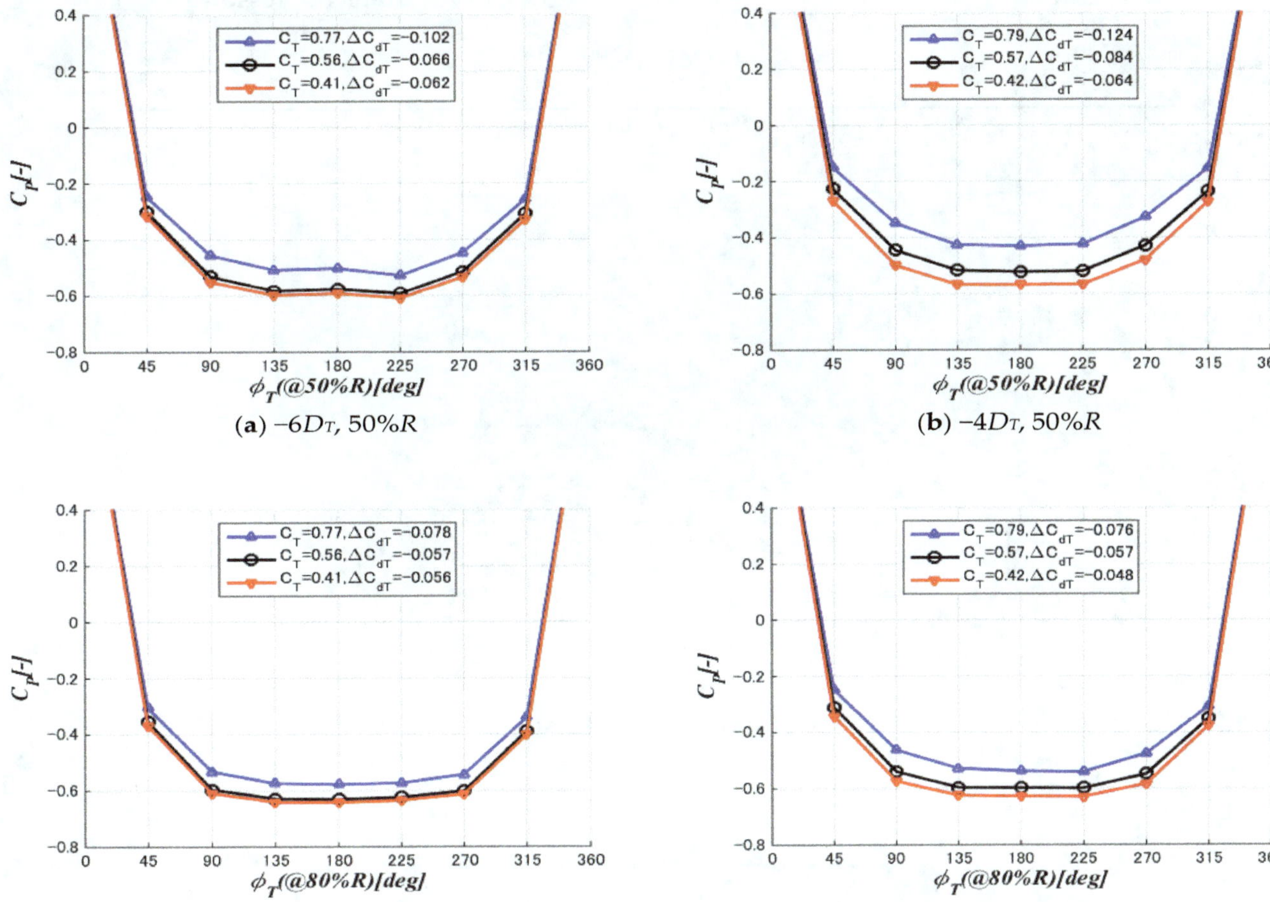

(**a**) $-6D_T$, 50%R (**b**) $-4D_T$, 50%R

(**c**) $-6D_T$, 80%R (**d**) $-4D_T$, 80%R

Figure 6. Distribution of the pressure coefficient on the dummy tower in typical cases.

4. Analysis

4.1. Outline

The relationship between the rotor thrust and the tower drag coefficient was determined by the proposed method described in Section 2. The model is generally the same as the one described in Section 3.3; however, the influence of the support was neglected as the support was located far downwind of the rotor, and this distance was fairly large compared to the support's diameter. In addition, the influence of the nacelle was neglected, as it does not affect the outboard sections, as discussed in this study.

4.2. CFD Outline

A CFD analysis was conducted for the rotor using ANSYS CFX [19] considering the k-ω SST turbulence model. The rotor was modeled by an ADM with uniform load distribution.

The simulation domain and the boundary conditions are summarized in Table 2. The model consisted of structured cells around the ADM with $y+$ is about 1 and unstructured ones otherwise. The total number of cells was approximately 50 million. The typical cells are shown in Figures 7 and 8.

Table 2. Simulation domain and boundary conditions.

Boundary	Distance from the Rotor Center	Boundary Condition
Rotor	$-D/140$~$+D/140$	ADM
Tower	Not available	Not available
Inlet	$-4D$	Wind speed (uniform)
Outlet	$+5D$	Ambient pressure
Bottom	$-2D$	Slip
Top	$+2D$	Slip
Side	$-2D$, $+2D$	Slip

"D": Rotor diameter.

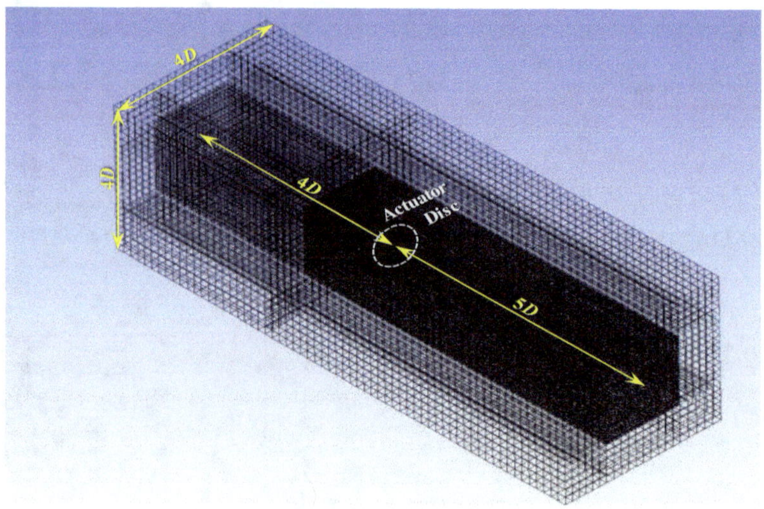

Figure 7. CFD domain.

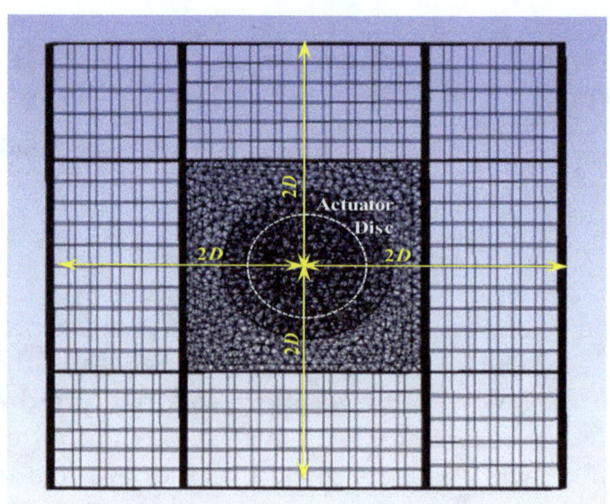

Figure 8. Cells around the rotor.

4.3. Analysis Conditions

- Wind speed (U_0): 6 m/s (uniform, steady)
- Yaw angle: 0 deg
- Thrust coefficient (C_T): 0.4–0.9 (each 1.0)

4.4. Wind Speed Distributions in Front of the Rotor

The wind speed distributions in typical conditions, $C_T = 0.4$ and 0.9, are shown in Figures 9 and 10. The circles in these figures denote the positions of the dummy towers at $-6D_T$ and $-4D_T$. The wind speed tended to decrease in front of the tower in general, and the wind speed was lower in the vicinity of the rotor. A comparison between the two conditions shows that the wind speed around the dummy tower decreased as C_T increased. A comparison between the two sections shows that the wind speed in front of the rotor was lower at 50%R compared to that at 80%R. The load was uniform on the rotor, and the inboard sections were considerably affected by the rotor.

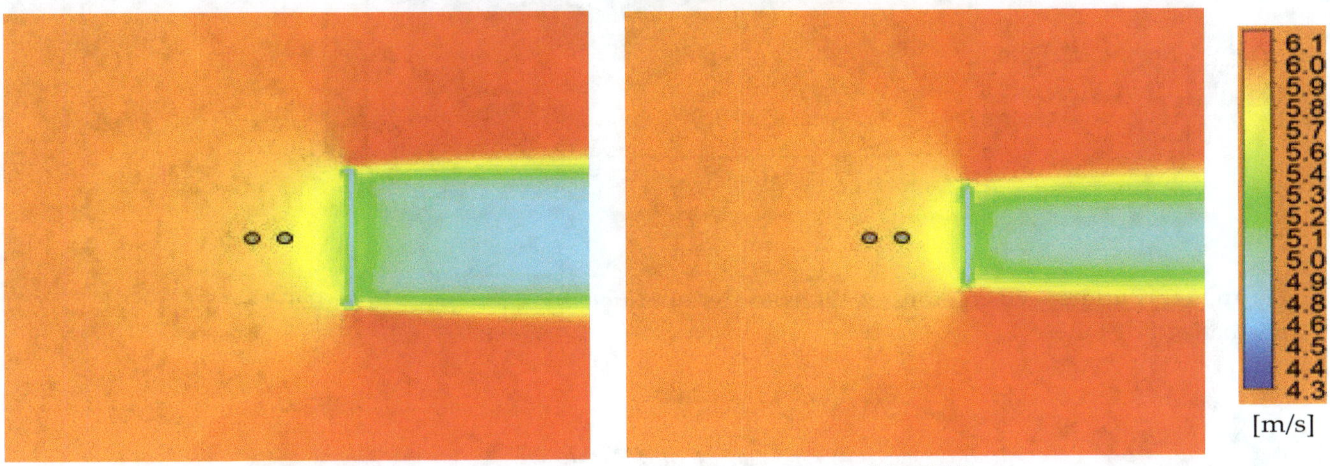

(**a**) 50%R (**b**) 80%R

Figure 9. Wind speed distribution in front of the rotor at $C_T = 0.4$, $U_0 = 6$ m/s. (The circles indicate the dummy tower positions at $-6D_T$ and $-4D_T$).

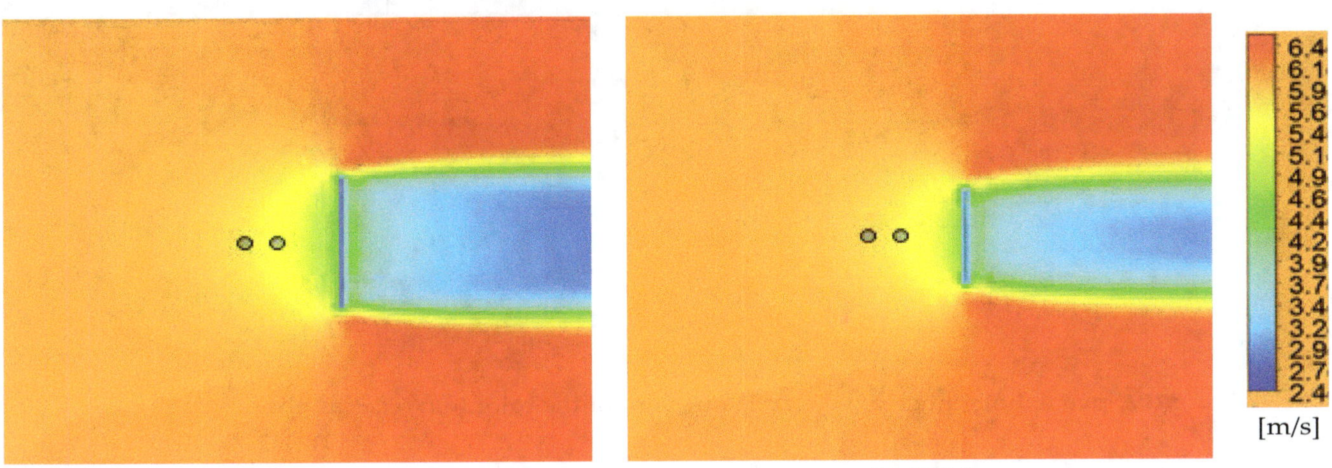

(**a**) 50%R (**b**) 80%R

Figure 10. Wind speed distribution in front of the rotor at $C_T = 0.9$, $U_0 = 6$ m/s. (The circles indicate the dummy tower positions at $-6D_T$ and $-4D_T$).

The distributions of the normalized wind speed μ_T and its differential $\partial\mu_T/\partial\xi_T$ with respect to the normalized distance ξ_T in front of the rotor are shown in Figures 11 and 12. The distributions at $C_T = 0.4$ and 0.9were identical to those in Figures 9 and 10 in the symmetrical plane. The top subplots correspond to the distributions at 50%R, and the bottom ones correspond to those at 80%R. The wind speed was lower in the vicinity of the rotor and decreased as the thrust coefficient increased, as mentioned above.

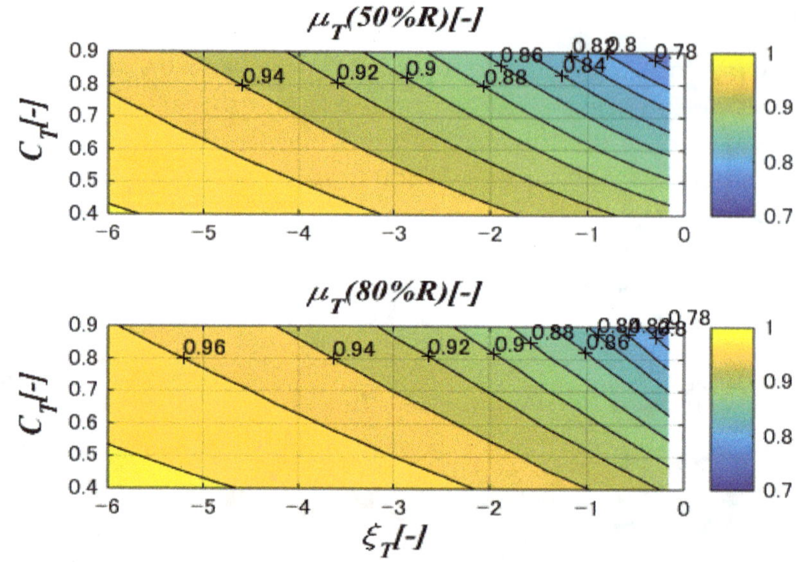

Figure 11. Wind speed in front of the rotor as obtained by CFD. μ_T, normalized wind speed, ξ_T, normalized distance.

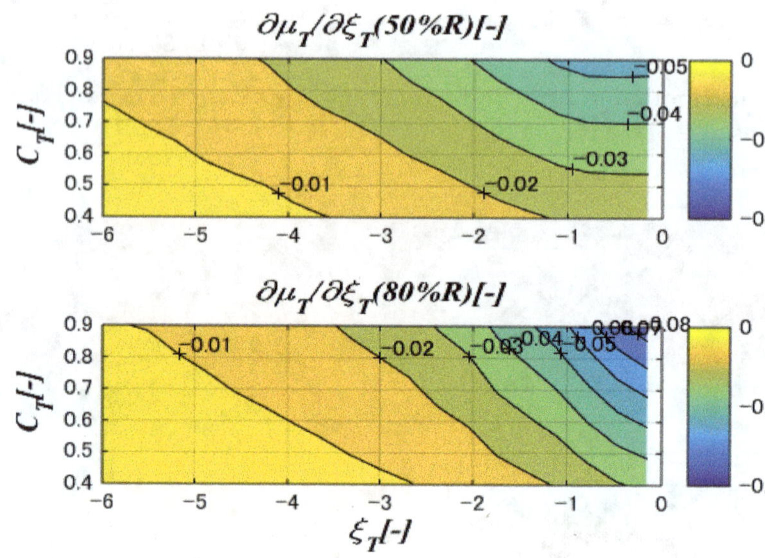

Figure 12. Spatial differential of wind speed in front of the rotor ($\partial\mu_T/\partial\xi_T$) as obtained by CFD.

4.5. Drag Coefficients of the Virtual Tower

The deviation of the drag coefficient of the virtual tower calculated by the present theory is shown in Figure 13. The x-axis denotes the virtual tower position, and the y-axis denotes the thrust coefficient. The diameter of the tower D_T was assumed to be 64 mm, as in the experiment. As indicated by Figures 11 and 12, the drag coefficient tended to be smaller as the rotor–tower clearance decreased and the thrust coefficient increased. The drag tended to be smaller at 50%R than at 80%R.

As shown in Equation (14), the rotor thrust-induced tower drag deviation consisted of the wind speed term and the pressure gradient term. Figure 14 shows the share of the wind speed term to the total deviation. It indicates that approximately 80% of the change in the drag wss caused by the first term of Equation (12), i.e., the decrease in the wind speed, rather than the pressure gradient.

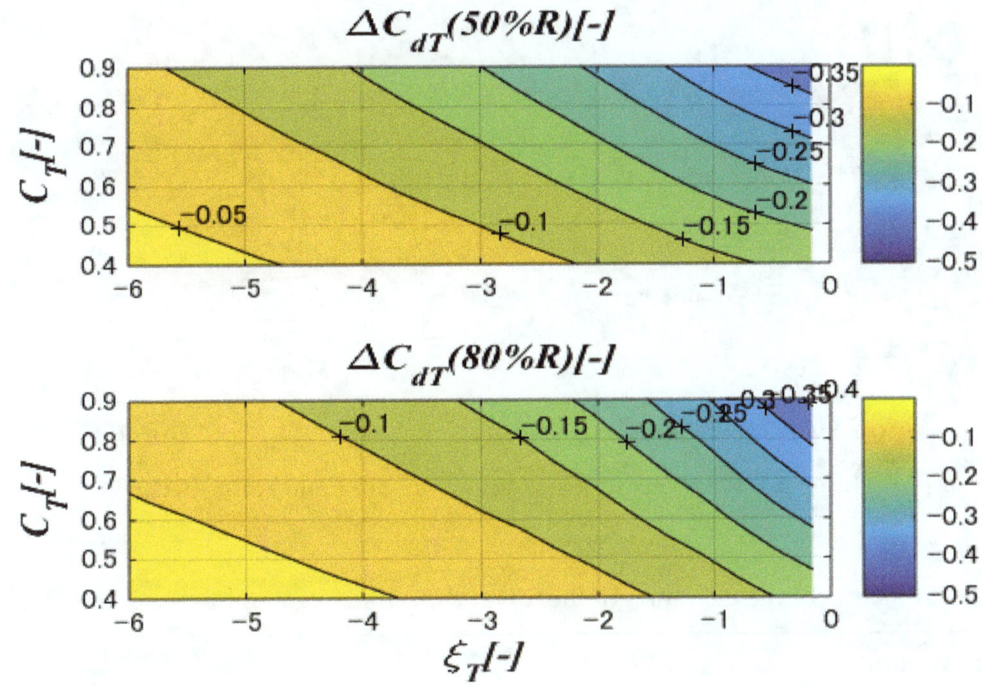

Figure 13. Relationship between the tower section drag and rotor thrust and the virtual tower position. ΔC_{dT}, drag coefficient.

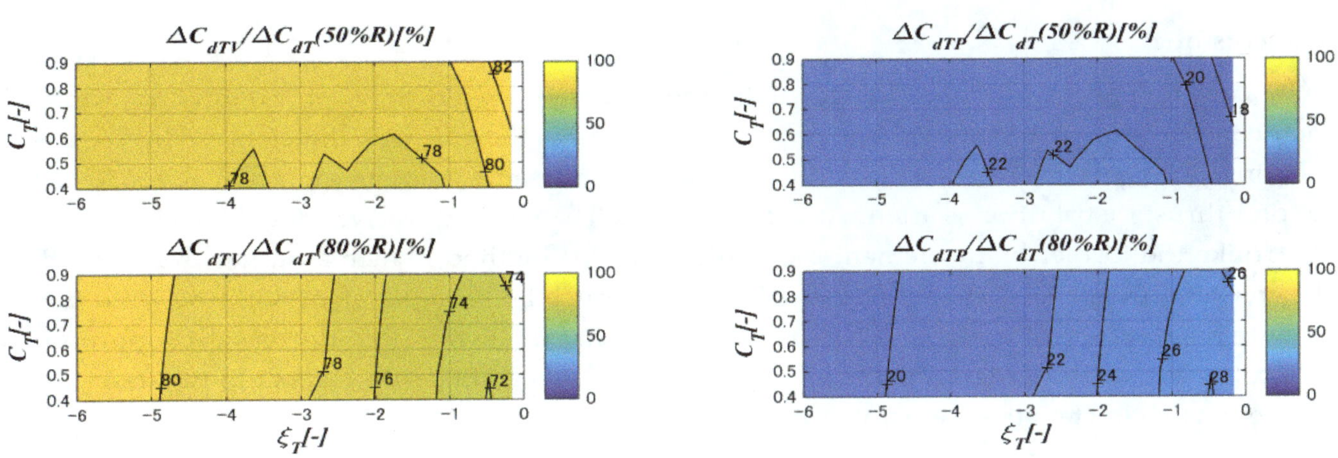

(**a**) Share of wind speed term C_{dTV}.

(**b**) Share of pressure gradient term C_{dTP}.

Figure 14. Share of the wind speed term C_{dTV} in the total C_{dT} tower section drag coefficient.

5. Validation

The thrust-induced drag deviations ΔC_{dT} to the thrust coefficient at $-6D_T$ and $-4D_T$ are shown in Figure 15. The lines are approximations for which the intercepts are at $\Delta C_{dT} = 0$ and $C_T = 0$ by definition. The ΔC_{dT} values for all the cases are shown to be almost proportional to C_T. The proportion factors are summarized in Table 3. The factors of the former BEM are zero, as the method does not consider the influence of the rotor thrust on the tower drag. On the other hand, the proposed method shows good agreement with the test. The drag coefficients tended to decrease as the rotor thrust increased. The drag coefficient at the 50%R section was smaller than at the 80%R section. Further, the drag coefficient decreased as the tower was placed closer to the rotor.

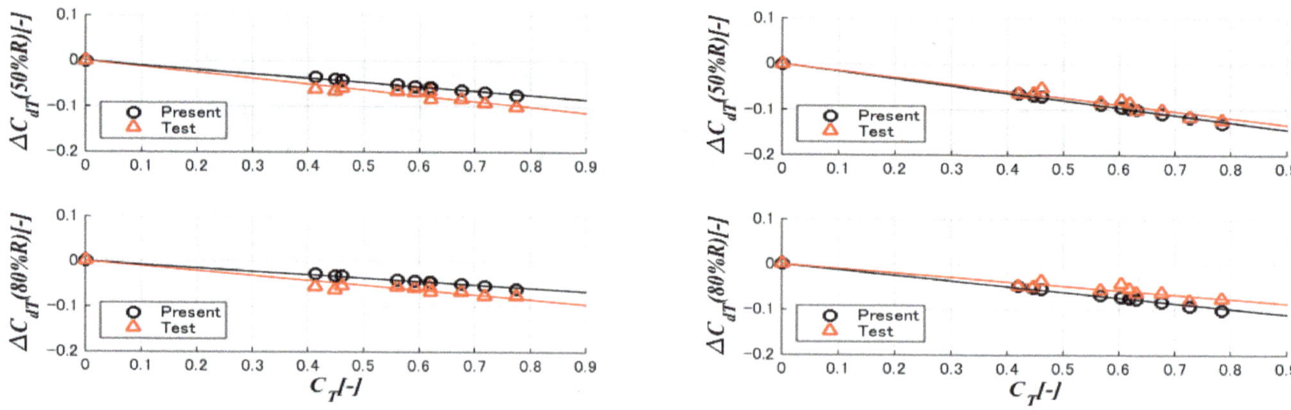

(**a**) $-6D_T$. (**b**) $-4D_T$.

Figure 15. Relationship between tower drag and rotor thrust.

Table 3. Fraction of the deviation of the tower drag coefficient to the thrust coefficient.

Dummy Tower Position	Tower Section	Test	Present	
		$\Delta C_{dT}/C_T$ [-]	$\Delta C_{dT}/C_T$ [-]	Deviation from the Test
$-6D_T$	50%R	−0.1270	−0.0937	−26.2%
	80%R	−0.1058	−0.0741	−30.0%
$-4D_T$	50%R	−0.1484	−0.1594	+7.4%
	80%R	−0.0961	−0.1225	+27.5%

6. Conclusions

A novel analysis method to calculate the average tower drag of downwind turbines, which considers the rotor-induced average tower drag coefficient, was developed. It consists of two terms, that is, the decrease in the wind speed and the pressure gradient caused by the rotor thrust. The rotor thrust distribution is assumed to be uniform. The method was validated by a wind tunnel test. Unlike the former blade element momentum (BEM) method, which assumes the tower drag to be constant, the proposed method demonstrates a much better agreement with the wind tunnel test, with an accuracy of up to 30%. The results show that the tower drag decreases proportionally to the rotor thrust. With regard to the sensitivity of the tower drag with respect to the rotor thrust, the following characteristics were noted:

- The drag coefficient was larger in the inboard section (50%R) than in the outboard section (80%R).
- The drag coefficient decreased as the tower was positioned closer to the rotor.
- Of the two terms influencing the deviation of the tower section drag, the effect of the decrease in wind speed was more dominant in leading to the decrease in the tower section drag.

By preparing the database of the relationship between the rotor thrust coefficient for the wind speed distribution in front of the rotor, the proposed method is expected to improve the accuracy of the load calculation in the BEM.

In future studies, the present model is planned to be extended to variable loads of wind turbine blades in succession.

Author Contributions: S.Y. organized the research, formulation, and validation. K.F. conducted the computational fluid dynamics simulations, and M.H. and A.T. conducted the wind tunnel test.

Acknowledgments: The authors express gratitude to Keiji Matsushima, Kimihiko Watanabe, and Yuji Ohya for their support for the wind tunnel test. The authors would also like to thank Masataka Motoyama and Omar Ibrahim for their support in the computational fluid dynamics simulation.

Nomenclature

C_{dT}	Tower drag coefficient
C_{dTO}	Tower drag coefficient with no rotor thrust
C_P	Power coefficient
C_p	Pressure coefficient of the tower
C_T	Thrust coefficient
D	Rotor diameter
D_T	Tower diameter
f_{XT}	Tower section drag
f_{XTO}	Tower section drag with no rotor thrust
p_0	Free stream pressure
p_T	Pressure at the tower position
R	Rotor radius
r	Station radius of the blade element
T	Rotor thrust
U_0	Free stream wind speed
u_T	Wind speed at the tower position
w	Thickness of the actuator disc
T	Longitudinal (or windward) position
C_{dT}	Deviation of the tower drag coefficient from that of the isolated tower
ΔC_{dTP}	Deviation of the tower drag coefficient by the effect of the pressure gradient
ΔC_{dTV}	Deviation of the tower drag coefficient by the wind speed
Δf_{XT}	Deviation of the tower section drag from that of the isolated tower
Δf_{XTP}	Deviation of the tower section drag by the effect of the pressure gradient
Δf_{XTV}	Deviation of the tower section drag by the wind speed
θ	Blade pitch angle
λ	Tip speed ratio
μ_T	Normalized wind speed at the tower position
ξ_T	Position normalized by the tower diameter
ρ	Air density
ϕ_T	Azimuth position around the tower (0 deg in front, clockwise around the vertical axis)
ADM	Actuator disc model
ALM	Actuator line model
BEM	Blade element momentum (method)
CFD	Computational fluid dynamics

References

1. Gipe, P. *Wind Energy Comes of Age*; John Wiley and Sons: Chichester, UK, 1995.
2. *Wind Turbines—Part 1: Design Requirements*; IEC61400-1, Ed. 3; International Electrotechnical Commission: Geneva, Switzerland, 2005.
3. Germanischer Lloyd. *Guidelines for the Certifications of Wind Turbines, Edition 2010*; Germanischer Lloyd: Hamburg, Germany, 2010.
4. Burton, T.; Jenkins, N.; Srpe, D.; Bossanyi, E. *Wind Energy Handbook*, 2nd ed.; John Wiley and Sons: Chichester, UK, 2011.
5. Jonkman, J.M.; Buhl, M.L., Jr. *FAST User's Guide*; NREL/TP-500-38230; NREL: Golden CO, USA, 2005.
6. DNV GL. *Bladed, Version 4.7*; DNV GL: Bristol, UK, 2016.
7. Zhao, Q.; Sheng, C.; Afjeh, A. Computational aerodynamic analysis of offshore upwind and downwind turbines. *J. Aerodyn.* **2014**, *2014*, 860637. [CrossRef]

8. Zahle, F.; Madsen, H.A.; Sorensen, N.N. Evaluation of Tower Shadow Effects on Various Wind Turbine Concepts. Available online: http://orbit.dtu.dk/files/3550506/ris-r-1698.pdf (accessed on 19 July 2018).
9. Matiz-Chicacausa, A.; Lopez, O.D. Full downwind turbine simulations using actuator line method. *Model. Simul. Eng.* **2018**, *2018*, 2536897. [CrossRef]
10. Wang, T.; Coton, F.N. A high resolution tower shadow model for downwind wind turbines. *J. Wind Eng. Ind. Aerodyn.* **2001**, *89*, 873–892. [CrossRef]
11. Yoshida, S.; Kiyoki, S. Load equivalent tower shadow modeling for downwind turbines. *J. JSME* **2007**, *37*, 1273–1279. [CrossRef]
12. Yoshida, S.; Kiyoki, S. Load equivalent tower shadow modeling. In Proceedings of the European Wind Energy Conference 2007, Milano, Italy, 7–10 May 2007.
13. Nagao, T.; Kato, T.; Shiraishi, T.; Yoshida, S.; Sugino, J. Development of advanced wind turbine systems for remote islands. *Wind Eng.* **2006**, *28*, 729–740. [CrossRef]
14. Yoshida, S. Performance of downwind turbines in complex terrains. *Wind Eng.* **2006**, *30*, 487–501. [CrossRef]
15. Yoshida, S. *Nacelle Yaw Measurement of Downwind Turbines in Complex Terrains*; Windtech International: Groningen, The Netherlands, 2008.
16. Saeki, M.; Sano, T.; Kato, H.; Owada, M.; Yoshida, S. Concept of the HITACHI 5MW offshore downwind turbine. In Proceedings of the European Wind Energy Conference 2014, Barcelona, Spain, 10–13 March 2014.
17. Yoshida, S. Combined blade-element momentum—Lifting line model for variable loads of downwind turbine towers. *Energies* **2018**, *11*, 2521. [CrossRef]
18. Research Institute of Applied Mechanics, Kyushu University, Boundary Layer Wind Tunnel. Available online: https://www.riam.kyushu-u.ac.jp/windeng/en_equipment.html (accessed on 10 November 2018).
19. ANSYS. *ANSYS CFX*; ANSYS: Canonsburg, PA, USA, 2006.

Permissions

All chapters in this book were first published in MDPI; hereby published with permission under the Creative Commons Attribution License or equivalent. Every chapter published in this book has been scrutinized by our experts. Their significance has been extensively debated. The topics covered herein carry significant findings which will fuel the growth of the discipline. They may even be implemented as practical applications or may be referred to as a beginning point for another development.

The contributors of this book come from diverse backgrounds, making this book a truly international effort. This book will bring forth new frontiers with its revolutionizing research information and detailed analysis of the nascent developments around the world.

We would like to thank all the contributing authors for lending their expertise to make the book truly unique. They have played a crucial role in the development of this book. Without their invaluable contributions this book wouldn't have been possible. They have made vital efforts to compile up to date information on the varied aspects of this subject to make this book a valuable addition to the collection of many professionals and students.

This book was conceptualized with the vision of imparting up-to-date information and advanced data in this field. To ensure the same, a matchless editorial board was set up. Every individual on the board went through rigorous rounds of assessment to prove their worth. After which they invested a large part of their time researching and compiling the most relevant data for our readers.

The editorial board has been involved in producing this book since its inception. They have spent rigorous hours researching and exploring the diverse topics which have resulted in the successful publishing of this book. They have passed on their knowledge of decades through this book. To expedite this challenging task, the publisher supported the team at every step. A small team of assistant editors was also appointed to further simplify the editing procedure and attain best results for the readers.

Apart from the editorial board, the designing team has also invested a significant amount of their time in understanding the subject and creating the most relevant covers. They scrutinized every image to scout for the most suitable representation of the subject and create an appropriate cover for the book.

The publishing team has been an ardent support to the editorial, designing and production team. Their endless efforts to recruit the best for this project, has resulted in the accomplishment of this book. They are a veteran in the field of academics and their pool of knowledge is as vast as their experience in printing. Their expertise and guidance has proved useful at every step. Their uncompromising quality standards have made this book an exceptional effort. Their encouragement from time to time has been an inspiration for everyone.

The publisher and the editorial board hope that this book will prove to be a valuable piece of knowledge for researchers, students, practitioners and scholars across the globe.

List of Contributors

Adrian S. Halifi and Sharidan Shafie
Department of Mathematical Sciences, Faculty of Science, Universiti Teknologi Malaysia, Johor Bahru 81310, Malaysia

Norsarahaida S. Amin
Department of Mathematical Sciences, Faculty of Science, Universiti Teknologi Malaysia, Johor Bahru 81310, Malaysia
Department of Mathematics, Faculty of Science and Technology, Universitas Airlangga, East Java 60115, Indonesia

Mikhail Sukharev
Department of Applied Mathematics and Computer Modeling, National University of Oil and Gas «Gubkin University», 119991 Moscow, Russia

Chao Wang and Liangyu Wu
School of Hydraulic, Energy and Power Engineering, Yangzhou University, Yangzhou 225127, China

Feng Yao
School of Hydraulic, Energy and Power Engineering, Yangzhou University, Yangzhou 225127, China
Jiangsu Key Laboratory of Micro and Nano Heat Fluid Flow Technology and Energy Application, School of Environmental Science and Engineering, Suzhou University of Science and Technology, Suzhou 215009, China

Juan Shi and Mengchen Zhang
Key Laboratory of Energy Thermal Conversion and Control of Ministry of Education, School of Energy and Environment, Southeast University, Nanjing, Jiangsu 210096, China

Nan Shao, Jijian Lian, Guobin Xu, Fang Liu, Heng Deng, Quanchao Ren and Xiang Yan
State Key Laboratory of Hydraulic Engineering Simulation and Safety, Tianjin University, No. 92, Wei Jin Road, Nan Kai District, Tianjin 300072, China

Mohamed F. El-Amin
College of Engineering, Effat University, 21478 Jeddah, Saudi Arabia
Department of Mathematics, Faculty of Science, Aswan University, Aswan 81528, Egypt

Usama Khaled
Department of Electrical Engineering, College of Engineering, King Saud University, 11421 Riyadh, Saudi Arabia
Department of Electrical Engineering, Faculty of Energy Engineering, Aswan University, Aswan 81528, Egypt

Abderrahmane Beroual
Ecole Centrale de Lyon, University of Lyon, AMPERE CNRS UMR 5005, 36 Avenue Guy de Collongue, 69134 Ecully, France

Younggi Park, Jaiyoung Ryu and Hong Sun Ryou
School of Mechanical Engineering, Chung-Ang University, 84, Heukseok-ro, Dongjak-gu, Seoul 06974, Korea

Wuyi Wan, Boran Zhang and Xiaoyi Chen
Department of Hydraulic Engineering, College of Civil Engineering and Architecture, Zhejiang University, Hangzhou 310058, China

Lei Shi and Hongwei Ma
School of Energy and Power Engineering, Beihang University, Beijing 100191, China

Lixiang Wang
Midea Corporate Research Center, Foshan 528300, China

Joon-Yong Yoon
Department of Mechanical Engineering, Hanyang University, 55 Hanyangdeahak-ro, Sangnok-gu, Ansan, Gyeonggi-do 15588, Korea

Choon-Man Jang
Department of Land, Water and Environment Research, Korea Institute of Civil Engineering and Building Technology, 283, Goyangdae-ro, Ilsanseo-gu, Goyang-si, Gyeonggi-do 10223, Korea

Seok-Yun Jeon
Department of Mechanical Engineering, Hanyang University, 55 Hanyangdeahak-ro, Sangnok-gu, Ansan, Gyeonggi-do 15588, Korea
Department of Land, Water and Environment Research, Korea Institute of Civil Engineering and Building Technology, 283, Goyangdae-ro, Ilsanseo-gu, Goyang-si, Gyeonggi-do 10223, Korea

Ahmed Hamood and Artur J. Jaworski
School of Computing and Engineering, University of Huddersfield, Queensgate, Huddersfield HD1 3DH, UK

Xiaoan Mao
Faculty of Engineering, University of Leeds, Woodhouse Lane, Leeds LS2 9JT, UK

Yueheng Song
School of Reliability and System Engineering, Beihang University, Beijing 100191, China

Jiming Ma
Sino-French Engineering School, Beihang University, Beijing 100191, China

Shengkui Zeng
School of Reliability and System Engineering, Beihang University, Beijing 100191, China
Science and Technology on Reliability and Environmental Engineering Laboratory, Beihang University, Beijing 100191, China

Qiaorui Si, Wenting He, Asad Ali and Shouqi Yuan
National Research Center of Pumps, Jiangsu University, Zhenjiang 212013, China

Gérard Bois
LMFL, FRE CNRS 3723, ENSAM, 8 boulevard Louis XIV, 59046 Lille CEDEX, France

Qifeng Jiang
Key Laboratory of Fluid and Power Machinery, Xi Hua University, Chengdu 610039, China

Shigeo Yoshida, Masahiro Hamasaki and Ao Takada
Research Institute for Applied Mechanics, Kyushu University, 6-1 Kasugakoen, Kasuga, Fukuoka 816-8580, Japan

Kazuyuki Fujii
Interdisciplinary Graduate School of Engineering and Science, Kyushu University, 6-1 Kasugakoen, Kasuga, Fukuoka 816-8580, Japan

Index

www.ingramcontent.com/pod-product-compliance
Lightning Source LLC
LaVergne TN
LVHW081017210926
848485LV00004B/885

* 9 7 8 1 6 4 7 2 6 6 7 8 3 *